TEAM AND COLLECTIVE TRAINING NEEDS ANALYSIS

To the memory of
Phillip Pike
and
Alan Huddlestone

Team and Collective Training Needs Analysis

Defining Requirements and Specifying Training Systems

JOHN HUDDLESTONE
Coventry University, UK

&

JONATHAN PIKE

CRC Press
Taylor & Francis Group
Boca Raton London New York

CRC Press is an imprint of the
Taylor & Francis Group, an **informa** business

CRC Press
Taylor & Francis Group
6000 Broken Sound Parkway NW, Suite 300
Boca Raton, FL 33487-2742

First issued in paperback 2017

ISBN 13: 978-1-138-09215-0 (pbk)
ISBN 13: 978-1-4094-5386-4 (hbk)

Visit the Taylor & Francis Web site at
http://www.taylorandfrancis.com

and the CRC Press Web site at
http://www.crcpress.com

Contents

List of Figures

List of Tables

About the Authors

Dr John Huddlestone is a Senior Research Fellow in the Human Systems Integration Group within the Engineering and Computing Faculty at Coventry University in England. His research interests include team training, training needs analysis, training methods and media, and aviation human factors. Current research projects include the human factors of future flight-deck technologies and single-pilot operations, and the team and collective training implications of future maritime unmanned systems concepts. His research has also included the specification and evaluation of multiplayer simulation systems and the evaluation of novel training media. Before joining academia, he was a Royal Air Force Officer. Working in the training specialisation, he was responsible for the analysis, design and delivery of a wide variety of training solutions in the aviation and engineering domains. He was a member of the Human Factors Integration Defence Technology Centre team that was awarded the Ergonomics Society President's Medal for their outstanding contribution to Human Factors research. He holds a PhD in applied psychology from Cranfield University, a Master's degree in Computing Science from Imperial College, London and a Batchelor's degree in Education from Nottingham Trent University.

Jonathan Pike is a freelance training specialist currently living in Perth, Western Australia. Between 2005 and 2014, while working in the Human Factors Department of Cranfield University and the Human Systems Integration Group of Coventry University, he conducted research for the Defence Science and Technology Laboratory under the auspices of the Human Factors Integration Defence Technology Centre and Defence Human Capability Science and Technology Centre. He was a member of the Human Factors Integration Defence Technology Centre team that was awarded the Ergonomics Society President's Medal for their outstanding contribution to Human Factors research. A visiting researcher at Coventry University, and a past visiting research fellow at Cranfield University at he holds a BSc in Biology from University College London and an MSc in Applied Computing Technology from Middlesex University.

Human Factors in Defence

Series Editors:
Dr Don Harris, Professor of Human Factors, Coventry University, UK
Professor Neville Stanton, Chair in Human Factors at the University of Southampton, UK
Dr Eduardo Salas, University of Central Florida, USA

Human factors is key to enabling today's armed forces to implement their vision to "produce battle-winning people and equipment that are fit for the challenge of today, ready for the tasks of tomorrow and capable of building for the future" (source: UK MoD). Modern armed forces fulfil a wider variety of roles than ever before. In addition to defending sovereign territory and prosecuting armed conflicts, military personnel are engaged in homeland defence and in undertaking peacekeeping operations and delivering humanitarian aid right across the world.

This requires top-class personnel, trained to the highest standards in the use of first class equipment. The military has long recognised that good human factors is essential if these aims are to be achieved.

The defence sector is far and away the largest employer of human factors personnel across the globe and is the largest funder of basic and applied research. Much of this research is applicable to a wide audience, not just the military; this series aims to give readers access to some of this high-quality work.

Ashgate's *Human Factors in Defence* series comprises of specially commissioned books from internationally recognised experts in the field. They provide in-depth, authoritative accounts of key human factors issues being addressed by the defence industry across the world.

Reviews of
Team and Collective Training Needs Analysis: Defining Requirements and Specifying Training Systems

This in-depth analysis of team training challenges addresses a highly complex and interactive issue which has long plagued program managers, operators, and trainers in every service. The complexities of modern weapons systems, the ever evolving threat and severe national budgetary constraints require that training systems be optimized for efficiency and harmony of effort. In their excellent work the authors have given the community of practice a methodology and framework to accomplish the former. Their book should be part of the kit used by every weapons system program manager and team training organization.

Rear Admiral Frederick L. Lewis, USN (Ret.), President Emeritus,
National Training and Simulation Association

This is an exceptionally comprehensive look at team and collective training. It provides a rare insight into the methodology of training needs analysis and how it can address the complexities beyond the individual level. The author's credentials are apparent and amply demonstrated in the worked examples and case studies. This is a definitive guide not just to those responsible for training and operational delivery but also those involved in R&D or procurement.

Lt Col (Retd) Guy Wallis, Principal Analyst,
Defence Science and Technology Laboratory

Training is expensive, but failing to train realistically is even more expensive. The text provides a masterly, logical approach to understanding, decomposing and integrating key variables involved in a complex world of weapons and training systems. Moreover, it enables the individuals and teams who deploy them to work together and build an accumulation of marginal edges sufficient to distinguish effective professional military performance from that of the enthusiastic amateur.

Professor Victor Newman, Knowledge & Innovation Management,
The Business School, The University of Greenwich

In an age where organisations now have to justify spending on complex training and exercising for teams and collective capabilities, this book is long overdue. It provides processes on the elements of team and collective training analysis, gives a toolkit for those involved with the acquisition of related training systems and, more importantly, is an essential guide for those who want to make their training better.

Commander Paul Pine, Royal Navy (Maritime Training Acquisition Organization)

Acknowledgements

The work presented in this book originates in part from work carried out under the Human Factors Integration Defence Technology Centre (HFI DTC) and the Defence Human Capability Science and Technology Centre (DHC STC) contracts, conducted on behalf of the Defence Science and Technology Laboratory, as part of the Chief Scientific Advisors Research Programme. Any views presented in this book are those of the authors and do not constitute the official views of the UK Ministry of Defence.

The development of Team and Collective Training Needs Analysis as a methodology is the product of eight years of research, and encompassed numerous projects and case studies. Any endeavour of that scale can only be successful with the support of many others. Whilst it is impossible to name everyone that assisted us, there are a number of people that deserve a particular mention. From BAE Systems, Dr Karen Lane (as Director of the HFI DTC and latterly, the DHC STC) did much work behind the scenes to ensure that our various research programmes came into being, whilst Dr Carole Deighton, in her technical assurance role, was an invaluable sounding board for our ideas. Dr Colin Corbridge from the Defence Science and Technology Laboratory (Dstl) was an enthusiastic supporter of the project, facilitating a number of case studies and challenges for us to overcome.

A number of military colleagues made a significant contribution to our work. Firstly, Lieutenant Commanders Tom Harrison, Paul Pine and Paul Newall, all Royal Navy Training Needs Analysis specialists, acted as sounding boards for our ideas and gave much pragmatic and constructive feedback throughout the project. Often cast as customers for our work, they gave generously of their time to support our progress and critically review our outputs. Colonel Hugh Russell (British Army) applied an early version of the methodology to a case study, and gave us invaluable feedback about the experience both of using the methodology and of interpreting our written guidance. Lt Rob Driscoll orchestrated and hosted our visit to US Naval Station Norfolk to observe a Fleet Synthetic Training Exercise whilst aboard USS Eisenhower. Lieutenant Colonel Bo Andersen (Danish Army) at the NATO Joint Warfare Centre arranged for us to observe a major NATO exercise that he was running. In some lively discussions, he shared many invaluable insights into the process of planning and delivering multinational exercises in an Alliance context. Finally, the staff of the Joint Training and Exercise Planning Staff at Northwood put up with our endless questions with good humour during the weeks that we spent under their feet whilst they developed and ran major exercises.

The production of the manuscript was a team effort in itself. Whilst any errors that you might spot are entirely our own work, many perished at the hands of the team that helped us through the process. Joseph Wallis brought his copywriting skills to bear on the first half of the book and provided sound advice on writing style. Kevin Bessel at BAE Systems tenaciously proofread the final manuscript. Jessica Onslow from Dstl navigated the manuscript safely through the rapids of the MOD Permission to Publish system and, along with colleagues Sarah Bowditch, Claire Ford and Ian Greig, reviewed the work and provided an invaluable critique. Mention must be made of Guy Loft, our Publishing Editor who encouraged us to write the book in the first place, and has patiently waited for it to evolve from urban myth into a tangible entity. Guy was ably assisted by Charlotte Edwards, who led us gently through the publication process.

Finally, we must thank colleagues, friends and family who have supported/ put up with us whilst we have been on what feels like an epic writing journey. We could not have done it without them.

Foreword

I have been variously involved with Collective Training throughout my Army career. As a junior MOD staff Captain in the early 1990s I typed the first drafts of the Army's *Compendium of Collective Training Tasks* – a huge tome that eventually begat the Mission Essential Tasks List (METL). I say 'typed' as at that time there was not a lot of thinking involved, just lots of listing. I also had much Collective Training done to me: on battlegroup, brigade and divisional exercises we busily did things, waited in the rain, then did more things and then we were considered to be 'trained'. We had certainly had experiences, we had certainly learned things but usually the connection between the two was obscure and perhaps incidental. As the Armed Forces moved from the Cold War 'training and waiting' footing through to the need to sustain a continuous operational footing through the Gulf War, the Balkans, Iraq and Afghanistan so the design and development of Collective Training developed, but not necessarily the way that we thought about that design and development.

There is a very human tendency where if we are convinced that we are doing the right thing and it doesn't seem to be working then we continue to do the same thing, but harder, in the conviction that it will eventually work. We operate the ignition, spinning the starter motor and flattening the battery and still the car will not start. Nor will it with no fuel. Armed with an otherwise perfectly good idea it is difficult to perceive how or why it might not be the right idea for that particular problem. And so it is for Collective Training.

General Martin Dempsey, now (2015) US Chairman of the Joint Chiefs of Staffs, made much of the need to be imaginative and adaptive in order to face a decentralized and networked enemy in his Campaign for Learning: Avoiding the Failure of Imagination 2010 Kermit Roosevelt lecture tour and following papers.[1] This means doing things differently. The established MOD Training Needs Analysis methodology is a good tool in the right context but it was designed and developed against individual training needs and methods. The Team and Collective Training Need Analysis (TCTNA) methodology presented here gives us the right tool to do things differently, to look at the problem through a different lens, thinking with imagination and being adaptive. Critically, this TCTNA methodology allows us to move away from thinking about individual skills in the collective space but to thinking about collective actions and outcomes. The sum of the parts is less than the whole. This is the key difference between individual 'Taskwork' and collective 'Teamwork'.

1 For example *RUSI Journal*, June 2010, vol. 155, no. 3.

We cannot strategically or financially afford to continue to do what we have always done before. In 25 years of military training design work I have been looking for a spanner that fitted this particular collective training nut, aware that the one I was using didn't fit. Here is one.

Col (Ret'd) Hugh Russell

Preface

When you stand on the quayside and look up at the USS Dwight D. Eisenhower, a US Navy Nimitz-class aircraft carrier, you cannot help but feel utterly dwarfed by it. It is simply huge. We found ourselves doing just that, on a brisk February morning in the US Navy base at Norfolk, Virginia, at the start of three days of observation of a Fleet Synthetic Training-Joint Exercise. All of the ships in the Eisenhower's task group were alongside and connected into a large common simulation environment, with numerous other US Navy and NATO assets and simulators participating. As it was a synthetic training exercise, only the warfighting staff aboard each of the ships would be participating with their associated command elements – a training audience of approximately 2,500 people networked together across both sides of the Atlantic. Of course, there were a few other bits and pieces: aircraft simulators on various sites, a Combined Air Operations Centre in yet another simulation facility, and a visiting Battlestaff from the UK who were roleplaying the higher command during the exercise. Should you think that sounded complex, you obviously haven't tried finding your way around inside a Nimitz-class aircraft carrier – that is an activity that needs a training course in its own right!

An event of such scale requires a correspondingly complex training and exercising organisation, and a great deal of planning, testing and rehearsal. We were about to meet the senior observer controller for the exercise who had a team of 50 instructional staff to observe and evaluate the training. The previous day, we had visited the training centre where we had met some of the 250 people who would be running the simulation and controlling the flow of events within the exercise. The point here is that team and collective training operates over a range of scales and at the top end can get very large and complex. Such complexity requires careful organisation. While team and collective training involves individual performance, it is embedded within a higher-level construct involving force integration, interoperability, coordination and communication.

To put this work into a little more context, when we started this work the global economic situation was putting significant pressure on defence budgets, global military operations were shining a spotlight on the need for effective team and collective performance, and the Royal Navy was in the process of acquiring a new class of aircraft carrier (the Queen Elizabeth class). At the same time, acquisition processes, including the training acquisition component, were under close scrutiny. The team and collective Training Needs Analysis (TNA) problem simply had to be addressed.

Our task was to try to solve the puzzle of how you carry out the analysis and design activities that constitute Training Needs Analysis for team and collective training. While much has been written on instructional analysis and design we found it difficult to articulate between the various theoretical approaches presented by different authors, and apply these models to the various exercises and events we observed as part of the research and analysis process. We characterised this issue as the 'archipelago of theories' problem. The problem was that Training Needs Analysis for team and collective training seemed to involve factors across a broad scope such as individual and team tasks, teamwork, command and control, task and training environments, team training approaches, instructional strategy and wide-ranging organisational and procurement considerations.

To attempt to try to fit the disparate elements of the puzzle together we started generating descriptions for the constructs we observed and read in the literature, in a format that was inter-relatable. We have tried to generate a modular toolkit where different aspects of the analysis can be inter-related within a unified framework. These descriptions and models form the first half of the book and provide a basis for the 'how-to' method of the second-half. We have deliberately put the theory and models at the front of the book, as we feel it helps to 'show our working', justify some of the synthesis of extant training research and provide a theoretical basis for some of the tacit best practice we observed at various training events. The models also provide some useful orientation for the experienced practitioner and essential context for the newcomer.

While we would not claim to have authored the final words on the subject, we hope at least to have provided some points of reference in the landscape. We hope that you find it a thought-provoking read and that the models and techniques that we put forward are of use to you. What we can say is that, when we have applied the methodology in a variety of different contexts, the outputs have been considered by the training and operational experts in the field to be sensible and useful. That at least is a start.

If you would like to contribute to the ongoing discussion about this research, please contact us as we would value feedback based around these models and their application.

We hope you enjoy the book.

John Huddlestone
(john.huddlestone@coventry.ac.uk)

Jonathan Pike
(jonathan.pike@tctna.com)

PART I
Underpinning Theory and Models

Chapter 1
Introduction

Background

An organisation's capability is delivered, almost without exception, by the team or set of teams that make up its structure. Consequently, effective team performance is critical to organisational success. A well-known incident highlighting the significance of team performance occurred on 3rd July 1988 when a United States Navy guided missile cruiser, the USS Vincennes, mistakenly shot down Iran Air Flight 655 over the Persian Gulf. Two hundred and ninety lives were lost. It was the air warfare team on board the Vincennes that had the responsibility for monitoring the airspace in the area where the ship was operating. This tragic event was attributed, in part, to failures in the interactions between members of the air warfare team on the USS Vincennes (Collyer and Maleki 1998). In the context of this book, the significant point about this incident was that the performance of the team involved depended on more than having skilled individuals in place; the critical issue was the way in which they worked together as a team. The United States Office of Naval Research responded to the incident by funding a seven-year research project, called Tactical Decision Making Under Stress (TADMUS), to investigate team performance. Notably, the improvement of team training was clearly identified in its objectives; such was the perceived significance of team training in relationship to team performance (Johnston et al. 1998). The purpose of this book is to provide a methodology for analysing team tasks to identify the critical aspects of team performance that need to be trained and to determine appropriate training solutions to enable teams to operate efficiently, effectively and safely.

Whilst few of us will find ourselves in a team involved in the decision-making process concerning firing a missile at an aircraft from a naval warship, most of us work in teams and encounter other teams both at work and in our everyday lives. As with the Vincennes incident, the issues that are involved in working successfully as a team are often highlighted either when things go wrong or when they go very well. Two such examples from everyday life came to light when the authors walked to a favourite restaurant for a lunch break whilst planning this book.

The first incident was observed as we walked to the restaurant. On the way we saw two bricklayers who were building a garden wall at the front of a house. The first few courses of bricks had been laid in a nice straight line and they were busy building the wall up to the required height. Since both could not work in the same place at the same time, they had made a plan to work at opposite ends of the wall and both work towards the middle. They had a piece of string stretched along the front of the wall as a guide so that the bricks could be laid level and in a straight line. As we were passing, the bricklayers were engaged in a free and frank exchange of views about how their task was progressing. The bricklayer at one end of the wall said to his colleague near the other end 'Your end of the wall isn't straight!', to which his colleague replied rather tersely as he looked along the wall, 'Well if you stopped leaning on the string it might help!' They had just demonstrated the need for communication, mutual performance monitoring and the provision of feedback between team members.

The second incident occurred at the restaurant itself. When we arrived our hearts sank, as a large party from a local company had arrived for a celebratory meal and had just ordered their meals. However, as we had always had good service in the past we elected to stay, hoping that we would be fed in a timely fashion despite there being a crowd ahead of us. Our dining experience was in the hands of two teams, the kitchen team led by the head chef and the front of house team led by the head waiter. In fact all went well. The large numbers of orders for some menu items meant that some items were no longer available by the time we were given menus. However, the kitchen team had communicated this to the front of house team and we were advised accordingly. Whilst the large party absorbed much of the teams' efforts, we were served hot, well-prepared food in a timely fashion, drinks were served promptly, and plates were cleared away soon after we finished eating. The large party appeared to receive a similar service. This suggested that both teams had a well-considered plan of action, that tasks were allocated effectively, and that communication and coordination was occurring both within each team and between the teams. The arrival of a large group dining together created a challenging set of conditions for the teams to work under. Not only did a large number of customers have to be served, everyone in the group dining together required their meals to be served at the same time. Equally, other diners expected to be served without delay. Therefore, the presence of a large group dining together causes the restaurant staff to experience a 'different sort of busy' to that of catering for a restaurant full of smaller groups of diners. The party of diners presented a more demanding requirement for synchronisation of team actions than that presented by simply having a restaurant full of smaller groups dining independently from each other.

Whilst these two examples are relatively small in scale, they serve to illustrate some of the complexities that are involved in carrying out team tasks. As team sizes increase and multiple teams become involved, these complexities amplify. Arguably, military organisations are faced with particularly demanding performance requirements and associated training requirements compared to those of civilian organisations. Whereas a civilian organisation tends to function in a clearly defined area of business in which it operates continuously, military organisations are required to be ready to undertake a wide variety of different tasks, of which many are only carried out when deployed on operations. The rest of the time they are in training. Furthermore, the exact configuration of the force required for a given operation will vary considerably depending on the nature of the operation itself. This complexity can be illustrated by considering the example of 1 Royal Irish Battlegroup as it prepared for and deployed to Iraq in 2003. The detail for this example is drawn from Colonel Tim Collins' account of his time in command of the 1st Battalion of the Royal Irish Regiment (Collins 2005).

In 2001 the Battalion had been deployed on peacekeeping duties in East Tyrone in Northern Ireland. On its return from Ireland in December 2001, the Battalion underwent training in the air assault role. After 10 months of training, the Battalion was deployed at short notice to an entirely different kind of task. They were to take part in Operation Fresco, manning Green Goddess Fire Engines to provide Fire and Rescue cover during the nationwide firefighters' strike in the UK during November and December of 2002. No sooner was this over, the Battalion received orders that it was to deploy to Iraq on Operation Telic as part of 16 Air Assault Brigade. At this point the Battalion was just over its established strength of 690. It was to deploy as the 1 Royal Irish Battlegroup with a strength of 1,225. This significant increase in numbers was due to the need to add additional units to the core Battalion to provide the required capability for the operation. These additional units included an artillery battery, an engineer squadron and 25 Air Naval Gunnery Liaison Coordinators from the United States Marines. The Battlegroup had less than a month to prepare for deployment. Colonel Collins was presented with a significant training challenge. Preparations included training at individual, team and collective (team of teams) levels. At the individual level, the soldiers had to be able to carry in excess of 100 lbs of equipment, survive in harsh conditions, and 'shoot straight … at night, under pressure, and when exhausted and even frightened' (Collins 2005: 100). At a company level, Colonel Collins' requirement was that the basic building blocks of the advance to contact with the enemy, the set piece night attack and meeting the engagement had to be perfectly understood and practised such that the troops knew the actions to be taken in any situation. They had to be able to carry them out with minimum of instruction as there would be no time to think in a battlefield environment (Collins 2005).

In the remainder of this chapter we first look at team and collective training systems to get an understanding of their nature and complexity. This is followed by a discussion about Training Needs Analysis as a construct and the need for development of a methodology specifically focused on team and collective training. The chapter concludes with a brief introduction to the Team and Collective Training Needs Analysis Methodology and some suggestions for alternative strategies for reading the rest of the book.

Team and Collective Training Systems

In this section we examine the nature of team and collective training and explore the complexities of the training systems required to deliver it.

Team and Collective Training

The structure of the training solution that Colonel Collins adopted in order to prepare his core infantry companies for deployment in the Battlegroup reflected the structure of the companies themselves. Figure 1.1 shows the generic structure of the infantry companies in the Battlegroup.

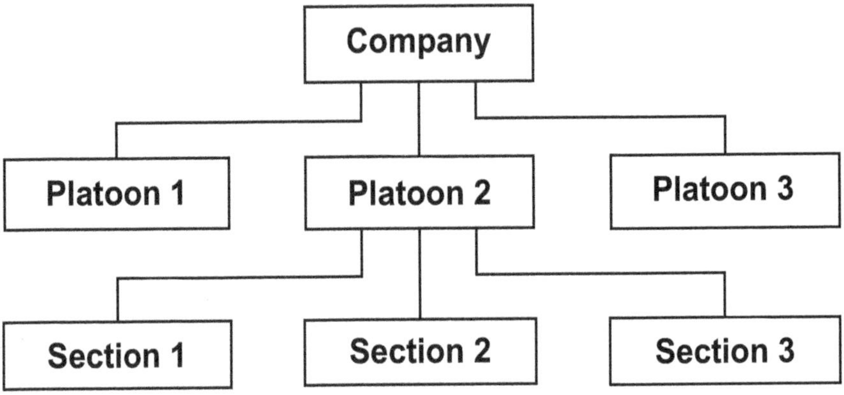

Figure 1.1 Infantry company structure

At the lowest level of aggregation are infantry sections. Each section is made up of a section commander, a deputy section commander and seven riflemen. The section commander and his deputy are responsible for the command and control of the section. The significance of this structure is that an infantry section can deliver significantly greater capability than can be delivered by nine infantrymen operating individually. For example, in order to neutralize an enemy gun emplacement the section can be split into two sub-teams: one team, led by the deputy section commander, to provide covering fire, whilst the other, led by the section commander, assaults the gun emplacement. At the next level of aggregation, three sections are combined to form a platoon under the command and control of a platoon commander and a platoon sergeant (the deputy). Platoon tactics, exploiting the coordinated action of three sections, enable more complex tasks to be undertaken. Similarly, at a company level, the employment of three platoons enables tasks of even greater scale and complexity to be undertaken. As the scale of aggregation increases, so does the command and control overhead. At a company level there is the company commander, his second in command, and a company sergeant major.

Training for each company began with individual training, with weapons training being a key component. This was followed by training at section level where section tactics were rehearsed. Once section training was completed, platoon level training was undertaken where the coordination of multiple sections undertaking platoon level tactics was practised. Finally, company level tactics were practised. Exercises took place both in the daytime and at night. This training sequence culminated in a live firing exercise on Sennybridge Training Area in Wales, conducted at company level. This exercise was supported by 105 mm light guns, 81 mm mortars and Milan wire guided missiles. Each of the companies in the Battlegroup had to advance under covering fire from medium machine guns and 0.5 inch heavy machine guns mounted on stripped down Land-Rovers (Collins 2005) and attack a wide range of targets.

The training solution which Colonel Collins adopted is representative of many collective training programmes in terms of the progression of training from individual to team and then aggregating teams together to train at more complex levels. Generally speaking the larger the team or collective organization then the greater the complexity of the tasks which it is required to undertake. This level of complexity is reflected in the complexity of the training system that is required to deliver suitable training.

Training System Components

The key components of a training system are shown in Figure 1.2.

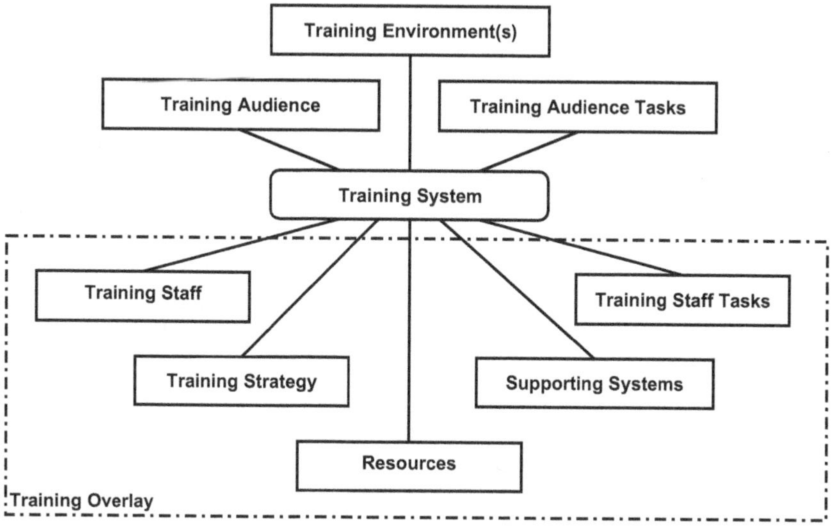

Figure 1.2 Training system components

Within a training system, the training audience undertakes a range of tasks in a training environment. This process is facilitated by the training staff undertaking their tasks in accordance with the training strategy and underlying methods, using a variety of training materials and assisted by supporting systems. In this book the term 'training overlay' is used to refer to the combination of the training staff, the tasks that they perform, the systems that they use to support the execution of their tasks, and the strategy and methods that they employ.

The complexities associated with each of the training system components for team and collective training include:

Training audience

The training audience may range from two to many thousands of people in number. As the size of the training audience increases, the range of ranks, trades and branches is likely to become more diverse. The natural processes that result in staff turnover (such as promotions, retirements, new people joining the organisation and personnel being allocated to new roles) mean that the experience levels within a team are dynamic. As a result, it can be difficult to characterise the input standard of a team or collective organisation prior to the start of training. Another confounding factor for team and particularly collective organisations is the availability of the training audience for training especially where different elements are drawn from different parts of the organisation and may have quite different demands on their time.

Training tasks

As the size of the team/collective organisation increases, so the complexity of the tasks which they undertake increases. Successful team/collective task execution depends on more than the cumulative effect of each team member executing their individual tasks; interactions between team members and between teams play a critical role. Integration of teams becomes a focus of training. In the military context, the team and collective training challenge is exacerbated by the ever-increasing complexity of contemporary warfare (particularly with the ever-increasing presence of asymmetric threats). This was characterised by Lt Gen Newton (Commander Force Development and Training, British Army) in his opening address to the International Training and Education Conference 2010 as a 'wicked problem'.

Training environments

The training environments for team and collective training are necessarily more complex and of greater scale than those required for individual training, because the task environment for each individual in the team/collective organisation has to be replicated. This is not to suggest that training environments for individual training cannot be complicated. A flight simulator for a single seat military aircraft such as the Lightning II (F-35 Joint Strike Fighter) would be a case in point. However, training Lightning II pilots to operate in a four-ship formation would require four aircraft or four such simulators connected together. On a similar theme, the British Army Combined Arms Tactical Training system, which enables the vehicle crews of an armoured battlegroup to train together in a synthetic environment, is made up of over 100 vehicle simulators networked together. The equivalent live training requires a large training area (typically of many tens or hundreds of square miles in size) and appropriate numbers of armoured vehicles, both for the training audience and the enemy forces.

The provision of suitable live training environments can be particularly problematic. As an example, the Royal Air Force has a requirement for aircrew to train in high altitude, hot environments. Such conditions do not exist in the UK so formations of aircraft have to travel to where such conditions can be accessed, such as in the United States. The provision of training areas to support joint exercises where land, air and maritime forces are exercising together can be challenging. In the UK there few places where land, air and maritime training areas are co-located.

The representation of weapons effects can be particularly challenging for military training. Live firing exercises, such as that undertaken by 1 Royal Irish before their deployment to Iraq, can only take place on specialist ranges and require a significant number of safety staff to oversee. However, such live firing can only take place against unmanned targets. A typical solution to this problem is to simulate the effects of weapons by such means as attaching lasers to weapons and having sensors attached to the equipment and vehicles of the participants which give an audible or visual indication of a hit. Such systems can also be enhanced by having data links from the components in the system so that position and event data (such as shots fired and hits) can be recorded for subsequent analysis and replay.

Training staff

The delivery of team and collective training requires a team of training staff, with a range of different skill sets. The number of staff supporting a large-scale, collective exercise may run into many hundreds of people with a wide variety of skills and backgrounds. As an example, the United States Navy Training Group Atlantic run large-scale synthetic training exercises for aircraft carrier task groups. These entail all of the ships in such a task group moored in the naval base in Norfolk, Virginia being connected into a synthetic training environment which provides synthetic inputs into all of the sensors on the ships and collecting data from all of the weapons systems. The training audience is comprised of all of the ships' warfare teams and the task force headquarters staff. Running the exercise requires in excess of 250 staff. Of these, approximately 50 are observer controllers who spend their time on board the ships, monitoring the training audience. These are drawn from an appropriate mix of specialisations which match the members of the training audience that they are monitoring (such as intelligence specialists, warfare officers and engineers). A similarly sized team is required ashore to actually run the exercise and control the synthetic environment to deliver an appropriate scenario. They are supported by a large team of role players who manoeuvre the synthesised opposing force ships and aircraft.

Training staff tasks

Training delivery typically necessitates the execution of instructional functions such as briefing, monitoring, evaluation and After Action Review (AAR) of a complex team undertaking a complex task in a complex environment. Significant effort is required to manage the training environment to ensure that events are controlled and timed to maximise the training effect. Furthermore, whilst individual training is typically analysed and designed once and then delivered many times, team and collective training events are often bespoke events with scenarios developed or adapted to meet the specific requirements of each team/collective organisation requiring training. Therefore, training staff are often involved in analysis and design as well as delivery and evaluation.

Supporting systems

Complex supporting systems are often required to support training staff tasks, such as systems for managing the training environment and for AAR. For example, the briefing, monitoring and debriefing of an armoured battlegroup exercise conducted at the British Army Training Unit Suffield in Canada requires a team of 20 to 30 training staff on the ground. They communicate via radio, both with each other and the team in Exercise Control who have a god's eye picture of what is happening on the exercise, generated from data collected from the weapons effects simulation systems. They are also equipped with laptops which have a feed from the same system, giving them a view of what is happening across the exercise. When a significant event occurs, such as when a noteworthy tactical engagement or manoeuvre takes place, they can also communicate with a team of analysts in Exercise Control who can record data for use in the subsequent AARs. Once the exercise is over, AARs can be delivered at different locations in the field with presentations which include replays of the tactical engagements downloaded from Exercise Control to mobile AAR theatres mounted on the trailers of large trucks. This saves the whole battlegroup having to recover to a central point for AAR then redeploy into the field for the next phase of the exercise.

Training resources

The conduct of team and collective training exercises can require a wide variety of resources, particularly when conducted in the live environment. For example, following an armoured battlegroup across a training area is not something that can be done on foot, so training staff travel in four-wheel drive vehicles. In addition, such exercises typically require the production of a significant amount of supporting training material. For the training audience this can include information about the scenario which they are going to experience in the form of background information and formal orders from which they may be required to produce plans of action. For the training staff this may include detailed information about the scenario including events described in a master scenario events list and assessment information.

Training strategy and methods
At a very high level the development of a training strategy and the selection of training methods for team and collective training may seem to be relatively simple, in so far as it will generally involve putting the team into a representative environment and getting them to carry out the task for which they are being trained. However, given the complexities of all of the elements of the training system described in the paragraphs above, there are typically a significant range of constraints, assumptions, risks and opportunities that have to be identified and taken into account when developing the training strategy and methods. Identifying all of the requirements in terms of the detail of what exactly has to be trained, determining how training should be decomposed (especially where multiple teams are involved), identifying all of the resource requirements, both for the training environment and the training overlay, and producing a solution that meets all of the constraints and exploits all available resources represents a significant undertaking.

Training Needs Analysis

Training Needs Analysis (TNA) is the name given to the systematic process of analysing training requirements, identifying possible training solutions to meet these requirements, and identifying the most appropriate solution for the organisation. In the United States in particular it is also referred to as Front End Analysis. TNA can be viewed as a subset of the processes in the Systems Approach to Training (SAT) model, from which it is derived.

The Systems Approach to Training

SAT, also referred to as Instructional Systems Development, has an established tradition in industry, government and educational settings for providing a framework to guide the systematic design, development and management of training and education courses. The North Atlantic Treaty Organisation (NATO) Training Group (1983) characterised the systems approach as a logical approach to problem solving with the following components:

a. defining the problem to be solved in the clearest possible terms;
b. considering every available method by which the problem could be solved, selecting and implementing the preferred method;
c. monitoring the effectiveness of the method adopted and incorporating modifications as required.

These components have been incorporated into a wide variety of SAT models over the last 40 years or more across the academic, commercial and military domains. Neil (1970) describes the development of a SAT model during the formation of the

Open University in the UK in the late 1960s. In the same era, Budget et al. (1970) describe the application of SAT in the Royal Navy. By 1980 over 60 different SAT models had been identified (Andrews and Goodson 1980) and the growth continues, including the development of contemporary models such as those proposed by Gagne et al. (2005) and Dick et al. (2005). Whilst the various models in existence vary in their degree of process decomposition and in the sequencing of process sub-components, the key SAT constructs can be illustrated by the generic model shown in Figure 1.3. Within SAT there are four main phases of activity or processes: analysis, design, delivery and evaluation. These are linked sequentially, starting with analysis.

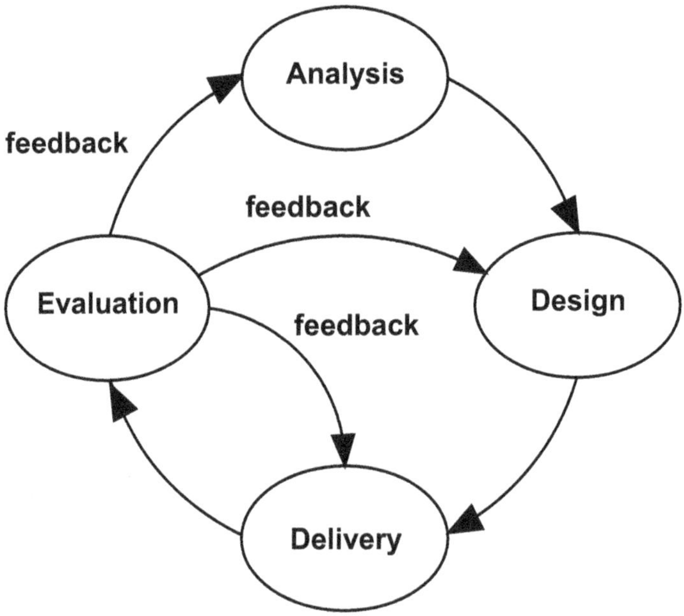

Figure 1.3 Generic SAT model

The analysis phase is concerned with the identification of the capabilities which the training audience should have on completion of training and the capabilities that they have on arrival. The gap between the two determines the instructional requirement and is typically expressed in terms of objectives which specify: the performance required, the conditions or environment in which the performance should be executed, and the level or standard of performance required (which informs assessment). For example, if a pilot wishes to gain an instrument rating, then one of the tasks that he or she has to be able to carry out is to fly an Instrument Landing System approach. In order to pass the instrument rating test this approach has to be flown to the required standards, which include flying within a specified speed tolerance and an angular deviation tolerance from the required flightpath. Such analysis also needs to identify constraints which must be considered during subsequent design and delivery such as cost, time, resource availability and the numbers and geographical distribution of the training audience.

The design phase follows on from the analysis phase and is concerned with the development of a course of instruction and assessment to meet the objectives identified in the analysis phase, taking account of the constraints that were also identified. Design can occur at a number of levels. At a high level, design is concerned with the determination of the overall strategy for instruction, which would typically include consideration of training methods and media, course structure and sequencing, and assessment strategies. The selection of appropriate training media is a critical step. For example, if an individual is to be trained to fry an egg, most of us would probably expect an egg, a frying pan and a stove to be involved. The related training method would most likely be a demonstration, followed by the opportunity for the student to practise. In this light it is interesting to reflect on the popularity of cooking programmes on television and reflect on whether the viewers ensconced on their couches watching the programmes turn into better cooks as a consequence of watching them. Certainly they get to see accomplished chefs demonstrating the appropriate procedures for preparing a dish with a high degree of skill, and they can see what the dish looks like at the end of the process. If they record the programme they can replay it to get step by step guidance as they try to produce the dish themselves. The one weakness in this is that when the final dish is presented, if they lick the television screen they get no impression of what the dish should taste like! At a lower level, design is focused on the development of the instructional and assessment materials for each lesson or instructional event. Consideration also needs to be given to instructor training requirements to ensure that the instructional team are capable of delivering the instruction as designed. Once design has been completed, training delivery can take place.

The final phase is evaluation. Evaluation can encompass both evaluation of the course outputs and the evaluation of the execution of the SAT processes themselves (Gagne et al. 2005). The evaluation of course outputs can include student performance assessments during the course, student feedback about the course, evidence of the skills taught being applied in the workplace following training, and evidence of the impact of the deployment of these skills on organisational outputs (Kirkpatrick 1998). The results of evaluation provide feedback into the analysis, design and delivery processes as shown by the feedback arrows in Figure 1.3. This closing of the loop should ensure that the system is self-correcting.

The benefits of applying such a systems approach to the development of instruction are commonly acknowledged to be: it focuses attention at the outset on the capabilities that students should have at the completion of the instructional process; the linkages between the stages ensure that the instruction that is developed is targeted specifically at delivering the required outcomes; and the systematic process is repeatable and auditable (Dick et al. 2005; Gagne et al. 2005; and Patrick 1992). Dick et al. (2005) note that whilst the approach has been widely adopted amongst educators, the greatest take up may be found in industry and military organisations. They attribute this to the premium that is placed in these environments on efficiency of instruction (which may be both in terms of cost and time) and the quality of student performance.

The Requirement for Training Needs Analysis

Having established that SAT provides a sound and widely accepted approach for the development and management of training, it begs the question as to why a subset of the analytical techniques should be extracted and separately identified as TNA? The answer to this question lies in the requirement for critical decisions that have to be made during the design phase of the SAT cycle related to expenditure and resource allocation.

Within commercial and military settings, the costs and resource requirements associated with development of training to support new systems that are being introduced or to support changes in business needs can be substantial. Typically these can include the acquisition of simulators and part-task training devices, infrastructure development (such as the construction of facilities to accommodate simulators, and associated briefing and AAR systems) and staff training. Consequently, it is essential that the ranges of options for delivering the required training effect are critically evaluated in terms of their capability to deliver the required output and their costs (both financial and logistic). Costs have to be determined for both the initial acquisition of the training system and its operation and updating throughout the lifetime of the system that it is supporting. It is this need to identify and critically evaluate training system options before significant training investment decisions are made that has led to the development of TNA as a defined subset of SAT processes, to support this critical decision point in training system acquisition. The mapping of TNA onto the SAT cycle is shown in Figure 1.4.

TNA embraces all of the elements of analysis, plus the higher level components of design sufficient to specify the requirements for a training system and evaluate putative options against these requirements. The importance of TNA is such that the United Kingdom Ministry of Defence (UK MOD) has mandated that a TNA should be conducted whenever the requirement for a training intervention has been identified to address a change in operational or business needs (UK Ministry of Defence 2014).

There is another practical aspect to the definition of TNA as a self-contained process in its own right. It is not uncommon for this analytical activity to be contracted out either to a main contractor that is supplying a new system to the organisation or to an independent company specialising in this type of analysis. For this purpose it is helpful to have a clearly defined procedure with distinct outputs against which a contract can be placed, and the outputs evaluated.

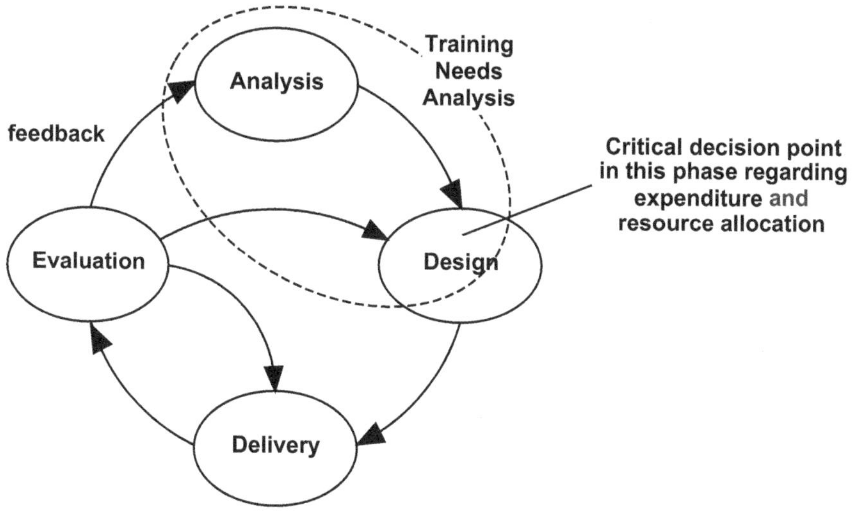

Figure 1.4 The mapping of TNA onto the generic SAT model

The Requirement for a TNA Method Targeted at Team and Collective Training

TNA as a construct has been exploited to great effect by organisations such as the UK MOD for over 20 years, particularly in the context of training to be provided for new equipment. The analytical methods that make up TNA are those that are employed in the SAT implementations from which they are derived. The main issue from the team and collective training perspective is that the analytical techniques that underpin contemporary SAT models, such as those detailed by Gagne et al. (2005) and Dick et al. (2005), and military equivalents such as the UK MOD Defence Systems Approach to Training Model (UK Ministry of Defence 2007), are aimed at individual training. They do not cater for the complexities of team and collective training systems as described above. Furthermore, whilst there is a significant body of research on team performance, including notably the outputs of the TADMUS programme funded by the US Navy (Johnston et al. 1998), few analytical techniques have been developed that could be exploited in the conduct of TNA for team and collective training.

The main developments have been in the area of team task analysis. Examples of new methods include Hierarchical Task Analysis for Teams (Annett et al. 2000) and Team Cognitive Task Analysis (Klein 2000). Hierarchical Task Analysis for Teams extends conventional Hierarchical Task Analysis with the inclusion of narrative descriptions of the communication and coordination required between team members to achieve each of the goals in the hierarchical description of the task. Team Cognitive Task Analysis extends the Critical Decision Method to include the capture of information about shared mental models and shared situational awareness. There has also been limited development of approaches to identify gaps in the capability of extant training environments. Examples include the Task and Training Requirements Methodology (Swezey et al. 1998) and Mission Essential Competencies (Alliger et al. 2003). These methods rely on ratings of the utility of the extant training environments by subject matter experts and do not address the specification of training environments for new systems with potentially new concepts of operation, such as the Queen Elizabeth-class aircraft carriers being acquired for the Royal Navy. The analysis of training overlay requirements does not appear to have been addressed at all.

Team and Collective Training Needs Analysis

The Team and Collective Training Needs Analysis (TCTNA) methodology described in this book has been developed to provide a TNA methodology that addresses the complexities of team and collective training. The overarching analytical framework is shown in Figure 1.5. The key points to note are:

- There are separate but interlinked stages which focus on the analysis of the team and collective task, the training environment and the training overlay. Each of these stages yields a comprehensive set of requirements for the training system.

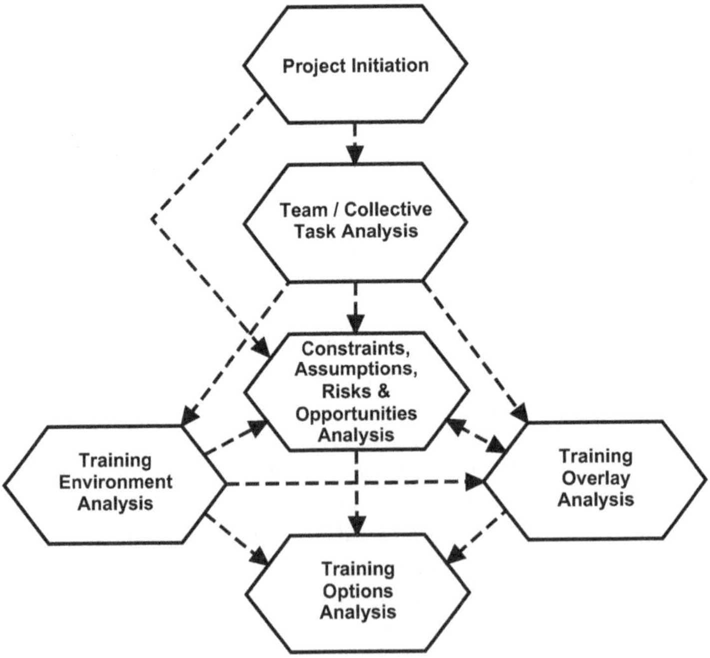

Figure 1.5 TCTNA analytical framework

- Constraints, Assumptions, Risks and Opportunities (CARO) Analysis captures all of the factors which impact on the development of a suitable training strategy and the viability and suitability of alternative training options.
- Training Options Analysis is concerned with identifying training options which can meet the requirements identified in the previous analytical stages and determining the optimal training solution for the organisation.
- The undertaking of a TCTNA is likely to be a complex task and needs to be set up correctly from a project management perspective, hence the project initiation stage which has to be completed before the analytical work commences.

In order to develop analytical tools to support each of these process steps, it has been necessary to develop a number of models to underpin the analysis process. Figure 1.6 shows the five models that underpin the TCTNA method and their relationship to each other.

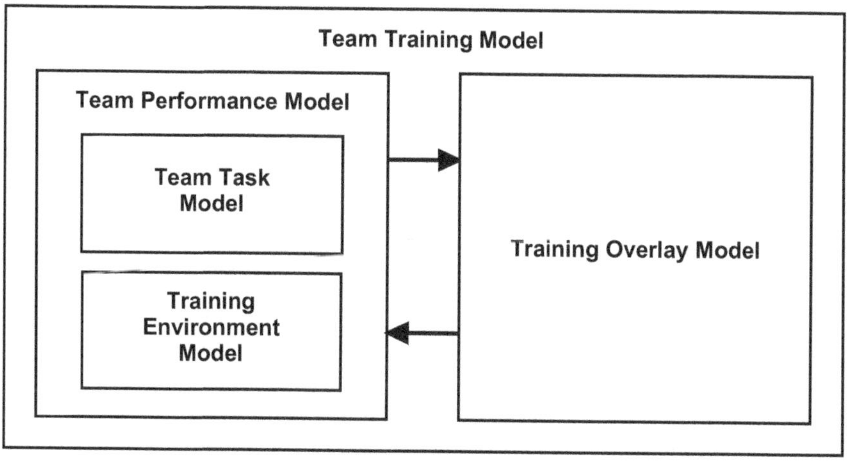

Figure 1.6 Models underpinning TCTNA

The purposes of these models are as follows:

- *Team Task Model* – captures the types of activities which a team engages in to undertake a task and the critical links with environmental conditions. This model provides the analyst with a 'shopping list' of activities and categories of environmental conditions to search for when analysing a task. This supports Team/Collective Task Analysis.
- *Team Performance Model* – shows that team task activity is linked to the environment and also captures the team member and team attributes that impact on performance. This supports Team/Collective Task Analysis.

- *Training Environment Model* – provides a decomposition of the training environment into components that are required to replicate the task environment for training purposes, each of which can be specified in detail, including fidelity requirements. This supports Training Environment Analysis.
- *Training Overlay Model* – provides a detailed breakdown of all of the elements of the training overlay in terms of tasks and resources that are required for training to be delivered effectively. This supports Training Overlay Analysis.
- *Team Training Model* – simply combines the Team Performance Model and the Training Overlay Model to provide a detailed view of all of the components that have to be captured in a TCTNA.

Approaches to Reading the Book

Figure 1.7 shows the structure of the remainder of the book. It is split into two parts: Part I (Chapters 2–6) covers the theory and models that underpin the TCTNA methodology; Part II (Chapters 7–13) presents the TCTNA methodology itself.

There are a number of different ways in which we suggest that you can work through this book:

- To gain a rigorous understanding of the TCTNA methodology we would suggest reading all of the chapters in order.
- If you want to get a feel for the methodology before delving into the detail, then you can read Chapter 7 to get the overview, then have a look at the case studies. This will give you a sense of the approach and an impression of what the outputs look like.
- If your interest lies in one particular aspect, such as the analysis of training environments, then we would suggest reading the chapter explaining the relevant underlying model, then reading the analysis chapter of interest and, finally, looking at how it has been applied in the case studies (reading Chapter 5, Chapter 11 and then the relevant sections of the case studies would provide a good understanding of Training Environment Analysis).
- To get a full appreciation of the view taken on the nature of team tasks, it is worth reading Chapters 2–4 in sequence. The team task model developed in Chapter 3 is an extension of the individual task model developed in Chapter 2. Chapter 4 then explores team coordination and adaptability in greater detail, and puts it into the context of task analysis.

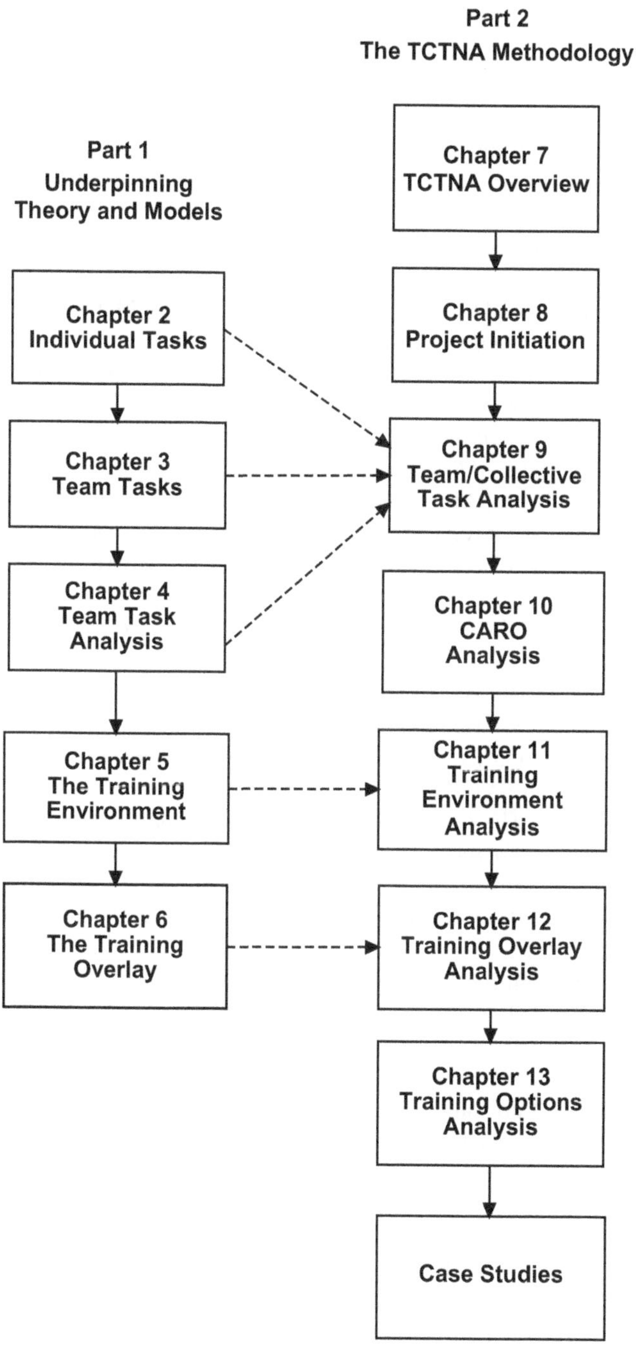

Figure 1.7 The structure of the book

References

Alliger, G., Garrity, M.J., Morley, R.M., McCall, J.M., Beer, L. and Rodriguez, D. (2003) Mission Essential Competencies for the AOC: A Basis for Training Needs Analysis and Performance Improvement, *Proceedings of the Interservice/Industry Training, Simulation and Education Conference*, Orlando, FL, December 2003.

Andrews, D.H. and Goodson, L.A. (1980) A Comparative Analysis of Models of Instructional Design, *Journal of Instructional Development*, 3: 2–16.

Annett, J., Cunningham, D. and Mathias-Jones, P. (2000) A Method for Measuring Team Skills, *Ergonomics*, 43(8): 1076–94.

Budget, R.E.B., Morse, S.L. and Stevenson, P.M. (1970) Applying a Systems Approach in the Royal Navy, in Romiszowski, A.J. (ed.) *The Systems Approach to Education and Training*. London: Kogan Page. 34–46.

Collins, T. (2005) *Rules of Engagement: A Life in Conflict*. London: Headline.

Collyer, S.C. and Maleki, G.S. (1998) Tactical Decision Making Under Stress: History and Overview, in Cannon-Bowers, J.A. and Salas, E. (eds) *Decision Making Under Stress*. Washington, DC: American Psychological Association. 3–15.

Dick, W., Carey, L. and Carey, J.O. (2005) *The Systematic Design of Instruction*, 6th edn. Boston, MA: Pearson.

Gagne, R.M., Wager, W.W., Golas, K.C. and Keller, J.M. (2005) *Principles of Instructional Design*. Belmont, CA: Wadsworth.

Johnston, J.H., Cannon-Bowers, J.A. and Salas, E. (1998) Tactical Decision Making Under Stress (TADMUS): Mapping a Program of Research to a Real World Incident – The USS Vincennes, in *North Atlantic Treaty Organisation Research Human Factors and Medicine Panel Symposium*, Edinburgh, 20–22 April 1998.

Kirkpatrick, D.L. (1998) *Evaluating Training Programs*, 2nd edn. San Francisco, CA: Berret-Kohler.

Klein, G. (2000) Cognitive Task Analysis of Teams, in Schraagen, J.M., Chipman, S.F., and Shalin, V.L. *Cognitive Task Analysis*. Mahwah, NJ: Laurence Erlbaum Associates. 417–30.

NATO (1983) *The Principles of the Systems Approach to Training*, NATO Training Working Group Publication No. 2, NATO.

Neil, M.W. (1970) A Systems Approach to Course Planning at the Open University, in Romiszowski, A.J. (ed.) *The Systems Approach to Education and Training*. London: Kogan Page. 59–67.

Patrick, J. (1992) *Training Research and Practice*. London: Academic Press.

Swezey, R.W., Owens, J.M., Bergondy, M.L. and Salas, E. (1998) Task and Training Requirements Analysis Methodology (TTRAM): An Analytic Methodology for Identifying Potential Training Uses of Simulator Networks in Teamwork-intensive Task Environments, *Ergonomics*, 41: 1678–97.

UK Ministry of Defence (2007) Joint Service Publication 822, Part 5, Chapter 3, Defence Training Support Manual 3, *Training Needs Analysis*, 2007.
——— (2014) Joint Service Publication 822, The Governance and Management of Defence Training and Education – Part 6, Defence Training and Education Capability Rules, Issue 1.1.

Chapter 2
Individual Tasks

Introduction

In the next few chapters we describe a set of models that underpin an approach for describing tasks, team performance and team task processes in a format useful for conducting TNA for teams and collectives as 'teams of teams'. The purpose of developing these models is to provide a theoretical basis for analysing and decomposing team and collective tasks. This task analysis approach enables linkages to features and characteristics of the environment within which the task is conducted, and to instructional tasks and broader training strategy.

Let us take the example of a pilot and co-pilot team being trained in an aircraft simulator by instructional staff. Task analysis describes what tasks the aircrew will be performing in operating the aircraft and diagnosing faults that may occur in flight. The environment in this situation is the simulator, which has a set of features or characteristics that may be presented to the trainees. Simulator features may include the replication of aircraft movement through a motion platform, or the ability to display visually a large number of real airports in different weather conditions and at different times of year. There are also a number of instructional tasks to be considered: aircraft faults must be selected and implemented on the aircraft simulator for the aircrew to diagnose and react to. Similarly weather conditions and airport location must be selected. These instructional tasks configure the environment and put it into a state which is suitable for trainees to react to. In this example, the simulator is a proxy environment for a real aircraft in flight. Some features of the real operational environment are preserved: the aircraft simulator may move, and the multifunction displays show appropriate values and behaviour. Other features of the operational environment are not preserved; in a simulator, if the crew crash human lives are not lost, and 'operating' a simulator does not consume jet fuel, but rather other resources such as electricity. Instructor tasks in this example go beyond the operation of the simulator; instructors are involved in assessing performance and supporting practice possibly through activities such as demonstration.

In a TNA context, analysis of the trainee team tasks is associated with the tasks instructors have to perform to establish and run training events that support trainee task practice. Task analysis is also associated with training strategy through related activities, such as the identification of assessment criteria for task processes and task products. Through the analysis and decomposition of tasks one can also accurately define which teams or individuals are involved in task performance, which leads to a functional definition of the scope of the training audience. In the case of the aircrew operating the aircraft and the aircraft simulator, the extent of the potential training audience is fairly clear-cut; we are dealing with the aircrew who will be flying the aircraft. In this situation we may need to represent air traffic control as external role-players to respond to aircrew queries and issue air traffic messages. This may be an additional instructional task that instructors need to take on and perform.

The determination of a potential training audience becomes more complex where team tasks are delivered by groups of units or platforms such as aircraft or ships. Here one must perform a task analysis first, and from that analysis determine who the personnel involved in the task are; rather than simply defining certain teams as constituting the training audience. A functional definition of the training audience is significant because it immediately mitigates early assumptions about the extent of the potential training audience. Unchallenged assumptions in this area can lead to the creation of stovepipe training environments which only cater to a subset of the actual collective training audience.

One of the considerations inherent within team and collective TNA concerns the unit of analysis and specifically the unit of performance. Are we looking at individuals or teams or both? Collectives are teams of teams, with teams consisting of a group of individuals. Within a team, individuals are clearly still conducting individual activity while the team itself is generating effects at a broader level. This means that team performance includes an element of individual performance, but clearly there is more to team performance than just simply the aggregation of individual performance. In the example of two individuals carrying a stretcher, both individuals must lift, and they must lift together and walk in the same direction (at the same speed) to perform this task. We can see that coordination and synchronisation of action is required even in simple team task examples. One might ask, by what processes or mechanisms is coordination and synchronisation of action generated and maintained? The question of how exactly the whole may become greater than the sum of the parts goes to the heart of the team training problem and touches on the critical aspect of teamwork as part of team performance.

This means that team TNA must capture (at some level) elements of individual performance embedded in team performance. The models of individual and team performance adopted for TNA should be clearly relatable such that the interface between individual, team and collective performance and training can be identified. Without the ability to relate analyses performed at the individual and team level one risks duplication of effort and a potential lack of coherency between concepts contained within teamwork and task measures. This is a significant issue in large procurement projects where multiple TNAs may have overlapping coverage.

The ability to relate between individual and team performance is also important when one considers that only individuals acquire new abilities in the areas of knowledge, skills and attitudes (KSAs). Learning processes are always addressed at the individual level, because learning is a change that occurs within the individual, based on their individual experience of a situation, even if that situation arises during participation in team activity. Teamwork is a complex concept which has been described in terms of team member attributes, interactions, processes, functions, strategies, behaviours, team properties, and products (Salas et al. 2005). Aspects of teamwork range from individually held KSAs that enable teamwork, for example communication skills or the attitude of supportiveness, through to observable interactions, behaviours or processes that occur between individuals such as communication, or performance monitoring (Salas et al. 2005). Teamwork concepts may also be taken to embrace elements of cognition such as situational awareness (Bowers et al. 1993), which may be shared or distributed (Salmon et al. 2009). Teamwork concepts may also describe the results or outputs generated by the team (teamwork products) which may include team characteristics such as cohesion, or task products such as the physical effects generated by the team (Annett 1997).

Unfortunately over time multiple studies and theories of teamwork have generated a proliferation of teamwork-related concepts with a lack of consistency in labels and definitions, leading some authors to question the utility of the teamwork literature (Salas et al. 2005). For this reason we anchor our teamwork-related concepts in specific areas, for example referring to team-task processes rather than using the broad umbrella term 'teamwork'.

The approach we adopt in the next two chapters is to suggest answers to three questions that we believe are highly relevant to training needs analysis as we move from an individual task to a team or collective task.

- Firstly, what is the relationship between tasks and the environment in which they are conducted? And how can this relationship be used in defining a model of individual performance of a task within an environment?
- Secondly, what changes occur when we move from an individual task to a team task? Answering this question looks at the types of activity that are seen within teams but are not seen in an individual context; describing how team task processes performed by multiple individuals are different from individual task processes performed by a single individual. In answering the second question we define a team task model, which is an extension of an individual task model. Within this team task model, a set of team coordination enablers are described, which we compare with the functions of command and control (C2) (Alberts and Hayes 2006).
- The third question is how to define a team performance model in accordance with the answers to the previous two questions. Team performance is of interest to TNA for a number of reasons; it identifies what must be addressed in team training and what must be supported in team practice. The construction of a team performance model is of relevance for TNA as it leads to the development of a framework of potential team performance *measures* forming the basis for what may be assessed in team training. Team performance measures while implemented in the delivery of team training, must be specified in the analysis and high-level training design phases contained within the TNA process.

Assessment strategy is closely linked with the specification of team performance and team performance measures, as making decisions of *what* to measure is closely allied to the *means* of measurement. Returning to our flight simulator example – our simulation environment should present data on the aircraft state which is relevant for assessment. Similarly it is not just aircraft position and speed we are interested in, but the flight path over time and other task critical variables. While this data may exist at an engineering level it is important that this data can be accessed and analysed by instructors who can apply judgement to it.

We are not just interested in the state of the environment over time, but also in being able to capture the interactions between the team. As an example, if we had a requirement to analyse communication frequency and content it would be a very poor simulator that did not offer us this data via a recording and playback facility. This impacts instructional strategy as instructors must have appropriate knowledge, skills, experience and supporting tools (and systems), and have access to the performance data on which assessment is based. Provision for data access may involve the instrumentation of training environments to support the capture of performance-relevant data, and catering for potential data playback to support post-performance debriefs or AAR. A simple example of this might be found on a rifle range – the environment is instrumented because targets have marked zones which have a denoted value. This performance data has to be accessed by instructors via a spotting scope or binoculars to enable judging, and the paper targets may themselves be used for debriefing students to point out errors that are suggested by the student shot grouping pattern.

A team performance model and associated team performance measures can inform the instructional content for the knowledge-based theory component of team training, and may also support internal team learning processes by helping to provide a shared mental model of team performance to team members. Finally as team performance is supported by a number of individually held KSAs, a clear articulation of the elements of team performance helps identify potential behavioural indicators of positive team performance, or areas for improvement. A shared model of team performance also helps instructional staff anchor their observations of team performance in appropriate and consistent areas. Twelve areas of potential team performance measurement are identified, and parallels with the findings of previous research within the teamwork literature are highlighted. These areas, identified from a synthesis of previous research in teamwork and command and control are:

1. individual team member coordination-based KSAs
2. individual team member communication KSAs
3. role assignment within the team
4. team level communication channel deficiencies
5. distributed perception processes
6. team situational awareness quality
7. goal generation processes
8. team goal quality

9. collaborative planning processes
10. team plan quality
11. leadership
12. team learning processes.

Tasks in the Environment

Tasks, whether individual, team or collective, can be viewed as occurring within a context or environment, where task performance acts to change the environment in some way. This statement applies whether we are boiling an egg or conducting strategic level military tasks. Figure 2.1 illustrates this view of the task of boiling an egg.

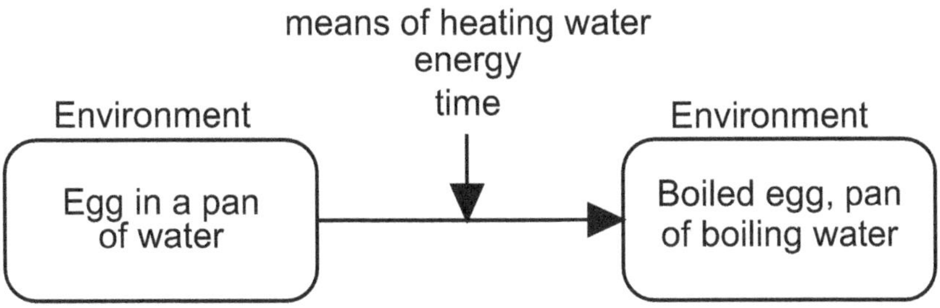

Figure 2.1 The task of boiling an egg

The task outcome is the state that the environment is in following, and generated by the completion of the task. Other synonymous terms for task outcome are task product, task effect or task output.

The term 'goals' is frequently used in task analysis (Kirwan and Ainsworth 1992: 1); we distinguish outcomes from goals on the basis that outcomes are environmental states whereas goals belong to individuals operating in the environment. As an example, the result of a lottery draw as a set of numbered balls appearing in a hopper is an outcome – an environmental state. Each individual that entered that lottery had their own specific goal which corresponded to the numbers that they held, which they wanted to be drawn. Goals as statements of intent held by individuals may be satisfied (or not) depending on whether they are enacted in the world as outcomes, and where we have multiple individuals we may have multiple goals. Robert Mager drew a similar distinction between goals and outcomes when he stated 'In other words, be able to describe specific outcomes that, if achieved, will cause you to agree that the goal is also achieved' (Mager 1972: 11). We discuss goals later in this chapter.

Task conditions or task inputs describe the state of the environment when we commence the task. These conditions may also be direct task triggers that cause us to undertake the task, as tasks may be reactive to a situation that presents itself. Task resources (usually in the form of equipment) are also an aspect of the environment before we commence our task. There is often interplay between task conditions and task resources which generates the context of the task. One example would be the relationship between weather (a condition) and clothing worn (a resource).

Task performance or task processes include the action parts of the task which act to change our initial environmental situation (conditions and resources) to task outcomes. The term 'initial' is relative to a specific defined timeframe. Task conditions and resources as well as describing the environment at the point we commence the task, also describe the dynamic elements of the environment which may impact on task processes, for example a gusting cross-wind will affect the process of landing an aircraft. In most tasks the environmental transformations generated by task performance are due to a combination of human decision making and technology.

Figure 2.2 summarises an input-process-output model of tasks; similar theoretical approaches can be seen in other task analysis methods such as the Precursor-Action-Results-Interpretation (PARI) method (Jonassen et al. 1999: 121). Tasks can be viewed as transformations of the initial environmental state through task processes to a task outcome, a subsequent (transformed) state of the environment. Task conditions and task resources impact on task processes and also comprise part of the initial environmental state.

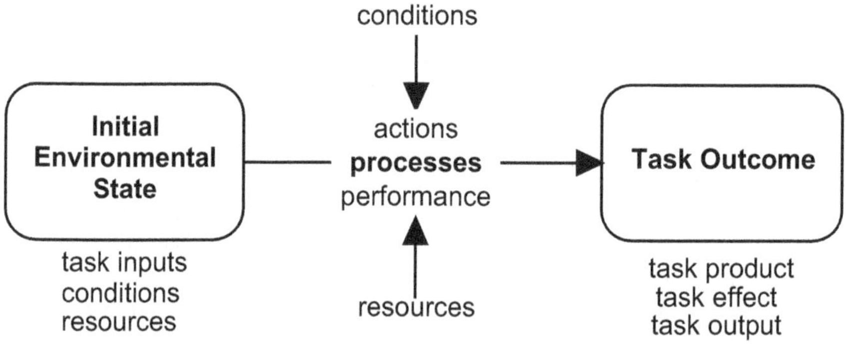

Figure 2.2 Tasks viewed as a transformation of the environment

The diagram in Figure 2.2, however, is not the whole story. While any particular instance of a task is based on a specific initial environmental state, all tasks of any complexity beyond boiling an egg are conducted under a range of conditions. This means that there may be a numerous and varied range of potential initial environmental states that need to be catered for within task performance. Take parachuting as an example – parachute jumps may be conducted from a range of altitudes, during the day or at night, with a range of equipment and a parachute which may or may not have been packed correctly.

This range of conditions describes a variety of potential initial environmental states for the parachuting task.

In the egg-boiling task, conditions such as the egg having gone 'off', it being broken or beyond the 'use by' date tend to lead to the task being abandoned. In the case of not having any water we could potentially fry our egg and so on, but since we have defined this task as the 'egg-boiling' task, an egg-frying task would be a different task leading to a different outcome state (a fried egg rather than a boiled egg). In these examples we see that the initial state of the environment and the presence or absence of resources impacts on task performance. We also see that when a particular type or means of environmental transformation is unavailable (egg boiling), an alternative transformation (egg frying) may be substituted as it generates an outcome that shares some properties with our ideal outcome.

Task Outcomes

While there are a number of potential initial environmental states for the parachuting task, the *intended* task outcome of the parachuting task is always the same – to have landed under a full canopy at the preferred landing spot, not having collided with any other objects on the way down, generally via flying into the wind and flaring the canopy at the appropriate moment. Success in this task always satisfies these criteria. Of course our *intended* outcome – what we would *like* to happen – is not the same as what *actually* happens (the actual outcome).

In the parachuting example to take our intended outcome statement 'under a full canopy' – not satisfying this criterion is the product of various accident situations:

- parachute malfunctions where the canopy does not deploy, deploys incorrectly, or both main and reserve parachutes deploy together;
- a low-pull (or no-pull) situation where the parachute does not deploy because the parachutist does not activate main or reserve, or either parachute deploys late;
- the parachutist separating from the parachute harness;
- mid-air entanglement or free-faller collision with parachutist open canopy.

To take our statement on 'preferred landing spot', not satisfying this simple criterion takes us into the richness of the world outside the drop zone: the runway, roads, barbed wire fences, trees, fields of staked crops and so on. Intended outcomes satisfy very restricted criteria, leaving a vast number of permutations free for unintended or suboptimal outcomes. Reading through the list of non-intended task outcomes above informs the topics for instruction for basic parachuting safety, imparting the knowledge and skills to prevent and manage these situations.

Between any initial environmental state and the task outcome we can define intermediate outcomes – in our parachuting example these would be states such as starting parachute deployment, parachutist under main parachute canopy and the points for commencing downwind, cross-wind and upwind sections of flight under canopy. These intermediate outcomes also describe potential accident nodes; the situation where a parachute malfunction occurs, where the parachutist initiates a cutaway procedure.

Figure 2.3 illustrates how a task generates a task outcome from a range of potential initial environmental states through task processes over time. The circled set of boxes in the centre of the diagram are the potential intermediate outcomes of the task.

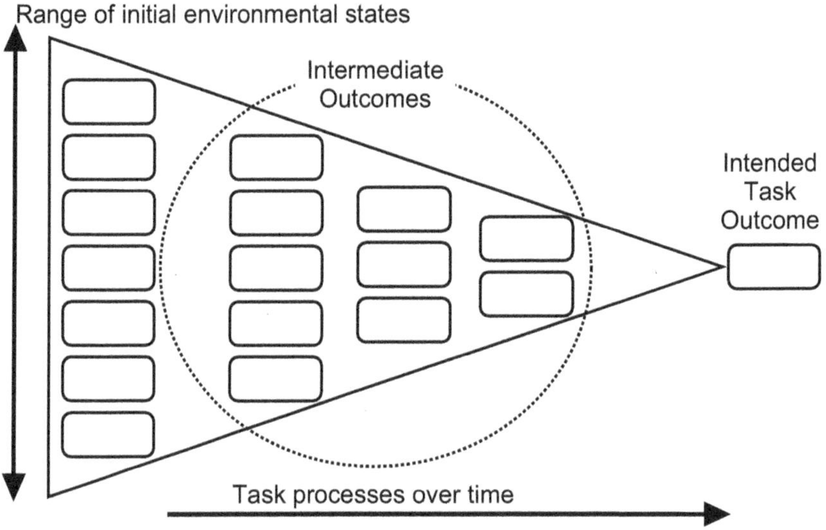

Figure 2.3 Idealised task conducted under a range of conditions

The discussion above centres on a description of a task from an idealised perspective, in other words it describes how the task is supposed to be performed, rather than how it is actually performed in practice. We have multiple intermediate outcomes, because our task may be conducted under a range of initial environmental states, and each of these multiple task trajectories to the intended task outcome may have unique intermediate outcomes.

If we look at the performance and assessment of a specific instance of a task under realistic conditions we see that the outcomes generated by the task are not always in alignment with our idealised task outcome. Going back to our parachuting example, not all parachute jumps go according to plan, within our range of intermediate outcome states we also have the emergency situations and procedures that flow from bad parachute packing or poor opening procedures, the reactions to high and low speed parachute malfunctions, two parachutes deployed at the same time in various configurations and so on. One of the functions of a training environment is to support the task under both idealised and emergency situations. Emergency task procedures invoked in reaction to these emergency situations are designed to return the individual to a non-emergency intermediate outcome, such as 'parachutist under [reserve] parachute canopy', by the most expeditious method.

If we take the example of target shooting, a range of factors such as the skill of the shooter, and environmental conditions (wind, distance, rifle setup) will impact on the performance of the task and on the task outcome generated. This is why in target shooting we generally see a grouping of shots rather than the bullets all going through the same hole in the target. The task outcome in this example (the grouping of shots) is determined by the task performance (that is, skill of the shooter) and the conditions of the task. The ability to recognise and minimise conditions that negatively impact performance can be considered in itself an element of an individual's skill when performing a task. In this way knowing how to zero the sights of a rifle or how to read environmental conditions to avoid shots in high wind will improve the accuracy of any shot taken.

Figure 2.4 illustrates an instance of task performance where a number of task outcomes may be generated.

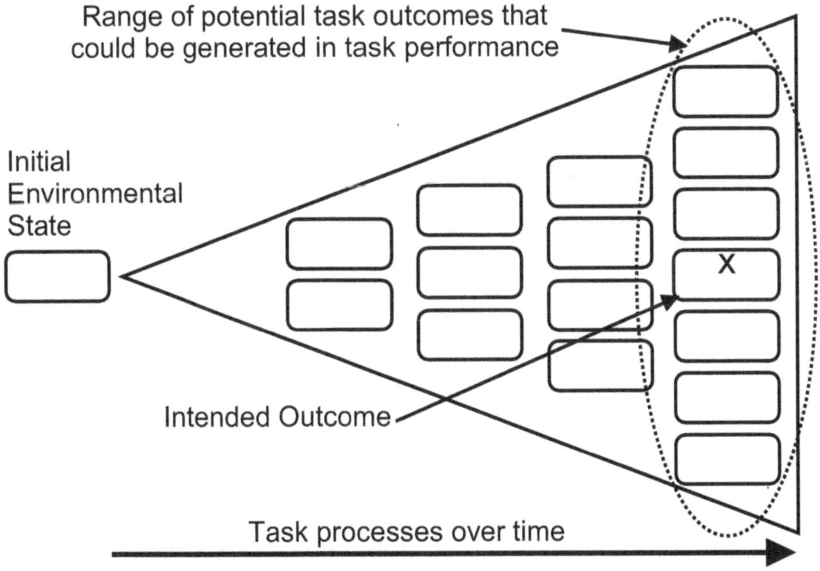

Figure 2.4 An instance of task performance or assessment generating a potential range of outcomes

The actual outcome generated in task performance or assessment may align very closely with our intended outcome (in the target shooting example hitting the highest scoring area of the target), or may be significantly at odds with it (missing the target altogether).

Intended task outcomes may be defined within a very narrow or very broad range of tolerance depending on the task and situation. In our egg-boiling task, a range of task outcomes from soft boiled to hard boiled are generally acceptable. In the parachuting example landing within 10 feet or 100 feet of the intended landing spot might be considered acceptable depending on the conditions.

When looking at the construction of performance measures whether for operational tasks (such as operational performance statements) or training objectives, we see that the commonly adopted format of 'Performance', 'Conditions' and 'Standards' first suggested by Robert Mager (1962) matches the areas previously outlined above. For example, an illustrative performance statement in parachuting might be 'Maintain stable heading in freefall for 10 seconds while conducting circle of awareness checks'

- 'Conditions' as an umbrella term describes: the factors relating to the initial state of the task, the resources available and the conditions which impact on task performance itself. All of these factors are performance-shaping factors external to the individual performing the task. In our example 'freefall' is the task condition.
- 'Performance' describes the task process; observable actions that are undertaken, normally with reference to best practice guidelines, policy or standard operating procedures. Here 'Maintain stable heading ... while conducting circle of awareness checks' describes the performance expected. Circle of awareness checks in this example include student altitude checks, horizon referencing and confirming the continued presence of instructor.
- 'Standards' statements, describe how one may judge the performance of a task against defined criteria – these may be outcome or process related. In our example '10 seconds [of stable heading]' is our standard. Outcome measures reflect the expected level of alignment between our intended outcome and the outcome actually generated in task performance or assessment. Process measures on the other hand reflect the expected alignment between predefined standards of a process (such as time taken to perform a procedure or number of attempts), and the actual task process as it is observed. Here '10 seconds [of stable heading]' is a process based standard.

The Task Environment

The combination of the two concepts illustrated above – the idealised task conducted under a range of initial environmental conditions and the specific instance of a task potentially generating a range of task outcomes – describes the task environment. The task environment defines what elements of interest from an environmental perspective are relevant for the task in question, and what must be supported for practice and assessment. Task environments must support the performance, conditions and standards required by the task. The task environment includes support for the range of initial conditions on which the task operates, and defines what must be supported to achieve a range of outcomes. The task environment also includes all of the task intermediate outcomes that can be generated in the course of the task (including potential errors and mistakes made), and all of the conditions and task related resources that impact on task processes. Lastly the task environment must support the implementation of standards that enable measurement of the difference between outcomes generated in actual performance and the standards set for expected performance. The implementation of standards also includes process as well as product (that is, outcome) measures of performance. Process standards include the task trajectory through the environment – in terms of which intermediate outcomes were generated, and process variables involved such as time and resources used.

The concept of the task environment is illustrated in Figure 2.5.

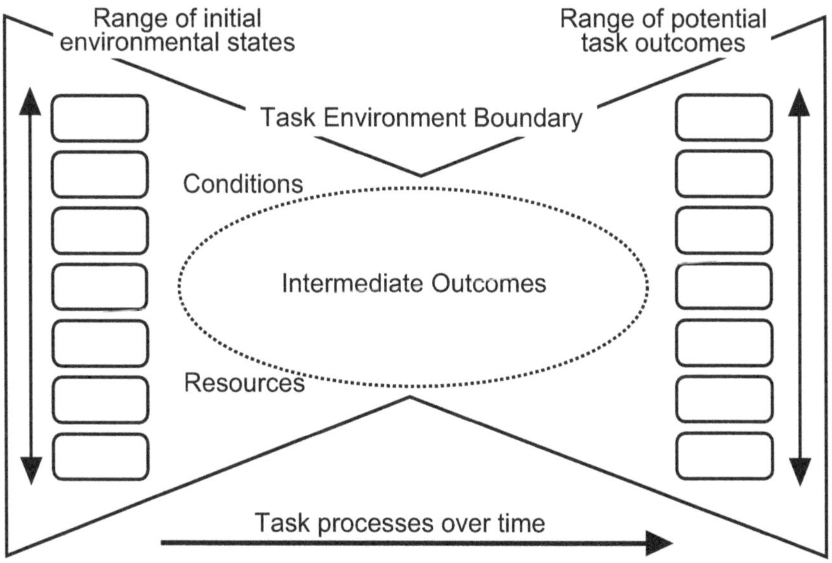

Figure 2.5 The task environment

Task Performance over Time

In the view of tasks that we are advocating, both the initial environmental state and outcome state (which includes task products or things seen to be 'created' by the task) are simply alternate states of the environment, with task performance through time as the transformations that move us from one state of the environment to the next. This idea is adapted from Roby who in his information transduction model of group activity on the task environment (Roby 1968: 13–22) explicitly modelled tasks as successive stages of cyclical activity within a task environment.

In this sense the terms 'initial environmental state' and 'outcome state' are simply labels for describing certain states of the task environment which are of interest to us. Figure 2.6 illustrates that through successive phases of activity, what was a task outcome in one phase of activity becomes an initial environmental state for the next phase of activity.

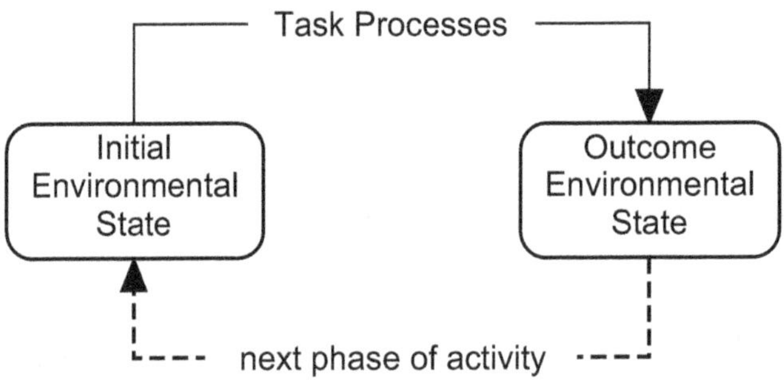

Figure 2.6 Phases of activity within a task

In this cyclical model of task performance we can represent the individual and his or her attributes as the agent conducting the task. In an individual task perspective we have individuals conducting tasks through individual task processes.

An Individual Performance Model

The relationship of individual attributes, the task environment (as represented by initial environmental states and task outcomes) and task processes describes a model of individual performance that is shown in Figure 2.7. The linkage of task outcomes to initial environmental states reflects task performance through successive phases of task activity as described above.

The construction of this model has some alignment with concepts expressed in the Input-Process-Output models of team effectiveness which were advanced by authors such as Hackman and Morris (1975), Tannenbaum et al. (1992) and the model of Command Team Effectiveness (NATO 2005). Because we are only addressing individual task performance in this model, group level factors are absent.

We have previously described initial environmental states, task processes and task outcomes, so our discussion centres on the relationship of individual attributes to other elements of the model.

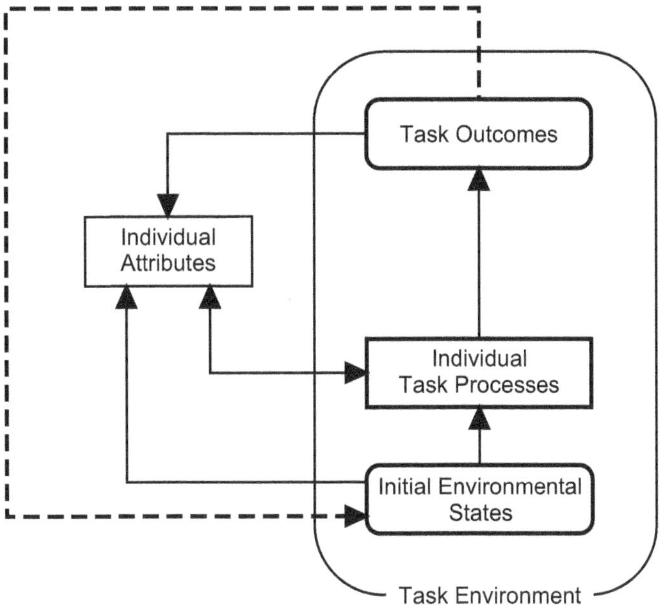

Figure 2.7 An individual performance model

Individual Attributes

In Figure 2.7 we see that an individual performing a task may be considered to possess a set of attributes which may impact on task processes. Some of these individual attributes are relatively stable and involve personally held knowledge, skills or attitudes (KSAs), which support task processes.

Other individual attributes are more dynamic, reflecting characteristics such as performance pressure, physical fatigue, tiredness, disorientation, hunger, stress, fear or frustration. These other non-KSA individual attributes have been described as 'emergent states' (Marks et al. 2001) representing 'member attitudes, values, cognitions, and motivations … constructs that characterize properties of the team that are typically dynamic in nature and vary as a function of team context, inputs, processes, and outcomes'. Other authors have identified these factors as resulting from acute stressors including 'time pressure, threat, impending failure, public scrutiny of performance' (Klein 1993) and can be seen frequently expressed in sportsmen and sportswomen in the course of performing tasks in competition. These emergent states or changes caused by acute stressors impact an individual's performance-related attributes which may then manifest in task processes that are undertaken by that individual. For example, a stressful situation may cause an individual's hands to shake which would then impact on psychomotor task performance.

The stressors that an individual experiences originate in the task environment shown in Figure 2.7, which may present as either initial environmental states (task inputs) or as task outcomes (task outputs). In the case of a sportsman or sportswoman taking a penalty, stress may manifest before the penalty, and either relief or disappointment following the taking of the penalty. In a similar way there is also interplay between task processes in the course of task performance and individual attributes. Task processes don't just generate transformations in the task environment, they also generate changes in the individual. Some of these changes may immediately manifest, other changes may appear over successive months and years.

The learning process is an example where task processes feed into individual attributes, which is why the practice and rehearsal of tasks is one stage in the development of KSAs. In a related manner, the successful completion of a task or conversely the making of mistakes as two types of task outcome may contribute to KSA development, as will the recognition of an initial environmental state as a form of task conditions. Conversely individual experience and skill (individual KSAs) will contribute to task performance processes that generate task outcomes, which is why we engage in training activity in the first place – to get better at performing tasks.

Before we can extend this individual performance model to a team performance model we must look at how the individual is linked to their task environment.

Individual Task Processes

So far we have discussed tasks from an environmental perspective where tasks and task processes are viewed as transformations of the environment. This is an environmental and action-centred view of tasks which describes the context within which tasks are conducted, the observable component of task performance and the results that are generated through task performance.

If one looks at an individual performing a task within an environment three types of task processes are apparent; sensing processes which furnish the individual with information from the environment, internal decision-making processes and doing processes which act to alter the external environment (Pike and Huddlestone 2007). Task actions involve both sensing and doing which are actively controlled by the individual. For example turning on a switch relies on the individual sensing the switch (either by sight or touch), and then effecting a change on the switch by pressing it (the doing portion). The stages of decision making may be characterised as sense, perceive (the interpretation of stimulus information), decide and act (that is, respond) (Gagne et al. 2005). Information processing models of cognition view these processes as being supported by the mutual interaction of sensory memory, long-term memory and short term working memory, under the control of attention (Ashcraft 2006: 52; Neisser 1976: 21).

An individual selects and receives stimuli (cues) either directly from the outside world, or through equipment (for example a radar display or a radio). The recipient receives this stimuli through a set of receptors (for example the eyes or ears) in an active process we characterise as sensing. The individual then performs some cognitive processing (decision making) on the received stimuli, which may result in physical action (doing). We describe the term as doing because sensing is also an action which is actively directed by the individual.

The recipient generates action through a set of effectors (the muscles of the body), which may deliver an effect directly (such as speech) or may be captured and transformed by equipment into another effect (such as pressing the 'transmit' button on a radio or pulling the trigger on a weapon). This effect is transmitted into the task environment and will generate a result (that is, a transformation of the environment). The quality or performance of this task outcome may be satisfactory or otherwise according to our predefined standards (in the shooting example hitting the target or missing it). Observing performance effects from the environment (that is, stimulus feedback from the effect) may lead to renewed or repeated action, ceasing action or a change to a new type of action. These three task processes – sensing, doing and decision making – are illustrated in Figure 2.8 in a simple information processing diagram.

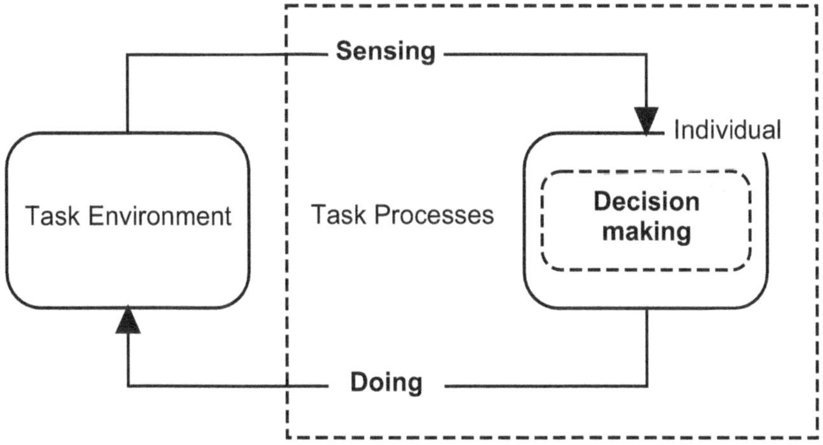

Figure 2.8 Individual task processes: sensing, doing and decision making

In our view of tasks as transformations, doing transforms the task environment and sensing transforms part of the individual's internal cognitive state. Clearly the process of decision making may also alter an individual internal cognitive state, for example if I have selected 'option a', rather than 'option b', that decision outcome will be defined and represented internally as an element of cognition before the associated physical action occurs.

Both sensing and doing as types of action or task process are tightly coupled with decision making in any instance of task performance, and all processes occur simultaneously under cognitive control. Both sensing and doing involve decision making, with doing actions being directly observable, whereas sensing actions are less observable in understanding what an individual might have seen or taken from a particular situation. In task performance we continually and simultaneously make action-related decisions, relating to what information we need to gather from the environment and what changes we need to effect in the environment.

Sensing as a form of action may be characterised by the type of stimulus received corresponding to the senses that receive the information. The stimulus type and corresponding sense organ are indicated in Table 2.1.

The first five senses are the most familiar to us; however, added to the senses that receive external stimuli, we have two senses that provide information on orientation and acceleration of the body. The sense of balance or equilibrium (the vestibular sense) is generated within the semi-circular canals of the inner ear, which enables us to perceive acceleration, and orientation of the body and to keep our balance. Kinaesthesia is the sense that describes the position of the body and the orientation of the limbs through stretch sensors in the muscles, tendons and joints (Blake and Sekuler 2006).

Table 2.1 Stimulus modalities and corresponding sense organs

Stimulus	Sense organ
Visual	Eyes
Auditory	Ears
Haptic	Skin/touch
Gustatory (Taste)	Tongue
Olfactory (Smell)	Nasal lining
Vestibular (perception of acceleration, orientation, balance)	Semi-circular canals of the inner ear
Kinaesthesia (perception of movement and orientation of the limbs)	Muscles/tendons/joints

For any training task there are a set of cues or types of stimuli that are required – for example in firing a personal weapon, observation of the target requires a visual stimulus, secondary cues related to system function may be visual, auditory (that is, hearing the weapon firing or hearing only a click) or kinaesthetic (such as recoil). These issues and issues relating to fidelity are explored in the training environment chapter (Chapter 5).

Decision-making Elements: Perceived Conditions, Goals and Plans

Just as tasks may be viewed from an environmental perspective as transformations of the external environment, a decision-making view of tasks sees them as transformations of the individual's internal cognitive environment. Some of these internal transformations are driven directly from the environment through activities associated with sensing, others less directly through other cognitive processes. In the representation of the task in the internal environment, a number of elements which comprise (internal) process and product can be considered to exist. These internal processes and products may alter when an individual undertakes a task, through interactions with the external environment and though internal interactions.

This is not proposed as a model of cognition but rather an abstraction of internal cognitive products and processes which we believe may have some value in trying to model individual decision making, and subsequently in defining team task processes. There are clearly a number of cognitive concepts which could be invoked or selected as being useful for representing decision making within tasks. We have selected three task elements for simplicity and because they appear to be directly associated with the environmental elements of tasks: initial environmental states, the task processes (sensing and doing) and task outcomes. The three decision-making elements we identify are an individual's perceived conditions, goals and plans.

- Perceived conditions describe the internal state that an individual holds relating to: what's happened, what's happening and what may happen in the outside world. This is not just the internal version of what an individual sees and hears, but is the product of a set of editing processes being applied to this environmental information. These editing processes include interpretation, selection, judgement and projection (anticipation of future events). All of these editing processes rely on recall of past events, general memory and a number of other cognitive processes such as planning.
- Goals define an individual's intention, what alterations the individual would like to see in the world, what that individual wants to happen or to prevent from happening. We draw a clear distinction between goals which we consider as properties that an individual possesses, and task outcomes which are things that have occurred and are properties of the external environment. If one considers the situation of a lottery, each individual participating is holding a goal that their selected numbers will be drawn, however there can be only one outcome. An intended outcome is an outcome which satisfies that individual's goals. This can also be described as a goal-related outcome.
- Plans describe how an individual's goals may be satisfied though the performance of actions. As such they describe the connection between the individual's perceived conditions and their goal. Planning (the activity that generates plans), creates a mapping between the currently understood situation and the desired future situation. Planning on its own, however, achieves nothing; for plans to generate effects in the world, they must be enacted.

Figure 2.9 suggests the relationship of the task environment to perceived conditions, goals and plans.

Each of these decision-making elements (perceived conditions, goals and plans) is held by the individual and can be considered as both internal product and process. For example goals are the information output of the continual process of goal generation and maintenance; these are illustrated by the small self-linking loops in Figure 2.9. Goals have a relative priority and importance – goal generation processes and maintenance processes will move the relative priority of these goals (that is, promote or demote goals) according to the demands of the situation.

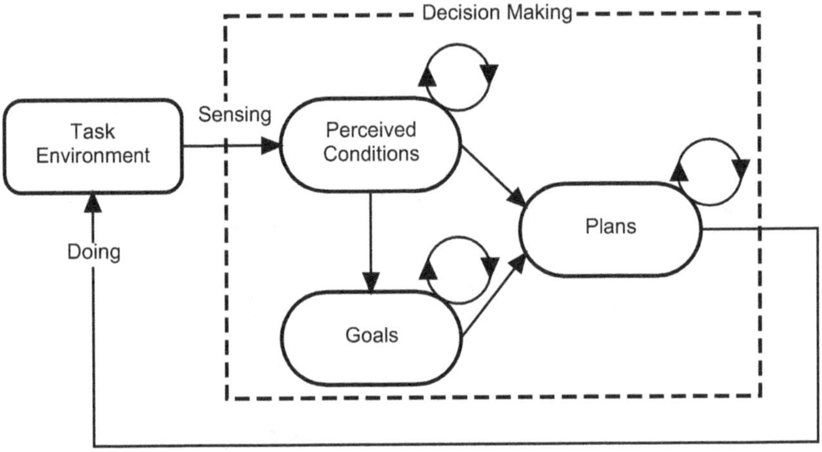

Figure 2.9 Perceived conditions, goals and plans

Decision making has a number of internal aspects: regarding sensing (what an individual chooses to focus on, or pay attention to in the external world), goal generation (what goals to generate and their relative priority) and planning (how resources and actions are selected and sequenced). The external face of decision making is the decision made, which is then enacted in the external environment. In the enacting of plans we can see the appropriateness of the plan to the situation and the adequacy of the plan in describing the task processes that are undertaken.

Referring to Figure 2.9:

- The task environment linkage to perceived conditions indicates that perceived conditions are in part (but not wholly) informed by information gleaned from the environment. Information that an individual gathers from the task environment may include information products containing specifically designed messages, such as signs, maps, heads up display information or illuminated captions.
- Perceived conditions linkage to goals indicates that goals may be reactively generated in response to environmental situations; if an individual

perceives that they are in a particular situation they may wish to extricate from or mitigate the situation. We have not indicated a connector from goals back to perceived conditions (that is, the opposite relationship from the one described); however. it may be that an individual's goals may have some bearing on what an individual perceives and this may explain some perceptual biases relating to individuals ignoring negative information. If established experimentally this link would demonstrate that individuals 'see what they want to do', in other words environmental stimuli may be selected or 'edited' to generate perceived conditions that allow goals to be satisfied through the enactment of plans.

- Perceived conditions and goals linkage to plans. Planning is predicated on knowing one's current situation and having a goal; goals being the description of intended outcomes. Plans describe the actions to be performed, the sequence that they are to be performed in, and under which perceived conditions. As actions may have dependencies such as the resources required, or particular team members having to be qualified to perform the task, plans may also describe these dependencies.

Goals

We have previously discussed task outcomes as being the state of the world at the completion of the task, with intended outcomes being the outcome that an individual wanted to achieve. Tasks are purposeful in that they are generally directed to achieving an end, just as one generally has a destination in mind when one sets off on a journey. Goals are descriptions of the intended outcome of a task, which is a description of a state of the world in the future that we would like to achieve. Goals exist internally as cognitive constructs and may also be recorded, stored and distributed as external information. For example, an individual goal of a win on the lottery (of more than a day's wage) or being on a holiday somewhere warm may be communicated as piece of information as well as being held cognitively.

The correspondence of goals and outcomes may be explained through a chess analogy. A chess player wants to win the game, and this has to occur through achieving checkmate which is a position that satisfies certain properties (the opponent's king is in check and cannot escape by moving, blocking or capturing the attacking piece). At the end of the chess game there are a huge number of possible chess positions each of which represents a task outcome. A large number of outcomes will satisfy the player's 'checkmate' goal criteria, other outcomes represent the player losing the game or a drawn or abandoned game. A goal, in this case 'checkmate', is an information description held by an individual which describes some properties of an intended task outcome. As goals exist as elements of information they are abstractions of real environmental situations, only describing certain desired environmental properties. For a goal to be realised, outcomes must be generated which share properties described by the goal. There is no reason why an individual cannot hold multiple goals relating to a situation – for example our chess player may want to win in 25 moves or less, or achieve checkmate using a rook.

Goals can be said to be satisfied when the properties of the goal are matched by the task outcome. If a child decides to build a sandcastle on a beach, they have gone through the process of deciding their goal (goal determination), enacting a set of actions in task performance while maintaining their goal and have generated a task outcome. At this point the task outcome can be referenced against the goal and the task concluded. The process of goal satisfaction is illustrated in Figure 2.10.

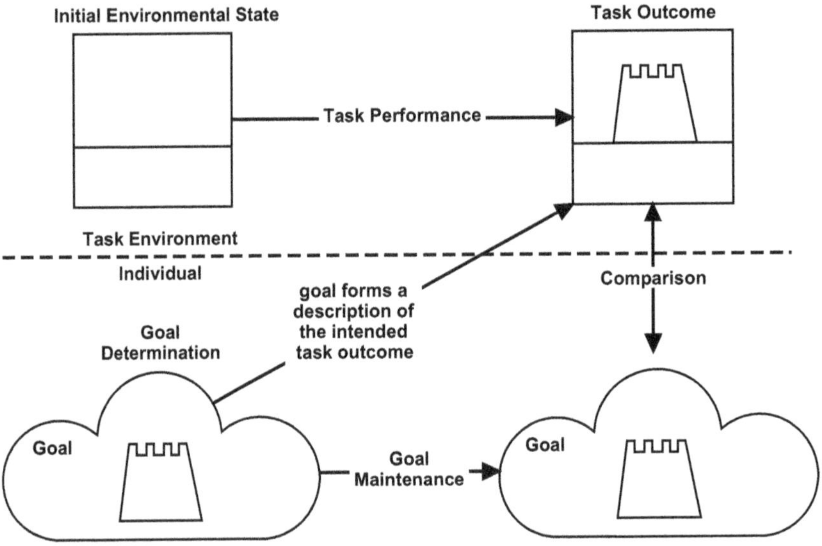

Figure 2.10 Goals and outcomes

We see goals in the instructional context, for example in assessment the trainee's task outcomes are referenced to a goal, which is in this case the assessor's goal of what constitutes a valid answer to the question.

Perceived Conditions

We characterise each individual's experience of the environmental situation that they are in as 'perceived conditions', with the world that is perceived being an individual's internal cognitive representation of the world.

Perceived conditions are of interest in TNA because individuals and teams must frequently make decisions based on less than perfect information in ambiguous and dynamic environments – where the consequences for error generate situations such as 'friendly fire' events. Friendly fire situations occur due to a mismatch between the real-situation (for example the armoured vehicles ahead are friendly) and the perceived conditions situation (armoured vehicles ahead are enemy). From a TNA perspective the perceived conditions relating to a task help us to understand which elements of the environment need to be dynamic and ambiguous to generate suitably realistic challenges for our trainees. Additionally understanding perceived conditions also helps inform how much background information on a situation may need to be provided to a trainee to generate a 'realistic' context. Perceived conditions are important as they are the cognitive task element that drives individual behaviour. It's not what's happening, but what we believe is happening that shapes performance.

Perceived conditions may drive goals directly in a reactive sense. As an example, if an individual is visiting the zoo and sees the lions escape from their enclosure, that individual might reasonably generate the goal of leaving the zoo as quickly as possible. Perceived conditions could also be 'delivered' indirectly to the individual as information on a sign at the zoo entrance 'Lions escaped, tickets half price'. Perceived conditions also link to plans, as plans describe actions that will be performed according to the context described in perceived conditions.

An individual's perception of an environmental situation as perceived conditions hopefully consists of more than a cognitive description of the world in its current situation – retrieval of memories of previous comparable experiences, and projections or anticipations of future situations will inform sense-making within the current environmental context.

We view perceived conditions as both process and product, a construct which is derived from the external world in its current state. As a concept it is closely related to the concept of situational awareness which embraces memory and projection and anticipation. Situation awareness (SA) may be characterised as product or both product and process (Stanton et al. 2005: 213). SA refers to 'the level of awareness that an actor has of the current situation that he or she is in' (Stanton et al. 2005: 213). Endsley (1995: 88) defines SA as 'The perception of the elements in the environment within a volume of time and space, the comprehension of their meaning, and the projection of their status in the near future'. Other authors have defined SA as product and process, with the perceptual cycle driving actor exploration of the world (Smith and Hancock 1995).

Figure 2.11 represents the relationship between perceived conditions and related concepts. It should be noted that this diagram does not represent all connections between entities but rather those relating to perceived conditions (in the sense that memory may feed plans and goals as well).

The importance of memory and projection/anticipation as they relate to perception of the current situation (as a part of perceived conditions) can be seen in the format of briefings that are given to trainees prior to training events to ensure that the appropriate context for task performance is generated. Such briefs may cover recent past events in the scenario setting, which represent a proxy for trainee memories, or past patterns of life information which represent proxies to help inform trainee projection or anticipation. In this sense we cannot expect our trainees to detect the unusual in the environment when we have provided no

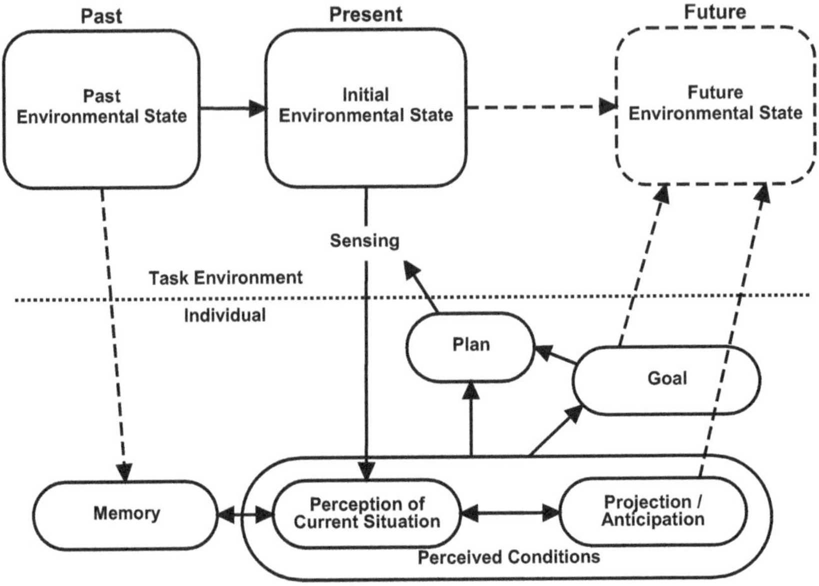

Figure 2.11 Perceived conditions

baseline data for what constitutes normality. Likewise in an AAR, an individual's anticipation and projection from a particular situation can be probed and discussed, which in turn helps expose and correct deficiencies in projection/anticipation and place events in proper context for future reference.

Perceived conditions do not have to be shaped purely by real-time updates from the environment but may be informed by information contained in briefs or online resources. Equally, things that were held to be true at one time can become inaccurate because of changes in the environment, suggesting that maintenance of perceived conditions in highly dynamic environments will be more challenging than in comparatively static environments. Perceived conditions also embraces the future states of the environment – anticipations which could be viewed as projected states of the external environment. These projections may become highly abstracted, and may go to the level of attempting to determine intent in a third-party actor, derived from our understanding about what the third-party actor understands about their environment, and what their goals may be.

Plans

Plans act as the bridge between goals and our perceived current situation as perceived conditions. Plans may be purely cognitive, such as driving a route by memory, or may exist as external information such as checklists, a project manager's Gantt chart, a chef's recipe, the instructions for the assembly of a piece of furniture, or a route drawn onto a paper map. Plans form the description for task processes that will be undertaken and the condition under which actions will be performed. In the simplest form a plan might have the following content: 'if it's raining, wear a rain coat', an alternative example could be 'assemble the wardrobe, then fit the wardrobe doors'.

Plans are formed in reaction to the integration of goals and perceived conditions. Planning as an activity and plans as the product of planning are predicated on having a goal or set of goals and having an awareness of the current situation.

Plans:

- describe the processes that move us from our initial environmental state, through a sequence of intermediate outcomes to our final task outcome;
- contain information about intended or potential activity which effects changes in the world;
- define information capture requirements; defining what we must understand about the shape of the world to proceed with our task processes;
- capture critical information requirements for tasks, the sensing requirements of tasks as well as the doing requirements of tasks;
- describe resources required to support tasks which may include time, or information that is provided to the individual rather than sensed as a task process.

Task processes described within the plan are invoked according to a set of criteria which describe either environmental conditions, or the intermediate outcomes which have been generated. A plan for getting up in the morning might be:

1. if it is 7am then get out of bed;
2. have breakfast;
3. have a shower;
4. clean teeth.

Here the implicit completion of one stage of the process then leads us to the next. We must conclude step 1 before commencing step 2. Most recipe books and furniture assembly plans omit this implicit plan information. Similarly there may be a set of assumptions built into the plan relating to the starting situation, or what resources are available. In the example of a plan for getting up in the morning there is an implicit assumption that I'm in bed in the first place, or in the case of furniture assembly that the piece of furniture is not already partially assembled (or missing a vital component).

When one looks at team and collective tasks in comparison to individual tasks the number of strands of parallel activity that are possible becomes larger. Teams can do many things simultaneously when compared to an individual, and team members may interact to generate the coordination of effort required to deliver team goals. The requirement to capture sequential and parallel phases of activity has implications for representational techniques that are used to record task analysis output; these considerations are discussed in Chapter 4 on task analysis methodology.

Integrating Environmental and Decision-making Views of Tasks

The importance of a cognitive view of tasks from a TNA perspective is that cognitive and decision-making processes and products underpin all observable performance and this is where the learning process occurs and what instruction as a design process addresses. In a team and collective context extending a decision-making view of tasks for individuals may be of some value for understanding team task processes. Task analysis to support TNA should attempt to align both the action-based and cognitive views of tasks, as well as individual and team based views of tasks and associated processes.

For each of the three elements of tasks that we have identified from an environmental perspective – initial environmental conditions, task actions and task outcomes – we suggest that cognitively-based internal equivalents could be defined which would have some value for TNA and in the subsequent stages of training design, delivery and assessment. Each individually held task element (perceived conditions, goals and plans) can be considered product and process. These task elements are summarised in Table 2.2.

Table 2.2 Environmental and individual task elements

Environmental task element	Individual task element	
	Product	**Process**
Initial environmental state	Perceived conditions	Generation and maintenance of perceived conditions. 'Sense making'
Outcome environmental state	Goals	Goal generation and goal maintenance
Actions	Plans	Planning, plan execution

Task outcomes have their cognitive equivalent in goals, which capture intentionality in an individual determining what particular end is to be achieved. For initial environmental conditions we have the cognitive equivalent of perceived conditions which reflect what information an individual has about their task environment. This information is selective and incomplete but is enhanced by the incorporation of elements experienced from past situations through memory and also incorporates aspects of projection which models projected future states of the task environment. Lastly we have plans which are the cognitive structures that drive task action processes. Combining the six elements of tasks together yields an expanded individual task model. This is illustrated in Figure 2.12.

Figure 2.12 is divided into a central decision-making environment surrounded by the external task environment. The outer triangle of the model reflects how

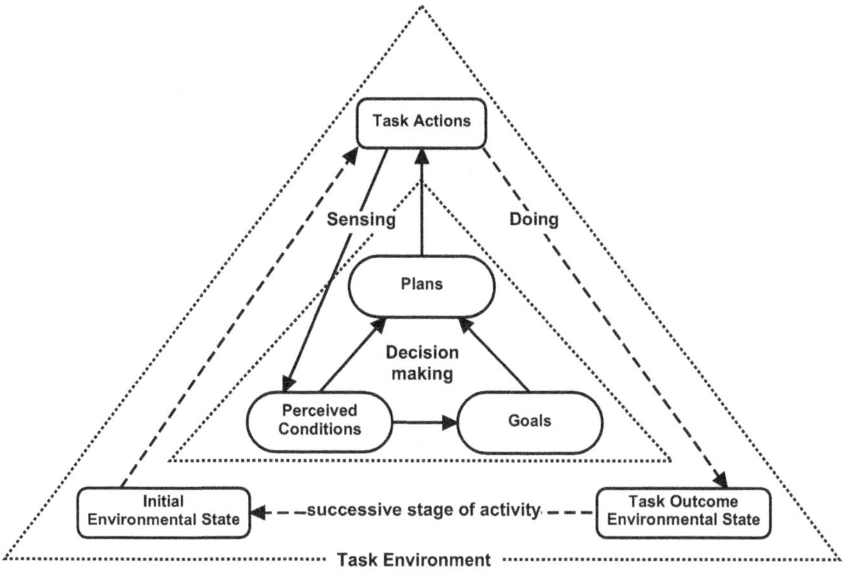

Figure 2.12 Individual task model

task processes change our initial environmental state to the environmental state of our intended task outcome. In each successive phase of task activity our previous task outcome may constitute the initial environmental state for the next phase of activity. As we have previously stated, the term 'initial' is defined with reference to a particular point in time.

We have not defined items such as short or long term memory, sensory memory or attention within this model. Within the inner triangle of the task model the function of recall is assumed to exist – and to feed perceived conditions, plans and goals – as are the other elements of cognition.

Within the internal triangle of perceived conditions, plans and goals we see perceived conditions selecting goals either directly as seen in Recognition Primed Decision-making (RPD) (Klein 1993), or contributing to the generation of goals through a more extended cognitive process of analytical decision making. Within the RPD model the 'recognition of typicality' has four subcomponents: plausible goals, relevant cues, expectancies and actions. Comparing this to the task elements within the internal triangle, cues and expectancies (projections) sit within perceived conditions. Actions are described by the plan construct and actions (or the effect of actions) are observed as perceived conditions. Anticipation of actions or the effect of actions may also be accessed via the recall of past plans. Finally, goals within RPD map to goals within the internal triangle.

Plans contain information of the actions that we may undertake, and are fed by both goals and perceived conditions.

Goals drive plans in that the goal is the cognitive representation of the outcome state of the task that the individual is trying to achieve. Perceived conditions feed into plans in that conditionality of execution of the task (for example an individual deciding to drive to work rather than take the train) is driven by what we perceive about the external environment (for example that the individual is running short of time, or that the station is closed). Perceived conditions may include information on the completion of intermediate steps of the task as intermediate outcomes (for example the individual realises when he or she has reached the train station).

In the case of tasks where the outcome of one task action may not be well established (because of environmental uncertainty) we have a requirement to initiate a task specifically to establish the state of the environment and therefore what the task outcome may be. Successive cycles of activity occur as the outcome environmental state of one cycle becomes the initial environmental state of the next.

In each of the corners of the triangle we have pairs of entities: initial environmental state and perceived conditions, goals and task outcomes, and plans and task processes (actions). In an ideal world these entities would all be in alignment with each other; however, in reality there may be considerable divergence between these task elements. To recap, we have defined six elements of tasks as being of interest from an individual perspective. These factors are summarised in Table 2.3.

Table 2.3 Task elements

Task environment element	Application in TNA and training design
Initial environmental states	Defines the range of initial conditions for the task which must be supported.
Task actions	Process based performance metrics and intermediate task outcomes that must be measurable. Conditions and resources needed for tasks. Standard operating procedures.
Task outcomes	Defines what the environment must support in task effects or products, critical errors and consequences.
Individual element	**Application in TNA and training design**
Perceived conditions	Cues for tasks, environmental stressors, scenario design. Reactive goal generation.
Plans	Task analysis, decomposition of the task, description of conditionality of execution, sequential and simultaneous task processes. Information and resource requirements for task processes.
Goals	Product based performance metrics – judgement of quality of outcomes.

References

Alberts, D.S and Hayes, R.E. (2006) *Understanding Command and Control*. Washington DC: CCRP Publication Series.

Annett, J. (1997) Analysing Team Skills, in R. Flin, E. Salas, M. Strub and L. Martin (eds) *Decision Making Under Stress: Emerging Themes and Applications*. Aldershot: Ashgate. 315–26.

Ashcraft, M.H. (2006) *Cognition*, 4th edn. Upper Saddle River, NJ: Pearson Education.

Blake, R. and Sekuler, R. (2006) *Perception*, 5th edn. New York: McGraw-Hill.

Bowers, C.A., Morgan, B.B., Salas, E. and Prince, C. (1993) Assessment of Coordination Demand for Aircrew Coordination Training, *Military Psychology*, 5(2), 95–112.

Endsley, M.R. (1995) Toward a Theory of Situational Awareness in Dynamic Systems, *Human Factors*, 37(1): 32–64.

Gagne, R.M., Wager, W.W., Golas, K.C. and Keller, J.M. (2005) *Principles of Instructional Design*, 5th edn. Belmont, CA: Wadsworth.

Hackman, J.R. and Morris, C.G. (1975) Group Tasks, Group Interaction Process, and Group Performance Effectiveness: A Review and Proposed Integration, in L. Berkowitz (ed.) *Advances in Experimental Social Psychology*, vol. 8. New York: Academic Press. 1–55.

Jonassen, D.H., Tessmer, M. and Hannum, W.H. (1999) *Task Analysis Methods for Instructional Design*. Mahwah, NJ: Lawrence Erlbaum Associates.

Kirwan, B. and Ainsworth, L.K. (1992) *A Guide to Task Analysis*. Boca Raton, FL: CRC Press.

Klein, G. (1993) *Naturalistic Decision Making: Implications for Design*. Crew System Ergonomics Information Analysis Center. Report CSERIAC SOAR 93–1. Available at http://www.dtic.mil/dtic/tr/fulltext/u2/a492114.pdf [acessed 18 August 2013].

Mager, R.F. (1962) *Preparing Instructional Objectives*. Belmont, CA: Fearon Publishers.

——— (1972) *Goal Analysis*. Belmont CA: Pitman Learning.

Marks, M.A., Matheu, J.E. and Zaccaro, S.J. (2001) A Temporally Based Framework and Taxonomy of Team Processes, *Academy of Management Review*, 26(3): 356–76.

NATO (2005) *Military Command Team Effectiveness: Model and Instrument for Assessment and Improvement*. NATO Research and Technology Organisation Technical Report TR-HFM-087.

Neisser, U. (1976) *Cognition and Reality*. San Francisco, CA: Freeman.

Pike, J. and Huddlestone, J.A. (2007) *Instructional Environments: Characterising Training Requirements and Solutions to Maintain the Edge*. Paper to the 29th Interservice/Industry Training, Simulation and Education Conference (I/ITSEC). Orlando, FL, December 2007.

Roby, T.B. (1968) *Small Group Performance*. Chicago, IL: Rand McNally.

Salas, E., Sims, D.E. and Burke, C.S. (2005) Is There a 'Big Five' in Teamwork?, *Small Group Research*, 36(5): 555–99.

Salmon, P.M., Stanton, N.A., Walker, G.H. and Jenkins, D.P. (2009) *Distributed Situational Awareness*. Aldershot: Ashgate.

Smith, K. and Hancock P.A. (1995) Situational Awareness Is Adaptive Externally Directed Consciousness, *Human Factors*, 37(1): 137–48.

Stanton, N.A., Salmon, P.M., Walker, G.H., Baber, C. and Jenkins, D. (2005) *Human Factors Methods: A Practical Guide for Engineering and Design*. Aldershot: Ashgate.

Tannenbaum, S.I., Beard, R.L. and Salas, E. (1992) Team Building and Its Influence on Team Effectiveness: An Examination of Conceptual and Empirical Developments, in K. Kelley (ed.) *Issue, Theory, and Research in Industrial/Organizational Psychology*. Amsterdam: Elsevier. 117–53.

Chapter 3
Team Tasks

Team Task Processes

In the previous chapter we described tasks and performance both from an environmental (the external) and individual actor or agent (internal decision making) perspective. In this chapter we extend these models applying them to team members performing a team task in a team context.

Teams have been variously defined. For example, Nieva et al. (1978), defined a team as: '*two or more interdependent individuals performing coordinated tasks towards the achievement of specific task goals*' whereas Salas et al. (1992) defined a team as '*two or more individuals, who have specific roles, perform interdependent tasks, are adaptable and share a common goal*'. In the Salas et al. (1992) definition, a team is defined as being adaptable and sharing a common goal. In a poorly performing team, team members may not necessarily be adaptable (or want to be adaptable) and may not necessarily share a common goal. This is why inadequate conflict resolution is quoted as a factor in poor teamwork (Baker et al. 2006). We define a team to be:

> a number of persons constituting a work group, assembled together for the purpose of joint action.

Team task processes are the processes that encompass all of the actions and interactions that the team undertakes in order to achieve the desired environmental end state (intended task outcome). Team task processes are different from individual task processes because we have more than one team member present in the task environment. This potentially allows for communication and interaction to occur between them. Team task processes include all of the individual task processes of sensing, decision making and doing – in a team context individuals still engage in these processes. Team task processes extend beyond these individual task processes to include other forms of activity that are only seen in a team context.

The relationship between team task processes and individual behaviours is that team task processes exist at the team level while behaviours are performed by individuals. Processes consist of behaviours linked or woven together that extend over time; the individual elements of these processes are behaviours. As an example, a conversation between individuals is a process that continues over time, comprising the individual behaviours of speaking and listening. The conversation stops at the point the individuals involved either stop talking or stop listening, and no one can be said to 'own' the conversation in the sense that a behaviour can be attributed to an individual. A similar example can be seen in the coordination of activity involved in running a team relay race – the race is a process that continues over time made up of individual contributing behaviours (running and baton passing).

Teams (and collectives) generate task outcomes through task processes that operate on a set of initial environmental states, so the concept of the task environment remains unchanged from the individual situation, except that the task environment in the context of a team is likely to be larger, more diverse and more complex than that found in an individual task.

Previous reviews of team performance literature (Bass 1980) have identified that an understanding of individual performance is not sufficient to understand team performance, and that team member interaction is critical to understanding team performance (Dyer 1984; Baker and Salas 1992). The question then becomes how to characterise team member interaction? One method of characterisation has been to describe interactions as either taskwork or teamwork.

Taskwork describes behaviours related to tasks performed by individual team members whereas teamwork describes behaviours related to team member interactions (Morgan et al. 1986). While providing two distinct categories (taskwork and teamwork), there are some behaviours within a team which would seem to fit both categories. As an example consider two individuals lifting a stretcher to carry an injured colleague to a place of medical treatment. Here both individuals are engaged in individual activity as taskwork, lifting their respective ends of the stretcher. These stretcher bearers are also engaged in teamwork in that they are interacting with each other by carrying the stretcher. If one team member drops their end of the stretcher the effect will be felt by both the casualty (who is no longer being carried), and by the other team member who will find walking (and carrying the casualty) much more difficult.

Such examples would necessitate an additional category that conveys the potential overlap between taskwork and teamwork. In the categorisation of taskwork and teamwork we therefore seem to need an additional category which captures how individual tasks may constitute a form of team member interaction. This third category (combining both taskwork and teamwork) could be characterised as coordination, a factor identified in a number of teamwork models (Dickinson and McIntyre 1997; Annett et al. 2000; Rousseau et al. 2006). Coordination describes how individual activity performed by team members generates task outputs which are beyond the capability of any single individual. In the stretcher bearer example, one team member on their own doesn't carry half a stretcher unless there is another team member to do their half of the task, at the same time.

Coordination and communication are highly related and co-dependant concepts. Communication supports coordination of activity; conversely, coordination supports communication. Communication doesn't operate effectively if team members don't acknowledge and reply to each other in a timely manner, or alternatively if team members all try to talk at the same time. In this sense communication must be coordinated according to certain rules or norms of behaviour if coherency of group activity is to be maintained.

Communication

In the situation of an individual working alone we characterised the task processes or actions as:

- sensing, which generates perceived conditions as information from the task environment;
- doing, which is how the task environment is changed by the task;
- the internal process of decision making.

In a team situation we have the potential for communication between individuals, as well as the task processes that we see in the individual context.

The model of communication below is an adaption of early work in communication theory by Shannon and Weaver (1949), who first proposed an engineering model of the communication process. Here the elements of transmitter, signal, channel, received signal, receiver and noise source in the Shannon-Weaver model are taken as comprising the task environment. Norbert Weiner introduced the principle of feedback into communication in the context of cybernetics (Weiner 1948). Other authors have elaborated on these ideas further with the introduction of double encode and decode stages for both the transmission of stimulus and transmission of response between two sources (Schramm 1955; Romiszowski 1988). In these communication models the environment is represented in a somewhat tacit manner – as 'medium' in Schramm's model of communication, and 'channel' in the Shannon-Weaver model.

The team task processes of communication and interaction within the task environment are illustrated in the example of two individuals having a conversation (speaking and listening) in Figure 3.1.

The outer loop in the diagram reflects the communication between team members through the task environment, the inner loop the individuals' interaction with the task environment. The communication loop is characterised by distinct 'transmit' and 'receive' phases in both individuals. In the example of verbal face-to-face communication between two individuals, the phases of communication are made up of speaking and listening, with the task environment being the air carrying the sound waves between them. In the situation of two individuals communicating by radio, the task environment includes the radios and electromagnetic waves that pass between radio sets. This model applies to both asynchronous and synchronous communication; in the case of email or correspondence by letter our individual's email or letter will reside in the task environment as a piece of information, and will be read by the recipient at some later time.

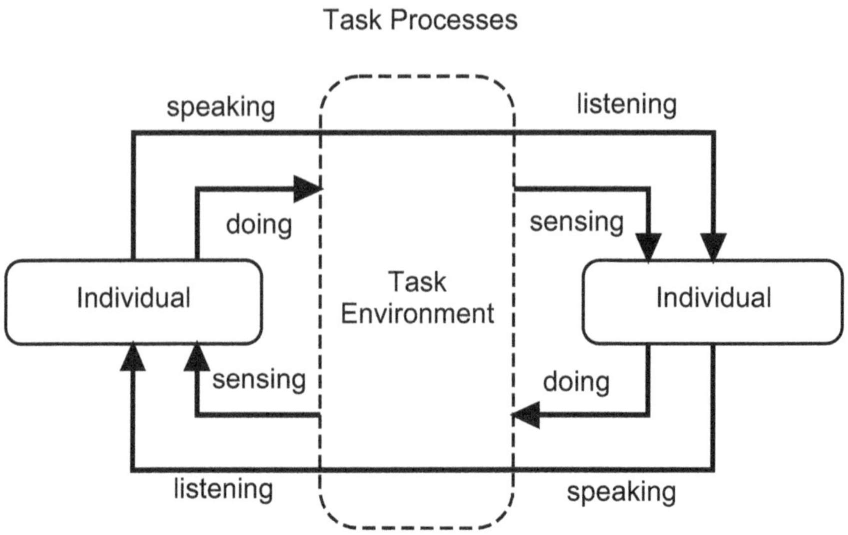

Figure 3.1 Team task processes, two individuals in a team communicating and interacting with the task environment

Referencing Figure 3.1, we can see that the communication process is a linked sequence of actions (that is, sensing and doing phases of activity), which allows connection and alignment between two team members through their task environment. Communication operates on a number of levels, the spoken words 'look at the red car' are vibrations in the air, which when heard and understood by other team members leads them to think of the colour 'red' and vehicle of type 'car'. This system works because all team members agree on what colour 'red' represents and what 'cars' look like. Communication is the transfer of information between team members, in this case a message 'look at the red car' from the first team member, to the second team member. This message passing involved the transformation of the second team member's internal state and may act as a cue for further action.

If one looks at the distinction between communication and interaction within a given environment the line can become blurred. Hoisting a red flag at a firing range area is one example, it is an effective means of communication providing everybody understands what the flag conveys. Flags were used as a two-way method of communication between naval vessels for hundreds of years, this communication depending on pre-agreed flag signal 'libraries' which related the flags flown to the associated piece of information or command being issued.

Coordination

Coordination is a concept closely related to communication; it is difficult to maintain coordination without communication, and in a variety of tasks it is difficult to maintain communication without coordination of activity. Coordination reflects how teams achieve transformations through activities that individuals on their own could not achieve. As one of the outputs of collective activity is the generation of information, poorly coordinated activity can compromise effective communication when the information that is required to be passed has not been prepared in a timely manner. Coordination of action exists at both the individual and the team level, and occurs between teams where we are dealing with collective organisations comprised of multiple teams.

A simple example of coordination within an individual task is the act of eating with a knife and fork, illustrated in Figure 3.2. The action of cutting a piece of steak relies firstly on the fork to pin the item in place and secondly, the knife to cut. The task outcome of having cut a piece of steak into two pieces relies on these actions being conducted concurrently in a parallel (that is, simultaneous) manner – neither action independently achieves the end result, or 'half' of the end result.

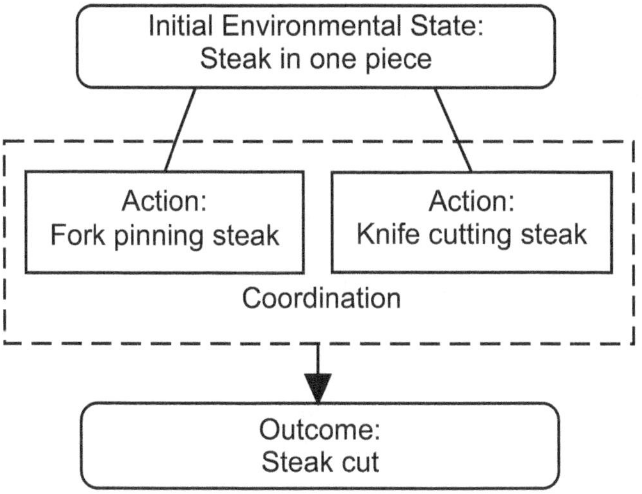

Figure 3.2 Coordination within an individual task

Coordination also occurs sequentially in the sense of 'passing the ball' in a team game. As a final example, a task like juggling involves coordination of both types – balls must be kept in the area (concurrent activity), through extremely well sequenced sets of catching and throwing (sequential activity). Communication and coordination are frequently related in a team setting as communication may be used as a mechanism to enable coordination. Coordination of communication establishes a dialogue between individuals where they will not be metaphorically 'all talking at the same time', nor sitting in silence waiting for the other team members to speak.

Coordination of action hopefully characterises activities that the team or collective organisation undertakes; this team activity generates effects in the wider environment. The presence or absence of coordination is observable in task outcomes providing an environment with appropriate properties in a suitable configuration has been defined. In this context communication is an enabler of coordination. In the next section we outline three team task processes (mediated through communication) that exist to support coordination of action.

Team Coordination Enablers

We have previously suggested that coordination of activity is our aim when developing team and collective capability, and that communication acts as an enabler to coordination. In this section we outline three team processes that support coordination in the highly dynamic and uncertain environments in which military teams may operate.

Within individual decision making we identified three task elements as being of interest from a training and TNA perspective: the constructs of perceived conditions, goals and plans. To recap:

- perceived conditions – refers to information that an individual perceives of the outside world;
- goals – refers to information on what individuals are trying to achieve (information describing future intended states of the environment);
- plans – refers to information that describes the sequential and parallel execution of task actions, including information and resources required and triggering environmental conditions.

While these elements may exist as internal cognitive products and processes, they may also exist as external pieces of information – such as a recipe book containing a set of plans for preparing and cooking food, the checklist procedure for an aircraft engine start, or the Air Tasking Order issued by the air component commander for a military operation.

When we move into a situation where multiple individuals or team members are performing team or collective tasks within a shared environment we have the potential for multiple sets of perceived conditions, goals, plans and actions all at the same time within a single external environment. If a team lacks coherency in its view of the situation, holds different goals (or disagrees on goal priorities) or has conflicting views on the best plan to achieve its goals, clearly this team is not going to perform efficiently. Therefore when we look at our team members we may find varying degrees of alignment between the various individually-held decision-making task elements. This principle may ripple upwards in levels of organisation, such that teams within a larger collective organisation may be well-aligned or not in these areas.

Some potential causes for team members holding different views of a situation are environmental. Things change in the outside world, and sometimes our view of the environment is not what it could be. Different team members may be observing the world from very different viewpoints, with different systems and may have quite divergent conclusions as to what the current environmental situation is. Communication may help maintain alignment between team members, but this can only occur if the mismatch between the constructs (perceived conditions, goals, plans) held by the individual team members is somehow identified. Conversely ineffective communication (or unwillingness of team members to communicate) will not cause an alignment of perceived conditions, plans and goals to occur within the team, even if the divergence of constructs is identified.

Misalignment or divergence may occur between team members' perceived conditions, goals or plans. In the context of perceived conditions, different team members may be observing different areas of the task environment, or may be observing the same area and drawing different conclusions as to what is going on, due to differences in past experiences. In the case of goals, team members may have different goals, or hold the same goals but with a different prioritisation regarding the order in which to achieve them, or which is more important. In the case of plans, our team members might have the same goal but disagree on the set of actions or resources required to deliver it.

Within a team, team members have requirements for different types of information – requirements for environmental information (that is, perceived conditions), requirements for defined goals and planning-related information. This does not imply that 'everybody needs to know everything' but rather that each team member according to their role and the situation will have different requirements; some of these information requirements overlap. If everyone did share everything it could potentially flood communication, cause information overload and effectively generate internal 'noise' which would mask important communication. The requirement for team task processes is to maintain alignment (that is, avoid conflicts) between the task elements that are shared or required to be shared within the team.

We have previously stated that the individual task elements may be considered as both processes and products, and that since the products are information they are capable of being communicated. The information expressed may be very simple or may involve complexities communicated in a more structured and formal manner. Klein (1993: 126), in discussing the development of team models, makes the comment, 'The idea is to take what we know about the way individuals think and bump it up one level as a model of teams'. This is a principle espoused within General Systems Theory, as Pask (1961: 14) comments, 'Further von Bertalanffy realised that when we look at systems … many apparently dissimilar assemblies and processes show features in common. He called the search for unifying principles which relate different systems, General Systems Theory'.

Our view is that team task processes mirror an individual's task element processes as expressed in Table 3.1.

In this view team task processes reflect internal task element processes. Just as an individual engages in planning when performing a task, so must a team engage in collaborative planning and ensure that the products of planning – the

Table 3.1 Team coordination enablers

Individual task element		Team task element	
Products	**Processes**	**Processes**	**Products**
Perceived conditions	Generation and maintenance of perceived conditions	Distributed perception	Team situational awareness
Goals	Goal generation and goal maintenance	Team goal generation and maintenance	Team goals
Plans	Planning, plan execution	Collaborative planning	Team plans

'team plan' – are communicated to team members. Through communication of the team plan, individual team members modify their own individual internal plans to achieve alignment with the team plan, facilitating team coordination. Each team task process (each of which involves communication) generates a corresponding team task product. The process of distributed perception generates team situational awareness, which in turn informs individual perceived conditions. The process of team goal generation and maintenance prioritises team goals – these goals then inform individually held goals.

Team task products exist as external information, whether the content of a face-to-face meeting, the written minutes of a meeting, a plan on the back of a napkin, an email attachment containing a list of objectives or formal documents or orders. Systems may support the update of information, such as overlays on printed maps or fully computerised information systems such as Link 16 – supporting a common tactical picture, displaying real-time positions of units. In these situations individuals reference the team task product and extract the information of relevance to them and their task.

Historically team members were involved in all aspects of these processes, for example in early fighter control operations, human observers on the ground would report the presence of enemy aircraft, including aircraft type, heading, number and altitude back to headquarters by telephone. The headquarters staff would then update the central map by pushing markers over a large physical map, while senior decision makers looking down on the map would then telephone airfields to scramble aircraft to intercept. In equivalent modern systems automation has replaced many of the non-decision-making processes, including automated communication between systems. This means that elements of technology have been inserted into what were previously purely human supported processes.

Team Task Model

Adapting the individual task model 'triangle' of task performance yields a team task model, which reflects team task processes.

The team task model, shown in Figure 3.3, is based upon some simple principles:

1. The purpose of a task is to have some effect upon the environment, or to build information on the environment to support subsequent change.
2. Goals describe the desired outcome of the task.
3. Actions are undertaken to attempt to achieve the desired effect of transforming the environment.
4. Actions generate task outcomes, these outcomes may be positive in that they are in alignment with the goals, neutral or negative.
5. Plans are descriptions of the actions that will be undertaken to achieve relevant goals.
6. Plans are predicated upon the perception of an initial state of the environment and a determination of the desired end state, the goal.
7. Actions described in the plan are enacted to generate outcomes which satisfy the goals.
8. External factors such as the weather and the actions of other agents (such as enemy forces and local population) also have effects on the environment that are not under one's control.
9. The effects of one's own forces' actions and other events on the environment are monitored and goals and plans are revised as required.

This model is consistent with well-established models such as Boyd's Observe, Orient, Decide, Act (OODA) loop model (Boyd 1987). Boyd's observation and orientation phases correspond to the perception and development of situational awareness (SA) activities in the determination of perceived conditions. Decision making in Boyd's model corresponds to the activities of goal generation and planning.

The Team Task Model shown in Figure 3.3 shows how team task processes can be decomposed into different types of activities. The model is composed of two nested triangles. The rectangular boxes of the outer triangle capture the actual state of the external task environment and the events and actions that are occurring within it. The initial and outcome environmental states are linked to show that in reality the environment is continuously changing and the whole model is in fact running as a continuous cycle. The end point of one task is the starting point for a successor task.

The rectangular boxes marking the points of the outer triangle represent the state of the environment and the events and actions occurring within it. The initial environment state is the initial set of conditions within which the task must be executed. The distributed action box captures the execution of actions

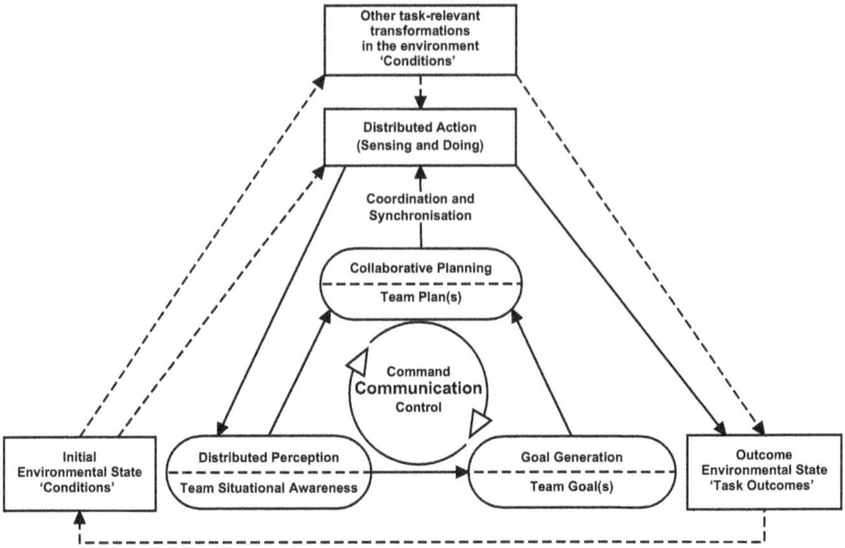

Figure 3.3 The Team Task Model

in accordance with the plan. The term 'distributed' has been used to reflect the fact that in a team or collective context actions are undertaken by multiple actors. Two types of action are identified, those which are intended to have an effect in the environment and those that are intended to develop SA (such as sending out a reconnaissance patrol). The top box captures the actions and events that occur in the environment which are outside the control of the team such as weather and enemy actions. The outcome environmental state is the result of the actions of the team and the other transformations that occur. The degree of mission success is the alignment between the collective task outcome (what has happened in the world) and our overall goal that describes what we were hoping to achieve.

The rounded boxes that mark the points of the inner triangle capture the team decision-making processes and products, which enable coordination and synchronisation of team activity. Coordination of team activity relies on aligned perception of the external environment as team situational awareness, aligned goals and aligned (that is, integrated) plans. Team task processes are shown in the top half of these boxes with the team task products in the lower half. In the centre of the model are the command and control functions, which provide the direction and coordination of all the processes, and communication, which is the core supporting process for team interaction.

A brief description of each component of the Team Task Model is provided below:

- *Initial conditions* – these are the initial set of conditions within which the task must be executed.

- *Dynamic conditions* – these are the actions and events that occur which are outside the control of the team and which may impact team task processes and actions.
- *Distributed perception* – distributed perception is the process by which the team analyses information about the environment, which may come from direct observation or other sources such as intelligence products, in order to build a picture of the current state of the environment and predict future events. Such predictions are typically informed by past experience. The product of the process is team SA. The terms team SA and distributed SA as described by Salmon et al. (2009) should be treated as synonymous, as they capture the fact that in a collective context, members of the collective organisation will each hold different elements of awareness of the environment, the critical issue being the unification of these views to inform planning and command and control. Team SA is a description of the situation as it is understood or assumed to be, which may differ from 'ground truth' (how things really are). The process is distributed because different team members may each have access to different sensor outputs and other information, and are likely to have had different prior experiences. Consequently, each team member is likely to form a different perception of the state of the environment. As such SA is also distributed across the team. The critical point is that effective distributed perception and team SA building depends on the communication of information between team members so that the right information gets to the right person in a timely fashion. This in turn requires team members to know who needs what information and when, and who might hold information that they need.
- *Goal generation* – goal generation yields the overall goal (for example overarching mission aim) and possibly a set of sub-goals. The goal is a description of the desired environmental state to be achieved as the task outcome. Goals may have to be revised during task execution. Task goals may be passed down from a higher level of command. In which case, goal generation typically focuses on identifying sub-goals that need to be achieved to meet the overarching goal.
- *Collaborative planning and team plan(s)* – collaborative planning yields a team plan which is a description of the set of collective actions to be undertaken to achieve the desired effect. Planning needs to include task and resource allocation across the team and capture how team tasks are to be coordinated and synchronised in time and space. When planning is undertaken by a team, with different team members having responsibility for the production of different parts of the plan, this activity itself also needs to be coordinated and the dependencies between different aspects of planning need to be understood to achieve this. This can only be achieved through communication between team members. The resultant plan then has to be communicated to all of the team members responsible for executing it, noting that not all team members may need to know all aspects

of the plan in detail. Planning can be both proactive and reactive. Reactive planning would include developing adaptations to the plan to deal with contingencies of the situation.

- *Distributed actions* – these are the actions executed in accordance with the plan. The term distributed has been used to indicate that in a team or collective context there are multiple actors undertaking simultaneous actions. Three types of action are identified: those which are intended to make a change in the environment (doing), those that are intended to inform perception and situational awareness (sensing), and communication. The coordination and synchronisation of actions in accordance with the plan may also require communication across teams.
- *Task outcomes* – these are the effects in the environment that are a result of the actions of the team and the other transformations that occur. The initial conditions and task outcomes are linked to reflect the reality of continuous and linked change – the whole model is in fact running as a continuous cycle.
- *Command, control and communication* – communication is the major interaction mechanism through which the functions of command and control are executed. In order to ensure that effort is directed effectively within the team, it is inevitable that communication between team members will be required. The key issue is what will the communication be about? If the team is to be effective then there will need to be a common understanding of who is generating and delegating goals, what information is required from the environment, who is doing what, with what resources, and how their activities are aligned. Task allocation and resource allocation need to be determined, and the coordination and synchronisation of actions need to be achieved. The team will also need to be aware of how it is progressing – it would be unfortunate if a team member noticed that something was going wrong but kept quiet on the basis that 'my end of the ship isn't sinking'. Therefore feedback based on performance monitoring within the team is necessary. Central to all of these is information sharing/coordination. These elements which support successful team working are well known and are captured in one form or another in most of the many teamwork models in existence.

The model is applied in the Team and Collective Task Analysis process by using the categories of processes, their connections to each other and to the task environment and the team interactions required to support them, as a shopping list for task processes and interactions when analysing a task.

Command and Control

The Team Task Model places command, control and communication at its centre – in this section we explore the relationship of command and control (C2) to the team task model. Command is concerned with the determination of intent (that is, the specification of goals), whereas control involves the shaping of the environment to deliver our goal. In the analogy of a central heating system in a house, command involves deciding how warm the room should be (that is, the setting on the thermostat), control is the means by which the warm room is generated (that is, the thermostat, the boiler and the radiators).

Alberts and Hayes, in their book *Understanding Command and Control*, point out that 'definitions of C2 are incomplete and potentially worthless unless a means is provided to measure existence or quality' (Alberts and Hayes 2006: 32). They put forward the view that the quality of C2 should be determined by how well C2 functions are performed by the organisation. The C2 functions identified are (Alberts and Hayes 2006: 35):

1. establishing intent (the goal or objective)
2. determining roles, responsibilities and relationships
3. establishing rules and constraints (schedules and so on)
4. monitoring and assessing the situation and progress
5. inspiring, motivating and engendering trust
6. training and education
7. provisioning.

Looking at each of these areas in more detail, allows one to compare established C2 concepts listed above with the team task model developed:

1. Establishment of command intent broadly matches the goal generation and goal maintenance team task process. In a multi-echelon team, a higher decision-making echelon will generate an overarching goal which may then need to be decomposed and distributed to lower echelon teams. In adversarial situations determination of command intent may include consideration of enemy goals. Alberts lists four points for evaluation within this: 'the existence of intent, the quality of its expression, the degree to which participants understand and share intent, and in some cases, the congruence of intent' (Alberts and Hayes 2006: 39).
2. Determination of roles and responsibilities involves the assignment of tasks (and responsibility to deliver task outcomes) to team members or teams. This is an element of planning within the team task model. Alberts identifies three potential quality areas: the completeness of role allocation, the establishment of required relationships and the level of understanding that team members have in the role that they have been assigned (Alberts and Hayes 2006: 41).

3. The C2 function of 'establishing rules, constraints and schedules' also fits within planning in the team task model, with quality measures involving the degree to which the plan is understood and accepted by team members.
4. The C2 function of 'monitoring and assessing the situation and progress' fits within the team task process of distributed perception. The team will be monitoring the effects it is generating in the external environment, and monitoring its level of awareness of the outside world (that is, whether critical information requirements are being met). The team will also be monitoring its own state and the state of its team members in the pursuit of their tasks. Measures of quality in this area include whether monitoring coverage is adequate and whether significant events are noted and reacted to.
5. Inspiring, motivating and engendering trust is a leadership-related C2 function which shapes the behaviour and interactions that are performed by team members. Team task processes may impact individually held KSAs, so leaders can shape behaviour by setting a good example and establishing a positive collaborative climate within the team. The view we adopt is that trust is established through meaningful interaction under realistic conditions rather than being 'trained' *per se*. Trust is one of a number of individual attributes related to positive team performance (Salas et al. 2005), with other examples being openness (Marks et al. 2001) or supportiveness (Gladstein 1984). Authors in leadership literature have identified ways leaders can influence team performance, including how individual team member KSAs may be positively influenced by leaders. Adair (1997) identifies eight leadership functions (broadly divided between task-related and team-related areas): providing an example, motivating, defining the task, planning, briefing, controlling, evaluating and organising. These functions are identified as part of his action-centred leadership model, which is commonly taught during leadership training in the British Armed Forces. There is overlap between leadership functions identified by Adair and C2 functions identified by Alberts and Hayes, which we would expect given that leaders perform a command and control role.
6. The C2 process of 'training and education' may be taken to include the establishment of internal team improvement processes, such as the structured sharing of knowledge that has been placed in context with team members given the time to reflect and discuss situations. More formalised approaches lead to 'lessons learnt' databases, the construction of internal systems for supporting team tasks and internal processes for inducting new team members into their roles. The focus for training may include communication, coordination and individual KSAs, all of which are covered within the models described in this chapter. In a group context, learning processes exist at both the individual and team level, with team learning processes involving elements such as discussion, debriefing, framing, socialisation, reflection, formalisation, rehearsal and application. Teams have an opportunity to engage in internal learning processes, which foster individual learning; this constitutes a unique team task process which spans multiple episodes of team task performance. Not unsurprisingly there are strong linkages between instructional strategy in a team context and team learning as a team task process.

7. Provisioning concerns the sourcing, allocation and distribution of resources within the team. This is an aspect of planning as plans describe not just how things are to be done, but who is performing what tasks and with what resources.

In summary we believe there is a good conceptual alignment between the team task model proposed and Alberts and Hayes' (2006) C2 functions, providing that we add 'team learning' to our existing list of team task processes that support coordination.

Team Performance

We define a team task as a task requiring the coordinated performance of more than one individual in a work group. A team task involves the performance of coordinated action between team members to a shared goal, generating a task product with specified value measures which may be evaluated in performance assessment. Coordinated action may be parallel (independent) or highly interdependent.

Having constructed a team task model we can now extend our model of individual performance (Figure 2.7) to team performance. We have identified coordination of action and communication of information as our focus for training at the team and collective levels, suggesting previously that deficiencies in team or individual level attributes would impact these two team task processes. Figure 3.4 summarises the key relationships between team task processes, communication and team coordination.

As we have suggested previously, task processes still act to transform the initial environmental state to the outcome state. The transformation achieved by the

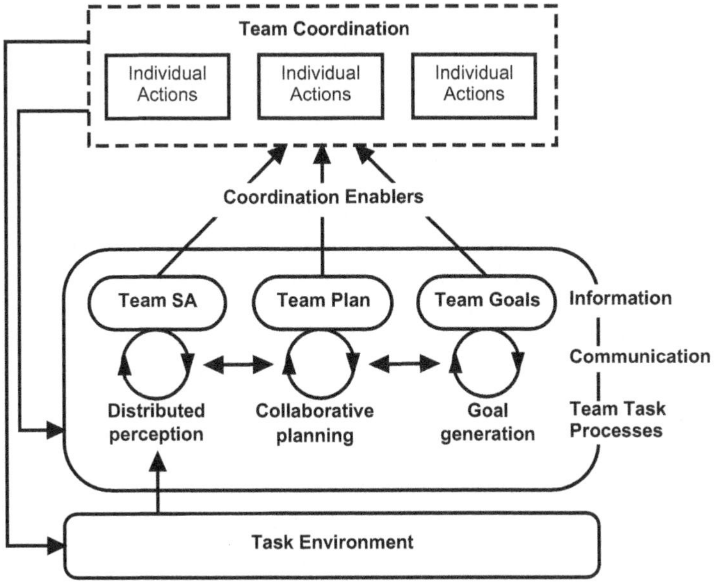

Figure 3.4 Contributing factors to team coordination

actions of the team depends on coordination of individual activity or actions which are distributed throughout the environment in both time and space. The individual actions are based on the products of team task processes (team SA, team plan, team goals), with these products based on the underlying processes of distributed perception, collaborative planning and team goal generation. Communication both distributes the products and supports the processes. Coordination of team activity also impacts on the communication of these products (that is, we need to ensure that team members have the appropriate information at the correct time) whilst also supporting the processes that generate and update team task products.

Our contention is that where issues in team performance exist it is due to a failure of coordination within the team, or individual performance issues. Coordination failures may originate due to individual performance issues, or a deficiency within team task processes. We would expect individuals with individual performance deficiencies to also be implicated in deficient team task processes. Put simply, personnel insufficiently prepared for their roles will exhibit poor individual performance *and* as a result cause deficient team task processes.

In team process assessment and measurement, process measures are generally considered diagnostic (Brannick and Prince 1997: 10). In the model of team coordination outlined in Figure 3.4, we would suggest tracing coordination issues back to the relevant team task process (distributed perception, team goal generation or collaborative planning). Team task processes of distributed perception, collaborative planning and goal generation are all supported by individually held KSAs which support the individual behaviours that together comprise teamwork.

Environmental conditions as characteristics of the external environment generate demands on teams that have to be mitigated by the team task processes with which the team engages. Where a situation may be said to be 'difficult' for a team to deal with, this difficulty stems from the mismatch of the demands of the situation and the capacity of the team to react to those demands and continue to generate the task outcomes required. For example in an environment that is dynamic, poorly understood, ambiguous or where past 'pattern of life' information is unavailable, the team task process of distributed perception to build team SA may be challenged and require great effort. These environmental factors have been identified as characteristics of naturalistic decision environments (Orasanu 1993), or team stressors (Cannon-Bowers and Salas 1998). These characteristics may challenge individuals and interfere with the team task processes that are being undertaken. The impact of naturalistic decision environment characteristics or environmental stressors on team performance is discussed in more detail in the task environment chapter, but in short we hypothesise that different environmental demands generate differing requirements and demands for the development of team task processes.

The model for team performance, Figure 3.5, is an extension to individual performance concepts, with the addition of team attributes (such as organisation or the allocation of roles within the team) affecting individual attributes and team task processes. Team attributes relate to how the team has been structured and how roles have been allocated within the team. As such they represent the structural aspects within which the team task processes operate.

In a team we have multiple individuals so now we have as many sets of individual attributes as we have individuals in the team. We also have a new set of individual attributes to consider which are those KSAs or competencies that support team task processes. As we have previously identified, team task processes are a superset of individual task processes, with the addition of communication, coordination and coordination enabling processes (distributed perception, goal generation, collaborative planning) in a team context. To this list we add team learning as a process which while not necessarily relevant for a single instance of team task performance becomes relevant for team performance over time.

Figure 3.5 summarises the relationship between key concepts.

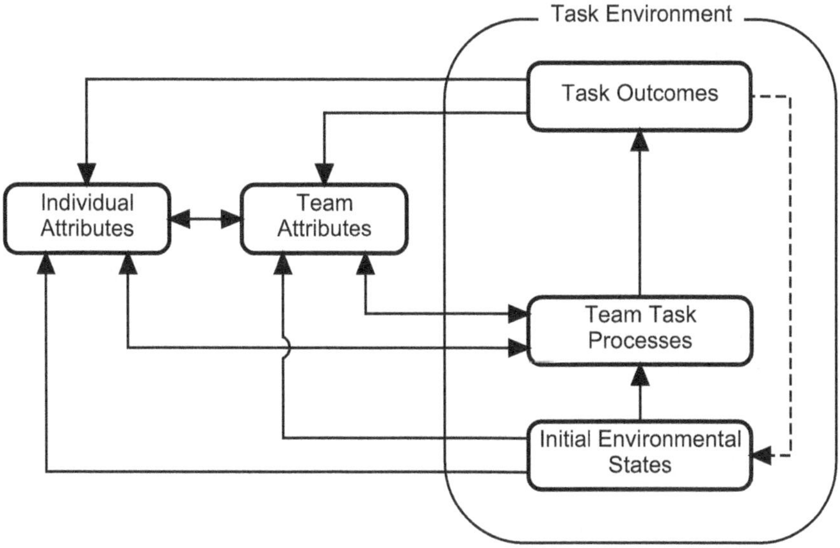

Figure 3.5 Team Performance Model

Team Adaption and Team Learning

Teams do more than simply perform tasks; teams adapt to changing conditions and challenges presented by the environment. In adapting, teams alter their task processes in reaction to their task environment to continue to generate the task outcomes that are required. The same principle of adaption applies in the individual context of performance, however the task processes in an individual context exclude coordination and communication, as well as the processes of distributed perception, collaborative planning and team goal generation. An individual performing an individual task is also divorced in an operational setting from any opportunity to engage in a team learning process.

Numerous authors have identified the requirement for teams to react to rapidly changing operational contexts (Rosen et al. 2011). This has been characterised as adaptive team performance or team adaptation, where what is being adapted is the team's performance processes (ibid.). The model of team adaptation advanced by Burke et al. (2006) has team adaptation (also known as team innovation and team modification) as the output of adaptive team performance. While we would agree that teams modify their team task processes in response to alterations in their environmental conditions, we would contend that the reason they do this is to modify their external environment. The implication is that when measuring team adaption we can do this both by monitoring the changes in team processes and by monitoring changes in the environment – by observing the outcomes or effects being generated by the team. A failure of adaption may be seen as a failure of the team to generate required task outcomes under changing conditions.

The impact of task-significant changes in environmental conditions is to require adaptive alterations in the following team task processes:

- *Coordination and synchronisation of actions may alter* – procedural coordination according to a predefined procedure or plan may be replaced by adaptive coordination. The implications for this shift are discussed in Chapter 4 – team task analysis.
- *Team distributed perception* – generates demands on the team distributed perception process as information on changes in the task environment must be distributed and shared within the team to maintain team SA.
- *Team SA* – changes in team SA may cause emergent team goals to be generated, team goals to be dropped or team goals to be re-prioritised.
- *Team goal generation* – changes in team goals may cause collaborative planning to occur which may include reactive planning and the dynamic reallocation of resources, information, actors and tasks.

All of these team task processes rely on communication in various forms and the generation of information products.

As well as adaption, teams may engage in learning processes simply through the performance of their tasks, independent of any instructional strategy we may implement – providing those tasks are supported in the workplace as part of 'day-to-day' activity. Team learning and team adaption are related and overlapping concepts, which are not synonymous. Teams may learn without adaption – such as the training of team Standard Operating Procedures through team drills. In these situations we do not necessarily want the team or team members improvising new innovative ways of doing things as this may cause confliction or accidents.

Teams may also adapt without learning, something innovative may be done in reaction to an unusual situation, but that adaption may not necessarily be socialised, shared, placed in context and reflected on. In these situations the team has adapted as a one-off to a situation, there is no guarantee that this adaption or innovation will be deployed in a future context. Finally we have situations where both adaption and learning occurs.

While adaption is often discussed in the context of a dynamic environment (Rosen et al. 2011), adaption could be viewed as simply the team managing the demands of the task environment. Adaption in non-dynamic environments might include having to operate in a wide range of potential situations (for example desert, snow, jungle and urban terrain), adaption in a relatively static but highly uncertain environment (such as found in low visibility environments) or highly stressful environments.

We have previously characterised learning as the acquisition of new KSAs that individuals previously did not possess; however, these are not necessarily the only outputs of learning. In an organisational context, teams may become reorganised or re-equipped after episodes of performance, new equipment is procured, new concepts developed, dangerous situations recognised or specific Tactics, Techniques and Procedures (TTPs) developed. Furthermore over time, team learning processes will become improved.

In this sense the outputs of the team learning process may impact individual team member attributes (KSAs, task-related expertise and experience, mental models, emergent states), team task processes and team attributes such as team organisation and the allocation of roles and responsibilities.

The Team Performance Model

The Team Performance Model (Figure 3.6) informs the conduct of Team and Collective Task Analysis. It also provides a breakdown of the nature of task outcomes, which is of relevance to both Team and Collective Task Analysis and to training strategy. Within this model (as part of Command and Control) is the representation of leadership as a specialised team task process which acts to modify team attributes such as team organisational structure and role assignment (as seen in Adair's 'organising' leadership function; Adair 1997). Leadership as a team task process also affects the individual attributes of other team members, for example in Adair's 'providing an example' and 'motivating' (ibid.). We also include team learning processes within Command and Control as a process relevant to task performance over an extended period of time.

The model shows how the team responds to cues from the task environment to produce outcomes as a consequence of carrying out team task processes. The details in the task environment and team task processes boxes are based on the constructs defined in the Team Task Model and the Training Environment Model respectively.

Task outcomes are categorised as being: intended outcomes, other outcomes or critical errors and their consequences. These are defined as follows:

- *Intended outcomes* – effects in the environment achieved by the team which are aligned with the team's goals.

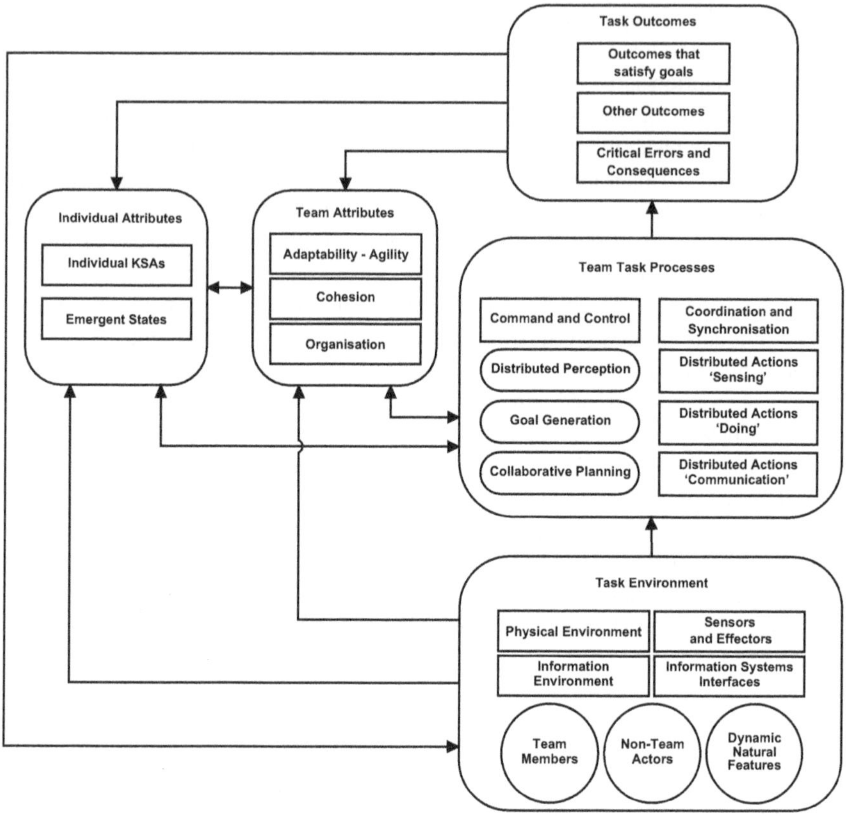

Figure 3.6 Team Performance Model – expanded

- *Critical errors and consequences* – outcomes that have occurred due to the team's actions which are undesirable, as far as the team is concerned. The identification of critical errors and consequences is of significance for determining whether training is required. Within Training Overlay Analysis a risk assessment is carried out to determine if the risk associated with the errors that would be likely to occur if the team were not trained is acceptable. If not then team training is deemed to be necessary.
- *Other outcomes* – the concomitant states that result from the pursuit of our goals, for example if we drive to our destination (our goal) the car will consume petrol. The consumption of petrol is not a goal, and it does not constitute a critical error or consequence, rather it is a neutral and concomitant consequence of our goal directed actions, and can therefore be characterised as an 'other outcome'. It is useful to identify other outcomes as they can highlight elements such as consumable resources that need to be captured in the analysis.

The model also shows the team and individual attributes that can impact on team performance.

Team Member Attributes include:

- *Individual team member KSAs* – relatively stable attributes possessed by individuals which may be developed in training. Given the complexity of team activity effective team interaction is facilitated by all of the team members having a shared understanding of how the team carries out its tasks. As such, there is a requirement for all team members to hold some underpinning knowledge about the team and how it operates. Such knowledge typically includes:
  - team/team member roles
  - team/team member capabilities
  - task models
  - team interaction patterns
  - doctrine
  - Standard Operating Procedures (SOPs)
  - TTPs.

  These need to be identified in the Team and Collective Task Analysis as they may need to be trained. Skills may include the requirement to operate and exploit systems to share information. Attitudes in a team context include developing a positive orientation towards the team, supportiveness and the opportunity to develop trust between team members.
- *Emergent states* – that an individual possesses reflects how individual performance in any given moment is influenced by being in the environment and undertaking the task. For example, the effect of incoming mortar fire may negatively impact on how well an individual can perform, whereas the successful completion of a task may have a positive effect. Cannon-Bowers and Salas' (1998) extensive observation of teams in training has led to the identification of the following factors as significant *environmental stressors* on individual and team performance:
  - threat/stress
  - performance pressure
  - high workload
  - time pressure
  - adverse physical conditions
  - shifting goals
  - high information load
  - incomplete and/or conflicting information
  - multiple information sources
  - visual overload
  - communication difficulties
  - ill-structured problems
  - rapidly changing and evolving scenarios (dynamic environments, uncertain environments).

Identifying the environmental variables that can give rise to these stressors ensures that the training environment supports an appropriately realistic and demanding environment for the team to operate within.

Team attributes that may impact coordination and communication include:

- *Adaptability* (also referred to as agility) – refers to the ability of the team to react to changing circumstances and is a much sought after attribute of military teams. It is promoted by presenting teams with a variety of novel problems to solve, which underlines the importance of capturing all of the appropriate environmental variables that can impact on a task when doing the analysis. Adaptability manifests itself in how the team responds to novel circumstances in terms of how it adjusts its planning and coordination in task execution.
- *Cohesion* – in contemporary research, cohesion is considered to be a multi-dimensional construct. Carless and De Paola (2000) identified three constructs which appeared to be at play in a work context:
  - *task cohesion* – the extent to which the team is united and committed to achieving the work task;
  - *social cohesion* – the degree to which team members like socializing together;
  - *individual attraction to the group* – the extent to which individual team members are attracted to the group.

  Of these constructs, task cohesion was found to be the one most strongly related to team performance. Whilst intuitively the notion of cohesion may have some resonance, its measurement is problematic. Typically, its measurement involves the administration of questionnaires to all individuals in a team. This type of approach is unlikely to be practical in a team or collective training context which has a strong task focus.
- *Organisational structure and stability of the team* – both how the team is structured and how it is located can impact on team performance. For example, a team that is widely distributed geographically is likely to face more challenges in communication than a team that is co-located in a single workspace. Training exercises can provide commanders with opportunities to experiment with and fine-tune team organisation. Organisational structure also includes how roles and responsibilities have been allocated within the team. However, team organisational structure cannot be 'trained' or 'measured' as such. That said, a skilled observer may be able to identify if structural weaknesses are impacting on team performance. For example, a convoluted decision-making chain may impede responsiveness.

Team Performance Measures

There are a number of implications of the Team Performance Model on the analysis, design and delivery of team and collective training. The following areas may be measured as contributing factors to teams delivering required outcomes:

- team task processes
- intermediate task outcomes generated
- individual KSAs supporting teamwork
- individual competencies or behaviours supporting teamwork.

The specification of measures of team performance for assessment is required in TNA because assessment requirements drive the requirements of both the environment that needs to be put in place, the events it supports and any instrumentation that be required to capture team performance. Training environments should be able to capture information on team and individual performance enabling diagnostic remediation by the instructional team. The question is therefore what to measure, how to record and what to playback?

Team performance measures are also required in TNA because training strategy relies on training staff being able to identify, describe and assess observed team performance. Training strategy relies on suitably experienced instructors who have requisite skills to enable team performance assessment, and TNA is the stage where this is specified. This is of particular importance where evaluation is based on team observation where the team working environment is not recorded, and individual teamwork behaviours (the observable component of a team member engaging in teamwork) are being observed and judged.

Finally, as previously identified in our Team Performance Model, team performance may involve individually possessed KSAs that support observable teamwork competencies and behaviours which are woven into team task processes. Identifying and describing these KSAs and competencies in the format of observable behavioural indicators forms part of what we may be trying to develop in our teams, and bringing the team's awareness to these factors may help support internal team learning processes.

The Measurement of Team Task Processes

Team coordination – the alignment of distributed activity over time to deliver goal-aligned team task outcomes – is a primary process measure that should be addressed in team and collective training. Coordination has been identified by numerous authors in teamwork literature (Denson 1981; Morgan et al. 1986; Dickinson and McIntyre 1997; Annett et al. 2000; Rousseau et al. 2006). Alongside and co-dependant with coordination is communication (Dickinson and McIntyre 1997: 21; Annett et al. 2000). Team coordination is co-dependent on communication as poor communication will impact on team coordination just as poor coordination will impact on communication (Annett et al. 2000). If individual actions are considered as pieces of a jigsaw, team coordination is ensuring that all pieces are fitted together appropriately and communication provides the mechanism for achieving this. If we view individual performance as a jigsaw piece, deficiencies in the shape of jigsaw pieces (that is, individual performance issues) may impact on coordination and communication. Deficiencies in communication and coordination may, however, occur because of team integration or team structural issues, which are not based in the individual (for example poor communication systems provision or design).

Team coordination cannot exist without team SA, team plans and team goals being aligned and appropriately distributed within the team. Communication is the mechanism that supports distributed perception, collaborative planning and goal generation processes. Table 3.2 identifies 12 potential areas of interest when observing a team performing a task. These are process-based measurement areas, which complement the product or outcome-based measures of team performance.

Table 3.2 Twelve potential team process measurement areas

Process issue	Coordination	Communication
Individual behaviour	(Area 1) KSA deficiencies	(Area 2) KSA deficiencies
Team level	(Area 3) Structural or role assignment deficiencies	(Area 4) Communication deficiencies
	Process	**Product**
	(Area 5) Distributed perception	(Area 6) Team SA
Team coordination enablers	(Area 7) Goal generation	(Area 8) Team goals
	(Area 9) Collaborative planning	(Area 10) Team plan
Leadership Processes	(Area 11) Leadership	
Team Learning Processes	(Area 12) Team learning	

Failures or errors in one process area (such as collaborative planning or communication) will propagate through the system degrading team coordination and in turn the collective task outcomes generated by the team. While one can characterise broad issues in, for example, communication between two team members, one could also go deeper and ask what the subject or content of the communication was. Was it planning related, or related to an observed change in the external environment, or a reassignment of a task to another team member? As such, while process-based measures have traditionally been considered diagnostic, the implication of a multi-layer system model is that a measure such as 'coordination' may not be granular enough. We may need to drill deeper to understand *why* exactly coordination failed. The implication for TNA is that training environments need to possess features to enable the capture of data in these areas, and instructional strategy needs to focus on these areas as a subject of training content and an area for potential assessment and debrief.

Prior authors have identified behavioural examples in various areas of teamwork. Table 3.3 fits teamwork skill dimensions and behavioural examples from Baker and Salas (1992), who adapted findings from Morgan et al. (1986). Here, teamwork skill dimensions relate to individual behaviour in a team context, and therefore correspond to area 1 and area 2 of the 12 areas within team process measurement.

Table 3.3 Teamwork skill dimensions and behavioural examples (Baker and Salas 1992, adapted from Morgan et al. 1986)

	Coordination	**Communication**
Individual behaviour	**(Area 1)** Cooperation – prompted another member on what to do next. Coordination – provided direction on what the team members had to do next. Adaptability – changed the way he or she performed a task when asked to do so.	**(Area 2)** Sharing suggestions or criticisms – asked if the procedure or information was correct when he or she wasn't sure. Communication – asked for specific clarification on a communication that was unclear. Acceptance of suggestions or criticism – thanked another crew member for catching his or her mistake. Team spirit and morale – discussed ways of improving team performance.

Table 3.4 illustrates a more extensive example of fitting prior team process research into the 12 area team process measurement construct. This example is based on the work of Rosen et al. (2011) who presented a summary of markers of team adaptability processes referencing work from Marks et al. (2001) and other authors (Kozlowski et al. 1999; Fleishman and Zaccaro 1992; Sutton et al. 2006; Burke et al. 2006; Stagl et al. 2006; Smith-Jentsch et al. 1998; McIntyre and Salas 1995; Mohammed and Dumville 2001; Ilgen et al. 2005; Edmondson 1999; Kasl et al. 1997; van Offenbeek 2001).

Table 3.4 Markers of team adaptability processes (Rosen et al. 2011)

	Coordination	Communication
Individual behaviour	**(Area 1)** *Coordination* • Articulate information about their status, needs and objectives as often as necessary (and not more) • Sequence team taskwork behaviours to minimise downtime within interdependent tasks • Synchronise teamwork behaviours without overt communication in high workload conditions • Resolve any conflicting task demands through role negotiation *Back-up behaviour* • Proactively assist fellow team members with task work • Redistribute workload and adjust resources to fit task requirements across all roles in the network, that is, workload balancing	**(Area 2)** *(Coordination related communication)* • Pass information to one another relevant to the task in a timely and efficient manner *(Back-up behaviour related communication)* • Provide feedback to fellow team members to facilitate self-correction *(Reactive conflict management related communication)* • Utilize negotiation or mediation strategies for conflict resolution *Team communication* • Clearly communicate problem definitions • Follow up to ensure that messages are received and understood • Acknowledge messages when they are sent • Cross check information with the sender, to ensure that the message meaning is understood • Articulate 'big picture' to one another as appropriate • Proactively pass information without being asked *Affect management* • Regulate problematic emotional responses
Team level	**(Area 3)** *Role differentiation* • Match member KSAs to subtask requirements • Explicitly state boundaries of responsibilities between team members	**(Area 4)** *Affect management* • Provide opportunities for development of social cohesion
	Process	
Team coordination enablers	**(Area 5) Distributed perception** *Affect management* • Identify problematic emotions that negatively impact team *Cue recognition* • Scan environment for cues that might influence the outcome of the mission • Rapidly detect problems or potential problems in their environment	

Table 3.4 Markers of team adaptability processes (Rosen et al. 2011)

	Process (cont)
Team coordination enablers (cont)	*Meaning ascription* • Explicitly define potential problems • Accurately assess the underlying causes of environmental changes • Categorize issues based on relevance to mission success • Map relationships between relevant information and mission success by identifying impact of cues on mission • Generate solutions to potential problems by drawing upon mental models of previously used courses of action *Mutual monitoring* • Observe team members' performance to identify errors • Identify unbalanced workload distributions *Systems monitoring* • Track internal systems resources such as personnel, equipment and other information that is generated or contained with the team • Track the environmental conditions relevant to the mission
	(Area 7) Goal generation *Goal specification* • Identify what needs to be accomplished, within what specified time frame at what level of quality • Explicitly define desired outcomes *Mission analysis* • Explicitly articulate the team's objectives • Discuss the purpose of the team in the context of the present performance environment
	(Area 9) Collaborative planning *Mission analysis* • Discuss how the available team resources can be applied to meeting the team goals • Agree on the methods and approaches the team should take to work towards its goal (for example low task conflict related to selection of performance strategies) *Goal specification* • Prioritize the required actions of the team for mission success *Strategy formulation: deliberate planning* • Explicitly articulate expectations for how a proposed course of action should unfold • Brainstorm plans and strategies for solving problems • Categorize, condense, combine and refine expectations into mutually agreed upon list of additional tasks and information requirements *Strategy formulation: contingency planning* • Explicitly articulate expectations for how a proposed course of action should unfold • Brainstorm plans and strategies for solving problems • Categorize, condense, combine and refine expectations into mutually agreed upon list of additional tasks and information requirements *Strategy formulation: reactive strategy planning* • Modify routine performance strategies in response to unexpected environmental or task changes

Table 3.4 Markers of team adaptability processes (Rosen et al. 2011) (*concluded*)

	Process (cont)
Team coordination enablers (cont)	**(Area 11) Team learning processes** *Recap: information search and structuring* • Filter and store the filtered information concerning past performance • Perceive cues and corresponding information concerning past performance • Seek information and feedback from other team members about past performance *Recap: review events* • Refer to past adaptation/performance cycles • Summarize key features of past performance episode *Reflection/critique: active listening* • Accept reasonable suggestions • Respond to other's requests for information • Avoid sarcastic and non-constructive comments *Reflection/critique: framing/convergent interpretation* • Frame the situation to provide the initial perception of the past performance episode. • Discuss errors *Reflection/critique: reframing/divergent interpretation* • Integrate and share views of causes of effective and ineffective past performance • Engage in 'what if' scenarios to reinterpret causes of past performance *Reflection/critique: strength/weakness diagnosis* • Discuss the consequences of their actions • Discuss how unintended consequences can be avoided • Discuss how an ineffective course of action or process can be revised *Summarize lessons learned: accommodation/integration* • Articulate plans for changes in future performance processes • Team members create new mental models by integrating viewpoints • Team members decide and act on new mental models

We have structured the team processes identified by the authors above into the respective team process measurement areas derived from the team task model. As part of the categorisation process we have generated titles in brackets to indicate where we see the behavioural indicators sitting within the team process measurement areas.

From a team training perspective the identification of team task processes (which are an expansion of individual task processes) leads to the implication of expanded KSA requirements to underpin these team task processes, over and above individual task requirements. The teamwork-related KSAs support observable teamwork behaviours. Effective teamwork behaviours may be described as teamwork competencies and may be referenced to specific team tasks or expressed in a generalised format.

The nature of the process-based assessment of team task behaviours (also referred to as behavioural assessment), as identified by Morgan et al. (1986) in their Critical Team Behaviours Form, leads to an appreciation of the importance of the observation of team task processes (Baker and Salas 1992). Within a TNA perspective this generates requirements related to both instructional systems for the recording and playback of trainee behaviour and communication for debrief, and within subsequent instructional design as SME input is vital in determining markers or indicators of effective team task process measures. If coordination and communication are co-dependent our training environments should be able to capture activity in both areas and view both comparatively. This then leads to the specification of requirements for the instructional environment to ensure that these factors may be captured in team practice.

In the model of team performance measurement we outline, while measurement of coordination, communication and process-based team coordination enablers is vital, one should not overlook the product based aspects of team coordination enablers. In teams relying heavily on information systems that distribute real-time information relating to team SA, team goals or the team plan, the information quality of these products (for example Link 16 datalink playback as part of the recognised air picture) may well be a useful indicator of team performance. In Command Post Exercises (CPXs), this may be a particularly pertinent form of measurement, as the focus of these exercises is the processes that generate information rather than the external physical transformations that these information products enable. This also reinforces the importance of maintaining views of both ground truth (as the information system that reflects 'real' reality, accessed by instructors and training exercise controllers), and the picture that the team is constructing (such as the Blue Common Operational Picture distributed through the command system). The way that information products may be analysed (offline, post exercise) may also provide an alternative to the presence of observers in the team exercise setting, capturing information on the process aspects of team coordination enablers.

Measurement of Outcomes and Intermediate Outcomes

An alternative and supplementary approach to the measurement of team task processes described above, which relies on the observation and evaluation of particular behaviours or events of teamwork during task performance, is to measure whether intermediate outcomes are being reached. This relies on a framework which can relate the intermediate outcomes (akin to project milestones) that have been generated by the team within a network structure (akin to a project management Gantt chart) which defines what the overarching outcomes are. The ability to instrument the external environment (or some suitable proxy synthetic environment) is beneficial, however teams generate effects in both their external environment, and through the processing, transformation and communication of information. Such information environments and internal communication systems/environments may be recorded or copied and played back in subsequent debrief if required. The terminal node on the network graph of intermediate outcomes is the final task outcome or task output.

Team output measures are referenced to task outcomes and their associated goal criteria. The task environment, as the external conditions and demands generated by the environment, is what drives team task processes. Team task processes act to mitigate the demands of the environment and keep the team 'on track', in delivering its required outputs. If in training we fail to replicate environments in the type of demands seen or anticipated 'on operations' we may be in danger of practising 'the wrong type of busy'. Similarly our training environments need to be adaptable to be able to present a range of possible initial environmental states. If goal criteria are omitted the description of the exact conditions that were in effect and the resources and manning state of the team may add useful information to what may otherwise be a simple binary outcome score.

Individual Behaviours Supporting Teamwork (Teamwork Competencies)

All team task processes which extend over the duration of the team task consist of a set of linked and interwoven behaviours expressed by members of the team. We have split teamwork supporting behaviours into two main areas; those related to communication and those related to coordination. Through coordination and communication we achieve integration within the team or collective organisation. As has been outlined previously, coordination is the activity and capability we are looking to develop and this is supported by the enablers of team coordination – a shared view of the situation, aligned goals and an integrated plan. These three coordination enablers are both process (communication) and product (information) based.

The list below offers a set of suggested behavioural indicators in the team task processes identified in the team task model. These indicators can be applied at the individual behavioural level (a team member displayed or did not display these behaviours). Alternatively these competencies can be applied more generally at the process level – did the observed process shared by the team over time have these attributes?

As part of the training overlay, instructors need to have skills to observe and debrief on these behavioural items. From an environmental perspective our environments need to be able to support the monitoring or recording of communication and activity that demonstrates behaviour in these areas.

Communication

Communication fundamentals:

- confirms communication channels function
- monitors communication
- confirms recipient identity
- confirms message has been received, follows up when necessary
- uses appropriate terminology, syntax and phraseology

- performs read-back to check message sent is that received to ensure alignment of information
- error-corrects bad information
- communicates to correct agent or agency by most appropriate channel
- communicates only when required to maintain clear channels of communication
- actively clarifies potential ambiguity in communication
- prioritises and organises communications appropriately
- does not overly/unnecessarily protect information
- provides feedback in team discussions.

Distributed Perception

Communication to maintain distributed perception in the team – communicates to maintain a shared understanding of situation and team performance:

- reports significant changes or unexpected features in the task environment
- reports blind spots or coverage gaps, actively seeks information
- clearly articulates level of uncertainty in the environment
- attempts to clarify uncertainty or ambiguity in the environment, checks assumptions
- discusses conflicting environmental information
- maintains and communicates multiple hypotheses on environmental situation concurrently
- shares views on projected changes in the task environment
- identifies opportunities offered by the task environment
- shares or requests views on own performance
- contributes to discussions on team performance and effectiveness
- reports errors
- reports progress against the plan including completion
- queries activity drifting off plan or tacit goal drift.

Goal Generation

Communicates to clarify, establish, prioritise, align and maintain team goals:

- shares or verbalises goals when appropriate
- engages in goal generation process including interpretation of higher intent through goal decomposition
- discusses relative merits of goals and priorities
- defines and effectively communicates goals to maintain team cohesion
- identifies and responds to goal divergence
- communicates potential critical errors or risks

- anticipates and communicates conflicting goals or priorities
- maintains the goal under adversity
- adapts to situation by switching or dynamically generating goals when opportunity presents (within command parameters)
- ensures team understands the goal(s) and priorities
- issues valid commands when required.

Collaborative Planning

Communicates to integrate and develop plans within the collaborative planning process:

- communicates during the collaborative planning process to ensure plan integration
- anticipates shared dependencies (information, resources, agents, tasks) in planning that may lead to conflict
- de-conflicts future activity as part of planning
- identifies unreachable or unrealistic goals as part of planning
- defines critical information requirements for tasks and distributed perception requirements for tasks
- defines and communicates plan-critical variables and values
- allocates resources and actors to tasks
- critiques, discusses and requests feedback on plans where relevant
- clarifies plan ambiguity, provides guidance
- communicates or issues the plan with appropriate plan coverage and detail
- communicates explicit coordination strategy.

Coordination and Communication in Support of Coordination

Enacts plans into action and engages in both procedural and adaptive coordination where necessary:

- takes the initiative and has a bias for action
- anticipates the task and role-based coordination requirements of others
- establishes acceptable coordination boundaries within the task
- actively and openly collaborates
- engages in supportive backup behaviour
- utilises knowledge of task and situation to define an explicit coordination strategy
- reassigns tasks and actors to best effect
- distributes workload throughout the team
- coordinates action with minimum communication overhead
- communicates when necessary to synchronise activity according to the plan
- engages in coordination activity where trust is required.

Individual KSAs Supporting Teamwork

Teamwork supporting KSAs are divided into two main areas, those related to either communication or coordination. Table 3.5 is adapted from Stevens and Campion (1994) with their original headings placed into the coordination or communication category. These teamwork supporting KSAs form part of the broader individual team member KSAs, and are potential topics for individual instruction to trainees in how to engage in effective teamwork.

Cannon-Bowers et al. (1995) initially compiled a list of required team competencies. Here the competencies are divided into Knowledge, Skills or Attitudes. This list has been subsequently elaborated by further research (Cannon-Bowers and Salas 1997: 47; Salas et al. 2001; Baker et al. 2003) (see Table 3.6).

Table 3.5 KSAs supporting teamwork (Stevens and Campion 1994)

(Coordination and Coordination enablers)
Conflict Resolution KSAs • The KSA to recognize and encourage desirable, but discourage undesirable, team conflict. • The KSA to recognize the type and source of conflict confronting the team and to implement an appropriate conflict resolution strategy. • The KSA to employ an integrative (win-win) negotiation strategy rather than the traditional distributive (win-lose) strategy. *Collaborative Problem Solving KSAs* • The KSA to identify situations requiring participative group problem solving and to utilize the proper degree and type of participation. • The KSA to recognize the obstacles to collaborative group problem solving and implement appropriate corrective actions. *Goal Setting and Performance Management KSAs* • The KSA to help establish specific, challenging and accepted team goals. • The KSA to monitor, evaluate, and provide feedback on both overall team performance and individual team member performance. *Planning and Task Coordination KSAs* • The KSA to coordinate and synchronize activities, information and task interdependencies between team members. • The KSA to help establish task and role expectations of individual team members and to ensure proper balancing of workload in the team.
(Communication)
Communication KSAs • The KSA to understand communication networks, and to utilize decentralized networks to enhance communication where possible. • The KSA to communicate openly and supportively, that is, to send messages which are: (1) behaviour- or event-oriented; (2) congruent; (3) validating; (4) conjunctive; and (5) owned. • The KSA to listen non-evaluatively and to appropriately use active listening techniques. • The KSA to maximize consonance between nonverbal and verbal messages, and to recognize and interpret the nonverbal messages of others. • The KSA to engage in ritual greetings and small talk, and a recognition of their importance.

Table 3.6 Essential team knowledge, skill and attitude competencies (Cannon-Bowers and Salas 1997: 47)

KNOWLEDGE COMPETENCIES	
Competency	**Definition**
Cue/Strategy Associations	Linking cues in the environment with appropriate coordination strategies
Shared Task Models/Situation Assessment	A shared understanding of the situation and appropriate strategies for coping with task demands
Teammate Characteristics Familiarity	Knowing the task-related competencies, preferences, tendencies, strengths and weaknesses of teammates
Knowledge of Team Mission, Objectives, Norms and Resources	Meaningful for responding to a specific team and task – when change occurs, knowledge must be adjusted to incorporate new team members and task demands
Task-Specific Responsibilities	Integrating task inputs according to team and task demands
SKILL COMPETENCIES	
Mutual Performance Monitoring	Tracking fellow team members' performance to ensure that the work is running as expected and that proper procedures are followed
Flexibility/Adaptability	Ability to recognize deviations from expected course of events to readjust one's own actions accordingly
Supporting/Back-Up Behaviour	Providing feedback and coaching to improve performance or when a lapse is detected; assisting teammate in performing a task; and completing a task for the team member when an overload is detected
Team Leadership	Ability to direct/coordinate team members, assess team performance, allocate tasks, motivate subordinates, plan/organize and maintain a positive team environment
Conflict Resolution	Ability to resolve differences/disputes among teammates, without creating hostility or defensiveness

Table 3.6 Essential team knowledge, skill and attitude competencies (Cannon-Bowers and Salas 1997: 47) (*concluded*)

Feedback	Ability for team members to communicate their observations, concerns, suggestions and requests in a clear and direct manner, without creating hostility or defensiveness
Closed-Loop Communication / Information Exchange	The initiation of a message by the sender, the receipt and acknowledgement of the message by the receiver, and the verification of the message by the initial sender
ATTITUDE COMPETENCIES	
Team Orientation (Morale)	The tendency to enhance individual performance through coordinating, evaluating and using task inputs from other group members while performing group tasks
Collective Efficacy	The belief that the team can perform effectively as a unit when given specific task demands
Shared Vision	Commonly held attitude regarding the direction, goals and mission of a team
Team Cohesion	The total field of forces that influence members to remain in a group; an attraction to the team as a means of task accomplishment
Mutual Trust	A positive attitude held by team members regarding the aura, mood or climate of the team's internal environment
Collective Orientation	The belief that a team approach is better than an individual one
Importance of Teamwork	The positive attitude that team members exhibit toward working as a team

The development and observation of the behavioural competencies associated with teamwork and the development of KSAs supporting teamwork are both subjects for initial instruction into teamwork, and must be supported through the instructional functions contained in the training overlay, which is discussed in Chapter 6.

The function that the training overlay provides in supporting and developing the team learning process, and methods for supporting and reinforcing the group's own ability to learn are also discussed in Chapter 6. This means we can develop the team's own ability to learn outside formal team/collective training, by developing the team's good learning habits, just as an individual can be taught 'study skills' in an individual context.

Having developed a model for tasks and team task processes within the team and external environment in Chapters 2 and 3, we now discuss the implications for task analysis in a team context for TNA.

References

Adair, J. (1997) *Leadership Skills*. London: Chartered Institute of Personal Development.

Alberts, D.S and Hayes, R.E. (2006) *Understanding Command and Control*. Washington, DC: CCRP Publication Series.

Annett, J., Cunningham, D. and Mathias-Jones, P. (2000) A Method for Measuring Team Skills, *Ergonomics*, 43(8): 1076–94.

Baker, D.P., Day, R. and Salas, E. (2006) Teamwork as an Essential Component of High-Reliability Organizations, *Health Services Research*, 41(4): 1576–98.

Baker, D.P., Gustafson, S., Beaubien, J.M., Salas, E. and Barach, P. (2003) *Medical Teamwork and Patient Safety: The Evidence-Based Relation*. Washington, DC: American Institutes for Research.

Baker, D.P. and Salas, E. (1992) Principles for Measuring Teamwork Skills, *Human Factors*, 34(4): 469–75.

Bass B. (1980) Team Productivity and Individual Member Competence, *Small Group Behaviour*, 11: 431–504.

Boyd, J.R. (1987) *Organic Design for Command and Control*. Available at http://www.ausairpower.net/APA-Boyd-Papers.html [accessed 20 February 2013].

Brannick, M.T. and Prince, C. (1997) An Overview of Team Performance Measurement, in M.T. Brannick, E. Salas and C. Prince (eds) *Team Performance Assessment and Measurement*. Mahwah, NJ: Psychology Press. 3–16.

Burke, C.S., Stagl, K.C., Salas, E., Pierce, L. and Kendall, D.L. (2006) Understanding Team Adaptation: A Conceptual Analysis and Model, *Journal of Applied Psychology*, 91: 1189–207.

Cannon-Bowers, J.A. and Salas, E. (1997) A Framework for Developing Team Performance Measures in Training, in M.T. Brannick, E. Salas and C. Prince (eds) *Team Performance Assessment and Measurement*. Mahwah, NJ: Psychology Press. 45–62.

——— (1998) *Decision Making Under Stress: Implications for Individual and Team Training*. Washington, DC: American Psychological Association.

Cannon-Bowers, J.A, Tannenbaum, S.I., Salas, E. and Volpe, C.E. (1995) Defining Team Competencies and Establishing Team Training Requirements, in R. Guzzo and E. Salas (eds) *Team Effectiveness and Decision Making in Organisations*. San Francisco, CA: Jossey-Bass. 333–80.

Carless, S.A and De Paola, C. (2000) The Measurement of Cohesion in Work Teams, *Small Group Research*, 31(1): 71–88.

Denson, R.W. (1981) *Team Training: Literature Review and Annotated Bibliography*. AFHRL TR-80-40, AD-A099. Wright-Patterson AFB, OH: Logistics and Technical Training Division.

Dickinson, T.L. and McIntyre, R.M. (1997) A Conceptual Framework for Teamwork Measurement, in M.T. Brannick, E. Salas and C. Prince (eds) *Team Performance Assessment and Measurement*. Mahwah, NJ: Psychology Press. 19–43.

Dyer, J.L. (1984) Team Research and Team Training: A State-of-the-Art Review, in F.A. Muckler, A.S. Neal and L. Strother (eds) *Human Factors Review*. Santa Monica, CA: Human Factors Society. 285–323.

Edmondson, A. (1999) Psychological Safety and Learning Behavior in Work Teams, *Administrative Science Quarterly*, 44: 350–83.

Fleishman, E.A. and Zaccaro, S.J. (1992) Toward a Taxonomy of Team Performance Functions, in R.W. Swezey and E. Salas (eds) *Teams: Their Training and Performance*. Westport, CT: Albex Publishing. 31–56.

Gladstein, D.L. (1984) Groups in Context: A Model of Task Group Effectiveness, *Administrative Science Quarterly*, 29: 499–517.

Ilgen, D.R., Hollenbeck, J.R., Johnson, M. and Jundt, D. (2005) Teams in Organizations: From Input-Process-Output Models to IMOI Models, *Annual Review of Psychology*, 56: 517–43.

Kasl, E., Marsick, V.J. and Dechant, K. (1997) Teams as Learners: A Research-based Model of Team Learning. *Journal of Applied Behavioral Science*, 33: 227–46.

Klein, G. (1993) *Naturalistic Decision Making: Implications for Design*. Crew System Ergonomics Information Analysis Center. Report CSERIAC SOAR 93-1. Available at http://www.dtic.mil/dtic/tr/fulltext/u2/a492114.pdf [accessed 18 August 2014].

Kozlowski, S.W.J., Gully, S.M., Nason, E.R. and Smith, E.M. (1999) Developing Adaptive Teams: A Theory of Compilation and Performance across Levels and Time, in D.R. Ilgen and E.D. Pulakos (eds) *The Changing Nature of Work Performance: Implications for Staffing, Motivation, and Development*. San Francisco, CA: Jossey-Bass. 240–92.

Marks, M.A., Matheu, J.E. and Zaccaro, S.J. (2001) A Temporally Based Framework and Taxonomy of Team Processes, *Academy of Management Review*, 26(3): 356–76.

McIntyre, R.M. and Salas, E. (1995) Measuring and Managing for Team Performance: Emerging Principles from Complex Environments, in R. Guzzo and E. Salas (eds) *Team Effectiveness and Decision Making In Organization*. San Francisco, CA: Jossey-Bass. 149–203.

Mohammed, S. and Dumville, B.C. (2001) Team Mental Models in a Team Knowledge Framework: Expanding Theory and Measurement across Disciplinary Boundaries. *Journal of Organizational Behavior*, 22: 89–106.

Morgan, B.B., Glickman, A.S., Woodard, E.A., Blaiwes, A.S. and Salas, E. (1986) *Measurement of Team Behaviours in a Navy Training Environment*. Technical Report TR-86-014. Orlando, FL: Naval Training Systems Center.

Nieva, V.F., Fleishman, E.A. and Reick, A. (1978) *Team Dimensions: Their Identity, Their Measurement, and Their Relationships*. Final Tech. Report, Contract DAHI9-78-C-0001. Washington, DC: Advanced Research Resources Organisation.

Orasanu, J.M. (1993) Decision Making in the Cockpit, in E.L. Weiner, B.G. Kanki and R.L. Helmreich (eds) *Cockpit Resource Management*. London: Academic Press. 137–72.

Pask, G. (1961) *An Approach to Cybernetics*. New York: Harper & Brothers.

Romiszowski, A.J. (1988) *The Selection and Use of Instructional Media*. London: Kogan Page.

Rosen, M.A., Bedwell, W.L., Wildman, J.L., Fritzsche, B.A., Salas, E. and Burke, S. (2011) Managing Adaptive Performance in Teams: Guiding Principles and Behavioural Markers for Measurement, *Human Resource Management Review*, 21: 107–22.

Rousseau, V., Aube, C. and Savoi, A. (2006) Teamwork Behaviours: A Review and Integration of Frameworks, *Small Group Research*, 37(5): 540–70.

Salas, E., Bowers, C.A. and Edens, E. (2001) *Improving Teamwork in Organizations: Applications of Resource Management Training*. Mahwah, NJ: Erlbaum.

Salas, E., Dickinson, T.L., Converse, S. and Tannenbaum, S.I. (1992) Toward an Understanding of Team Performance and Training, in R.W. Swezey and E. Salas (eds) *Teams: Their Training and Performance*. Norwood, NJ: Ablex. 3–29.

Salas, E., Sims, D.E. and Burke, C.S. (2005) Is There a 'Big Five' in Teamwork?, *Small Group Research*, 36: 555–99.

Salmon, P.M., Stanton, N.A., Walker, G.H. and Jenkins, D.P. (2009) *Distributed Situation Awareness*. Aldershot: Ashgate.

Schramm, W. (1955) How Communication Works, in Wilbur Schramm (ed.) *The Process and Effect of Mass Communication*. Urbana, IL: University of Illinois Press. 3–26.

Shannon, C. and Weaver, W. (1949) *The Mathematical Theory of Communication*. Urbana, IL: University of Illinois Press.

Smith-Jentsch, K.A., Johnson, J.H. and Payne, S.C. (1998) Measuring Team-related Expertise in Complex Environments, in J.A. Cannon-Bowers and E. Salas (eds) *Making Decisions under Stress: Implications for Individual and Team Training*. Washington, DC: APA Press. 61–87.

Stagl, K.C., Burke, C.S., Salas, E. and Pierce, L. (2006) Team Adaptation: Realizing Team Synergy, in C.S. Burke, L.G. Pierce and E. Salas (eds) *Understanding Adaptability: A Prerequisite for Effective Performance within Complex Environments*. Amsterdam: Elsevier. 117–41.

Stevens, M.J. and Campion, M.A. (1994) The Knowledge, Skill, and Ability Requirements for Teamwork: Implications for Human Resource Management, *Journal of Management*, 20(2): 503–30.

Sutton, J.L., Pierce, L., Burke, C.S. and Salas, E. (2006) Cultural Adaptability, in C.S. Burke, L. Pierce and E. Salas (eds) *Advances in Human Performance and Cognitive Engineering Research.* Oxford: Elsevier Science. 143–73.

van Offenbeek, M. (2001) Processes and Outcomes of Team Learning, *European Journal of Work and Organizational Psychology*, 10(3): 303–17.

Weiner, N. (1948) *Cybernetics: Or Control and Communication in the Animal and the Machine.* Cambridge, MA: MIT Press.

Chapter 4
Team Task Analysis

Recap

In Chapter 2 the perspective that individual tasks can be viewed as transformations of the environment was developed. Task processes change initial conditions into intended outcomes satisfying goal criteria held by individuals. Sensing, decision-making and doing processes were identified as contributing to task performance. In Chapter 3, team tasks were considered with individual task processes being supplemented by team task processes. Team task processes include coordination and coordination-enabling processes (distributed perception, goal generation and collaborative planning). The distribution mechanism for these information building and sharing processes is communication. The effectiveness of team task processes directly informs the transformations that the team may make in the external environment in pursuit of the goals that the team wishes to achieve.

The model of teamwork constructed in Chapter 3 is an extension of the information processes that an individual engages with when transforming their external environment. This approach follows previous work treating teamwork as a form of distributed cognition (Macmillan et al. 2004). The team coordination enablers were suggested to be processes and products which kept individuals and their activities aligned and synchronised, enabling the team to manage the demands of their environment while continuing to generate coherent team output. A number of individual competencies were suggested as supporting these coordination-related processes. The competencies were grouped into: (1) communication, (2) distributed perception, (3) goal generation, (4) collaborative planning and (5) coordination.

The other process that teams may engage in is the team learning process, which is social and shared. The team learning process may be considered as a form of team task process or teamwork, and may be a subject of team instruction, just as individuals may be taught 'study skills and techniques' such as mind-mapping (Buzan 1974) or thinking skills (De Bono 1982), independent of their specific subject domain. We have the opportunity to train teams to get better at training themselves while engaging in team tasks, whether as part of team training, as participants in training exercises, or as part of routine day-to-day activity. The team processes exist concurrently with individual task processes. Just as an individual may learn on their own, they may also learn by participation in team learning processes such as through contribution, discussion and reflection of shared team task experiences.

Team task analysis forms a critical part of the team and collective training needs analysis process and enables the design of appropriate situations and events in the environment for the trainees to react to. Other complementary training approaches which are non-task specific include instruction and practice opportunities for developing teamwork competencies through generic teamwork training, or by instructional approaches that aim to develop the team's self-learning processes such as team member self-critique and reflection as part of the After Action Review (AAR) approach. These team training approaches are addressed in Chapter 6 on the training overlay.

Introduction

Poulton (1957) quoted in Romiszowski (1999: 462) drew a distinction between 'closed' and 'open' tasks when looking at sports training. Closed tasks involved a reaction to a stable environment, whereas open tasks were required when an environment was unpredictable and changing necessitating continual task adjustment. Wellens (1974) identified the importance of planning skills in an industrial context; these skills necessitate a situation-specific response and have been referred to as 'productive' skills (Romiszowski 1999: 462). Productive skills involve applying principles and generating procedures, plans and strategies, whereas 'reproductive' skills involve following standard procedures or plans. Romiszowski continues that 'We may conceptualise a continuum of "reproductive-to-productive" skills as a basic model for the analysis of skills'. The distinction between reproductive and productive skills is also a valuable input when considering the training overlay, as Romiszowski comments: 'Perhaps one of the weaknesses of much early work on skill acquisition has been the concentration on simple movements or on sequences of simple repetitive steps'. The distinction between reproductive and productive skills, which originated in the physical or psychomotor skills domain, has been extended into other domains of skilled activity: cognitive, psychomotor, reactive (self-management) and interactive (social) skills (Romiszowski 1999: 463). In this chapter we extend the distinction of reproductive and productive skills into team tasks and draw some implications for team task analysis as part of the TNA process.

We characterise team reproductive tasks as '*team task procedures*' and team productive tasks as tasks involving '*team task planning and execution*' to avoid potential confusion. In planning and execution tasks, the team generates a plan for the situation which is then enacted through the process of execution; in contrast with procedures the task plan is contained within its description, and understood by the team members in advance. Procedures can be viewed as 'set pieces' which have been designed to deal with particular environmental situations predicted to occur in the operational task environment. As such procedures are 'packaged' for a particular environmental context, or range of contexts or scenarios. Where the environment is too complex for pre-packaged solutions to be effective the team must engage in planning activity to decide what course of action to adopt.

The complexity of the team task environment may necessitate that *adaptive coordination* is used, either to supplement or replace *procedural coordination* which is based on a predefined procedure or plan. Adaptive coordination may be required within both team task procedures and within tasks involving planning and execution, and is supported by the coordination supporting processes outlined in Chapter 3. These coordination supporting processes and their corresponding products maintain the team's shared mental model of task and situation.

Adaptive coordination may involve:

- recognition of significant changes in the environment that impact current or planned activity
- the modification of plans
- generation of emergent goals
- management of new coordination demands and emergent constraints such as emergent deconfliction of activity
- reassigning actors, tasks, resources and information dynamically
- new modes of communication.

Adaptive coordination occurs in response to changes in environmental conditions, and levels of information and uncertainty (Klein and Pierce 2001; Gorman et al. 2010). We cannot be prescriptive about what strategies teams may adopt in response to environmental challenges. As Vicente (1999: xiii) comments, 'In complex sociotechnical systems, like nuclear power plants, there is rarely one best way of achieving a particular goal. In other cases, the operators would deviate from the procedures because the desired goal would not be achieved if the procedures were followed. ... A small change in context might require different actions to achieve the very same goal'.

This view has implications for the process and product of team task analysis. Clearly procedures and plan-based team tasks are important, however, we should be mindful of situations which necessitate adaptive coordination rather than procedural coordination. These situations characterise the '*adaptive coordination conditions*' and convey part of what must be captured in scenario specification. In short, we must describe the team procedures and plans, and then define the relevant environmental factors or events which cause adaptive coordination to be exercised. The adaptive coordination conditions embrace factors identified as *characteristics of naturalistic decision environments* (Orasanu and Connolly 1993) and *environmental stressors* (Cannon-Bowers and Salas 1998) and help define training environments with appropriate types of complexity. These factors were described to the authors by one RAF AWACS (Airborne Warning and Control System) crew member as characterising 'the right type of busy' (RAF 8 Squadron officer personal communication). We speculate that specific types of environmental challenges require specific coordination-supporting processes. The implication being that we could generate 'the wrong type of busy' by generating environments which are complex but in the wrong way. As an example we could generate an environment which is very dynamic, but not very ambiguous – this would generate a certain type of workload in a team and require certain types of teamwork to compensate for the environmental demands. A concern here is if our operational task environment requires teamwork in a less dynamic but more ambiguous situation, our team may not be as prepared as desired as these coordination supporting processes have not been sufficiently developed.

Adaptive coordination conditions are environmental factors that necessitate adaptive coordination to replace or supplement procedural coordination. This may apply to procedural tasks, and either (or both) of the planning and execution phases of planning and execution tasks. Adaptive coordination conditions may originate from internal team factors such as performance issues within the team or poor communication. A team may have to engage in adaptive coordination just to manage its own performance deficiencies. For example a team executing a plan may fail to achieve an objective due to performance issues, requiring the planning team to replan the activity. However, there may not be time within standard planning cycles to achieve this so adaptive planning may be required. Similarly, a planning team may produce a substandard plan that necessitates adaption in the team executing it to compensate for the plan's shortfalls. A consideration for design of training exercises is that teams when introduced together may well self-generate 'fog and friction' (endogenous adaptive coordination conditions) solely through mutual interaction. In this situation the instructional team may not need to generate additional external events or complexities in the training environment at the start of the exercise. In this case, the trainees generating a bad plan or self-structuring a poorly designed organisation may suffice and require no instructional input.

Adaptive coordination may also be required where one team changes its procedure for engaging in planning. For example, in a military headquarters activity is structured according to a cycle – receiving higher-level intent and information from sub-units, planning, briefing and orders dissemination. This cyclical activity or 'battle rhythm' connects lower-echelon command or force elements and higher command entities to the headquarters. Modification of the senior-level (operational level) headquarters planning cycle may necessitate adaptive coordination in lower-echelon (tactical) teams who receive, interpret and enact higher-level decision-making outputs. The team members engaged in tactical planning will need to modify their team planning procedures, as they engage in the detail of planning, and simultaneously managing task execution (NATO JWC (Joint Warfare Centre) officer personal communication). Analogous processes can be seen in business such as in budgeting cycles; a lower-level department cannot set a budget or engage in detailed planning until the higher-level team has determined the allocation of funds for the year.

An inherent part of team task analysis is *scenario specification*. As we have seen, team tasks may be conducted under a wide range of conditions – scenario specification describes which set of task conditions are best selected to generate challenges for task procedures and planning and execution tasks. Both team task analysis and scenario specification feed into the specification of the training environment through the description of features or characteristics that are required. The training environment must support the superset of features required by each scenario, though not necessarily concurrently. This is illustrated in Figure 4.1.

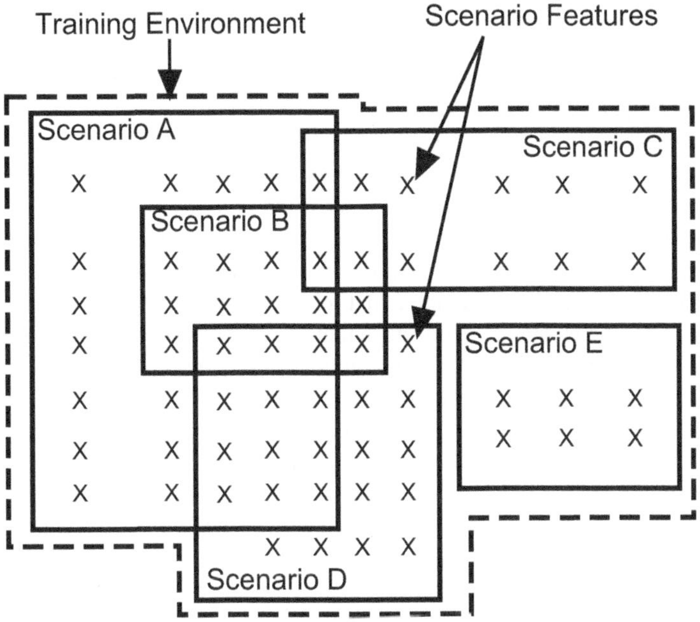

Figure 4.1 Scenarios and the training environment

Scenario specification helps ensure that simulation environments are populated with appropriate scenarios and content, can support a suitable range of events, have appropriate fidelity, have linkages to other simulation systems and specify the functionality required for the user to generate and modify the initial library of scenarios.

In this chapter we look at team task analysis building on the concepts outlined above, and establish connections through scenario analysis to the next chapter on the analysis of the training environment.

Task Analysis as Part of TNA

Task analysis can be defined as 'the study of what an operator (or team of operators) is required to do, in terms of actions and/or cognitive processes, to achieve a system goal' (Kirwan 1992: 1). As actions are conducted in a context or situation to generate a particular end state, task analysis may include description of the environment within which the task occurs (the conditions of the task) and the task team processes (coordination and coordination supporting processes) that are involved.

The process of task analysis is critical to TNA in general and therefore applies to TCTNA because it:

- identifies which team members are involved in performing actions or conducting activities such as communication, and are therefore are involved in the task. Team members involved in the task may comprise part of the training audience.
- describes the range of initial conditions, resources and intermediate outcomes as the features of the environment which the trainees must react to, or the situations that the trainees must be allowed to generate. These factors are the features of the environment which are relevant to the task.
- identifies the environment features that the trainees have to react to, one can start identifying the inputs that instructional staff have to generate.
- defines the sequence and dependency of actions that generate team outcomes. A description of these actions and their dependencies form part of the task's plan in defined task procedures.
- describes the outcomes and potential critical errors and consequences of the task, which may lead to subsequent tasks being performed.
- identifies 'other outcomes' which are goal neutral but may impose resource burdens on the organisation performing the task (for example consumption of fuel, ammunition, equipment servicing requirements).
- defines the goal criteria, by which we gauge whether outcomes are satisfactory. The definition of goal criteria drives the instrumentation requirements for the training environment, as we must have some means of recording and measuring outcomes generated against goal criteria to support the assessment process.
- offers the opportunities to capture team task processes such as coordination, the coordination enabling processes and communication.

There's a distinction to be drawn between the *process* of analysis and the notation used in the *product* of analysis. For example Gantt charts and network diagrams are two notations that can derive from the Program Evaluation and Review Technique (PERT) (Malcolm et al. 1959). PERT is a statistical project management analysis technique which seeks to derive the critical path of a project – the sequence of activities on which the timely completion of a project depends. The process of analysis seeks to derive this critical path information from the information that is available and by estimating durations of stages of activity. This critical path information is then conveyed graphically via a Gantt chart or network diagram to enhance understanding within the project team. In this example the requirement for analysis (critical path derivation), the process of analysis (PERT) and the output notation (Gantt charts and network diagrams) can be seen as discrete elements.

The task of making porridge or boiling an egg could be equally well decomposed by task analysis and represented as a Gantt chart, a network diagram or a hierarchical notation such as seen in Hierarchal Task Analysis (HTA) (Annett et al. 1971) or in Work Breakdown Structure (WBS) (described in US Department of Defense Standard MIL-STD-881C, 2011) format. The key concerns are firstly, why is the analysis being conducted and secondly, to what purpose is the product of analysis being put? In TNA we are seeking to describe the essential features of the task in the process of analysis and in a manner capable of being related to the task environment and the training overlay. Choice of presentation format for this analysis is based on how well these concepts can be communicated and interrelated. For example, Gantt charts are widely used to capture activity conducted by large numbers of people and their interdependencies. Other forms of notation have their own specific strengths. Different forms of notation may also be used through the process of analysis. For example one could start with a hierarchical decomposition technique, which may then act as a precursor for a subsequent network or time-like notation, as seen with a WBS feeding Gantt chart preparation.

In the next two sections we discuss two characteristics of team tasks – tasks that involve planning as a discrete phase of activity within the task, and the types of coordination used within team tasks: procedural coordination and adaptive coordination.

Task Analysis in Procedural and Planning-execution Tasks

In procedural tasks thc team is following a predefined plan, in planning and execution tasks the plan is generated by the team for the specific situation they are in, and is therefore unique. This difference has implications for task analysis and scenario specification. Procedural tasks are based on a set of conditions that are assumed to exist at the commencement of the task. These conditions include the triggering or initiation criteria for the task. For example the task of changing a tyre on a car is a reactive procedure – it is triggered by something happening to one of the pre-existing car tyres, and follows a clearly defined flow of activity. The defined flow of activity in changing a car tyre is shaped by the constraints of the environmental system; wheel bolts must be loosened before the car is jacked up and weight is removed – otherwise the wheel will spin, the old tyre must be removed before a new tyre can be put in place and so on. Task actions transform these conditions to generate intermediate outcomes and task outcomes. Task actions are sequenced according to the predefined plan inherent within the procedure, and this defined plan is one of the features of the task that is described in task analysis. Assessment of procedural tasks is relatively straightforward in that we can measure how the task was conducted against the procedure and whether the outcomes generated satisfied our goal criteria as described in the standards of performance associated with the task. This is illustrated in Figure 4.2 on the left-hand side of the diagram.

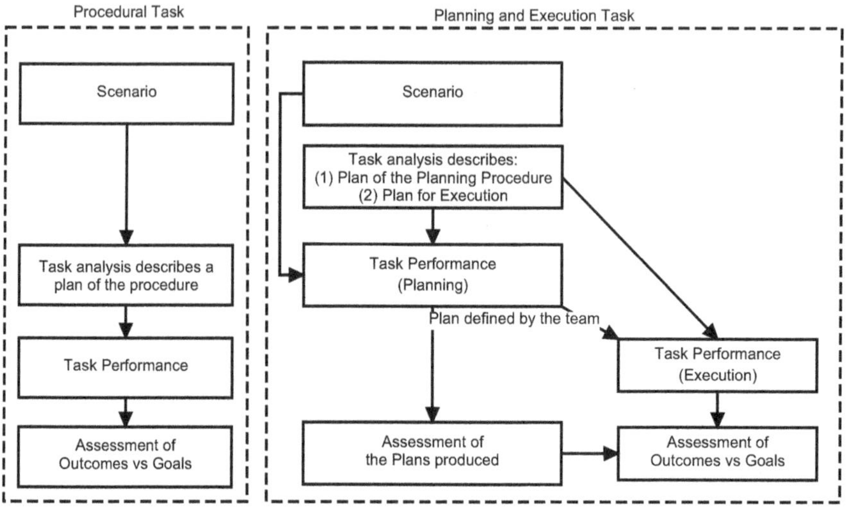

Figure 4.2 Task analysis of procedural tasks and tasks involving planning

In contrast to procedural tasks, team tasks which involve planning and execution do not have predefined plans which can be captured in task analysis in the same manner. After all the output of the planning activity that the team engages *is the plan of the task* as generated in the specific scenario in question. Task analysis in planning and execution tasks can describe the procedure for task planning and the procedure for task execution in a general sense but not the specific plans themselves. The plan for planning and plan for execution procedures are generic (that is, are always applicable in some sense) and context-free (that is, assume no specific initiation conditions). Due to these characteristics we need to create suitable scenarios of appropriate challenge, to contextualise the task and enable these planning procedures to be enacted in a meaningful way by the team.

Assessment of team performance planning and execution tasks takes on a somewhat different form from procedural tasks, because we can assess the plans produced by the team, as well as assessing the performance of execution. This is seen in CPXs where only planning is being trained, and tactical exercises such as some live flying exercises where only execution is being trained. Assessment of planning involves consideration of the scenario context and the suitability of the plan to the situation; this involves Subject Matter Expert (SME) knowledge and interpretation which has an implication for the support of instructional and assessment functions within the training overlay. Assessment of execution of the plan may be somewhat more procedural in the sense that actions performed may be referenced to the plan, but this is not the whole story. Changing conditions may render plans that have already been made either irrelevant, futile or impossible. In these situations the existing plan may be changed, amended or discarded, what matters in team task execution is that outcomes of suitable quality are being delivered, not that a redundant, inappropriate or outdated plan is being followed in a procedural manner.

Team procedures can be defined in task analysis with the flow of sequential and parallel activities of multiple team members according to a shared plan. This procedural coordination may be predefined and describes how we expect or anticipate the task to be performed. However, many collective tasks are at such a level where there is no predefined set procedure or plan to follow – the plan must be *generated* by the team for that particular situation, and this plan subsequently executed. In large organisations there is frequently a separation of team activity between teams involved in planning and teams involved in plan execution.

Procedural and Adaptive Coordination

Activity may be considered to be coordinated in two ways within team tasks. Firstly, we have *procedural coordination* which defines how individual activities are conducted according to a pre-defined plan. Secondly, *adaptive coordination*, where the team members must innovate to manage the environmental demands of the situation.

In a team context both types of coordination are important to team performance depending on task and situation. Expert teams may exhibit and modify the type of coordination due to the demands faced (Entin and Serfaty 1999). A related concept to procedural coordination is that of 'standardisation' defined as 'the degree to which task activities are specified in detail and standard operating procedures are established to direct behaviour' (Grote et al. 2010; van de Ven et al. 1976). Standardisation implies that procedural coordination will be required, however, some task conditions may necessitate adaptive coordination to supplement procedural coordination.

Procedural coordination may be conducted with limited communication because the individuals within the team are all aware of their respective activities and how these activities fit together. Coordination involving limited communication has been defined as implicit coordination by Entin and Serfaty (1999), who make a distinction between coordination involving communication (explicit coordination) and coordination not involving communication (implicit coordination). While the concepts of procedural and implicit coordination are related we suggest that procedural coordination *may* involve communication as team members offer and share information regarding task status, achievement of outcomes, fulfilment of goal criteria and the external situation. This information can be taken as the team 'thinking aloud'; it may facilitate error correction and maintain the teams' shared awareness of their situation, performance (distributed perception), intent (goals), future activity (plan) and current actions (coordination of action). As such, communication may support and improve the quality of procedural coordination – which could otherwise be considered as implicit coordination. In this sense communication during procedural coordination may not be an overhead but rather an enabler and error-correcting support process within team performance.

Communication may also modify or support personal attributes such as stress experienced while not necessarily impacting procedural communication *per se*. As an example, consider a rock climber who climbs into a section of rock face where she is not directly observed by her belayer below (and on whom her safety depends). The climber may continue to communicate verbally with the belayer as well as interacting by pulling on the rope as she climbs. The climber pulling on the rope causes the belayer to pay out rope in response. If as the climber continues to climb, she does not receive responses from the belayer below, she may hesitate, or become concerned and task performance may be impacted. In this example we can see that a task that *may* be performed with only implicit coordination (climber pulls rope, belayer feeds rope), can be supported by communication to maintain what has been characterised as a shared mental model (Entin and Serfaty 1999) or team 'common ground' (Klein and Pierce 2001). In this example two safety-critical variables which form part of the 'common ground' are: the amount of rope paid out (required for climbing with a safety implication if the climber falls) and the status of the climber (steadily climbing and managing the task, or barely hanging on and about to fall). This maintenance of a shared mental model or 'common ground' may impact team task performance through the modification of individual team member attributes such as emergent states or individual stressors, as described in the team performance model in Chapter 3.

Not unsurprisingly the type of coordination used (procedural vs adaptive) is related to the features of the task environment and what it requires, supports and demands. As an example, we cannot plan for every eventuality the environment may generate. Vicente (1999) in the preface (p. xiii) makes the comment that 'It is very difficult to write a procedure to encompass all possible situations' while discussing procedural compliance versus flexible adaptation in nuclear power plant management. Shepherd (2001: 63) illustrates this point with an event table for determining operator actions for a chlorine balancing task within a chlorine production facility. The event table has 32 rows each yielding a decision output for different system states, where the system is modelled as four resource-based binary variables (that is, a particular resource is either available to the team or not).

Other environmental situations may generate extreme workload to a point where team members simply don't have the time (or the mental resources) to communicate and perform the task. The flip side of the coin of not having capacity to communicate is not having the capacity to listen, or preferring not to listen – where team members 'switch off' and solely concentrate on their task. Anyone who has turned down the volume of the car stereo while reversing into a tight car parking space has experienced this. In situations where workload impacts on capacity for communication we must rely on the team members' shared mental model of the task, either based on the task plan (procedural coordination), or based on a complex set of flexible task principles established by prior experience. The efficiency of the type of coordination adopted is related to workload (Thordsen 1998 cited in Klein and Pierce 2001) – where adaptive coordination may work at low and medium levels of workload, procedural coordination is generally better suited in dealing with high workloads.

If the potential environmental demands of a situation can be predicted to some degree, for example an aircraft engine failure on take-off, then a predefined contingency plan or emergency procedure may be prepared for it. However, it is neither possible nor practical to attempt to describe every possible contingency of every situation we could feasibly plan for. For example, with the Apollo 13 incident where the crew faced an explosion and loss of oxygen on-board, there was no explicit predefined plan for dealing with that specific emergency situation that the crew faced. The crew had to engage in teamwork; innovate, identify their situation, communicate with mission control, generate some attainable goals, form a workable plan and coordinate activity. The crew had to manage the environment in the immediate timeframe, while simultaneously planning for the next hour and the timeframe beyond. All of these activities involved communication. These tasks, while involving adaptive coordination at a macro level, also involved procedural coordination within sub-elements (such as establishing system states, testing sub-assemblies) in forming part of the solution.

The distinction between procedural and adaptive coordination has implications for scenario specification. Scenario specification can be used to define the range of operational task conditions, from routine or expected situations requiring procedural coordination, to conditions that necessitate adaptive coordination.

Scenarios requiring either procedural or adaptive coordination can be linked together according to an elaboration strategy – where trainees are successively exposed to more challenging task examples (that is, more difficult task conditions). This supports the learning process by enabling trainees to use prior experience and knowledge in approaching unfamiliar situations. A common approach within military skills acquisition is 'crawl-walk-run'.

'Crawl-walk-run' is described in *A Leader's Guide to Lane Training* (US Army document TC 25-10, 1996). Lane training is 'a process for training company-size and smaller units on one or more collective tasks (and prerequisite soldier and leader individual tasks and battle drills)' (TC 25-10 1996: 5). Section 1-5 (TC 25-10 1996: 14) outlines the lane training concept and 'crawl-walk-run' – this is reproduced in Table 4.1.

Crawl-walk-run is implemented differently in different organisations and lacks a formal definition (and as a result means different things to different people); as a result not all crawl-walk-run training will epitomise these features above. However, 'perfect practice', reinforcement, overlearning, increase in speed of performance, and efficient procedural coordination and communication are characteristics of this form of instructional method which is widely used to teach Tactics Techniques and Procedures (TTPs) and drills (that is, procedural tasks). The instructional approach is procedural and this matches these tasks which are viewed as procedures.

Table 4.1 Crawl-walk-run

Phase	Description
Crawl (Explain and demonstrate)	The leader describes the task step-by-step, indicating what each individual must do.
Walk (Practice)	The leader directs the unit to execute the task at a slow, step-by-step pace.
Run (Perform)	The leader requires the unit to perform the task at full speed, as if in combat under realistic battlefield conditions.

As an adaptive coordination complement to 'crawl-walk-run' perhaps we should look to implement 'adjust-adapt-innovate'. Adaptive coordination may require minor *adjustments* to predefined procedures or plans, where an unusual combination of conditions forces some alternations in task execution. More significant *adaptions* might require new sequences of activity to bypass hurdles or manage constraints, or might require the abandoning of a plan and a stage of re-planning 'on the fly'. At the most demanding end of the adaptive continuum we may require *innovation* – where, for example, new goals are generated, resources matched to new tasks, novel sequences of activity performed or teams completely reorganised. To support the development of adaptive coordination our instructional approach needs to be adaptive as well. An adaptive instructional approach is less focused on the task as a detailed procedure but rather views the task as a means of shaping conditions to a goal, where information may be less than perfect. Instructional approaches might include not briefing the task in too much detail, highlighting critical environmental conditions that necessitate alternation of the plan, focusing on the development of alternatives and decision making, allowing non-safety critical mistakes to occur, and accepting that a bit of confusion is part of the learning process. Clearly our team has to have the basics grasped; however, problem-solving approaches in other skills areas have yielded positive results as reported by Schaab and Moses (2001: 12) of the US Army Research Institute.

An adaptive coordination approach could be supported within a simplifying conditions method (Reigeluth 1999: 442) that takes whole team task practice on an elaboration sequence from 'follow the defined procedure' within simplified, routine conditions, through to more complex situations which require increasing levels of adaptive coordination. This approach may avoid some of the issues outlined by Klein and Pierce (2001) who described methods on how to develop *un*-adaptive teams; these factors include 'Use the crawl, walk, run method of training', and 'Use part-task training methods'. Klein states that while development of adaptive teams is possible, it is unlikely given the characteristics of the teams' parent organisation and the traditional training strategies enacted by them (Klein and Pierce 2001). We develop these ideas in Chapter 6 on the training overlay.

We would suggest that training for both procedural and adaptive coordination is required, and that procedural coordination training should precede adaptive coordination training. Adaptive coordination has its foundations in procedural coordination; the team plan and the team members' shared mental model of task and situation. The danger with overreliance on procedural coordination is that the environment and task (outside training) may not support this, whereas the team's ability to adapt through engaging in adaptive coordination has potentially infinite applications.

Teams need to be able to engage in both procedural and adaptive coordination, and both types of coordination need to be trained. The ability to engage in adaptive coordination is to an extent dependant on procedural coordination, for example if one is learning martial arts one would learn to punch, kick and perform locks and throws (procedural coordination) as part of learning to fight – which involves adaptive coordination to manage the demands of the situation and the opponent faced.

The Necessity of Adaptive Coordination in Team Tasks

The complexity of the environment in which team tasks operate means we can rarely run everything exactly to a rigid predefined plan. Team tasks are 'bigger' in an environmental sense than individual tasks as they operate on a greater range of initial conditions and generate more intermediate outcomes, team plans are bigger than individual task plans and teams have a higher potential for simultaneous action performance.

Field Marshal Helmuth von Moltke the Elder observed that 'No plan of operations extends with any certainty beyond the first encounter with the main hostile force' (Hughes 1993: 45). Later Moltke broadens and amplifies this comment by suggesting that the plan may not even extend this far. 'To the calculations of a known and an unknown quantity, one's own and the enemy's will, also come factors that escape all foresight: weather, sickness, railroad accidents, misunderstanding and delusions' (Hughes 1993: 46). Sun Tzu makes similar comment in the *Art of War*, in Chapter 6 'Strengths and Weaknesses', in the section entitled 'The image of water': 'There is no hard and fast rules on good strategy. It has to be as versatile as water which changes course according to the ground … Be ever ready to change strategy according to the changing conditions of the opponent' (Chung, 1991: 69). An alternative translation of the same passage (Cleary 1988: 113): 'So a military force has no constant formation, water has no constant shape: the ability to gain victory by changing and adapting according to the opponent is called genius'. The authors, while observing military collective training, have heard participants make comments to the effect that 'if one doesn't have a plan one doesn't know what to change'. One could assert that this is due to the complexity of the task environment and the fact that in conflict things are especially problematic; the enemy has a plan and is deliberately effecting changes in the environment that impact our plan. In a complex, reactive environment adaptive coordination is a necessity. A plan if nothing else references a point of departure for subsequent adaption. A team's aspiration to turn within the enemy's decision cycle (Boyd 1987) might be rephrased as the ability of the team to engage in adaptive coordination, to shape the external environment in such a way to undermine the enemy's plan whilst supporting our own, more rapidly than the enemy.

Dwight D. Eisenhower echoes Moltke's comments

> I tell this story to illustrate the truth of the statement I heard long ago in the Army: Plans are worthless, but planning is everything. There is a very great distinction because when you are planning for an emergency you must start with one thing: the very definition of "emergency" is that it is unexpected, therefore it is not going to happen the way you are planning. (Eisenhower 1957)

Nixon (1962) reports Eisenhower as making the comment that 'In preparing for battle, I have always found that plans are useless but planning is indispensable' (Nixon 1962). One might ask the question what does the *process of planning* generate apart from the team plan, and what value might it offer the team? One answer might be found in the distributed cognition view of teamwork – in the shared mental model or understanding that the team now has of the situation and task, as a result of engaging in the planning process. By engaging in planning the team not only generates a plan, but also a shared mental model of the situation and task.

According to the team task model developed in Chapter 3, the mental model that the team shares includes information on:

- the situation, their level of awareness of the environment and its features that constrain desired actions
- constraints as environmental inhibitors to action (team situational awareness)
- the aim (team goals)
- action coordination and synchronisation requirements and opportunities
- communication opportunities and challenges
- the affordances, or team capabilities as defined by the entities (for example units, teams, actors), resources and information available to the team, team member capability and experience.

There are strong correspondences between the areas above that we might expect to be discussed within the team as part of planning, and that could be discussed from an instructional standpoint within the AAR process described in Chapter 6.

The complexities, uncertainties and hostile acts of adversaries within the environment can clearly frustrate team plans and therefore procedural coordination. In response the team can remain committed to the plan, re-plan and/or engage in adaptive coordination. Where a plan is only sketched in the broadest outline terms, a degree of adaptive coordination is to be expected. For team task analysis a conundrum is that while defining procedural coordination may result in a large complex plan of activities and dependencies, it does relate to an instance of a team task under conditions as they are currently understood to be (for a current instance of tactical or operational planning); or as they are envisaged to be in the future (for the design of TTPs). Since adaptive coordination is likely to be required how can we supplement a procedural task analysis approach to make statements about adaptive coordination?

In situations requiring team adaption we may have an environment with a high degree of complexity and variability (many uncontrolled variables), and teams may approach problems in different ways (the teams could be considered to have a high degree of freedom (Klein and Peirce 2001)). As a result, teams may adopt a range of strategies which may be equally successful in solving the problems faced through attaining goals in different ways. In this situation we are unable to define exactly what strategy or process the team might adopt, or be proscriptive about it from an assessment or evaluation perspective, however we do know that effective teamwork is what distinguishes high and low performing teams (Salas et al. 2005). This said, each route the team selects to success (or failure) will have a different cost in terms of time and resources, and may generate 'other outcomes' along with the goal-aligned outcome. These factors may be evaluated as part of instructional feedback and assessment.

One task analysis approach in support of a description of adaptive coordination is to *define task conditions that have significant impacts on procedural coordination within the team task and therefore necessitate adaption.* In other words, make the team form a plan, and then make them change it during task execution. Once these conditions are put in place, the team has to engage heavily in the team coordination processes identified in Chapter 3. Task conditions or scenario injects have been described as 'roadblocks' in the context of team perturbation training (Gorman et al. 2010). Extending the idea of roadblocks in task scenarios as defining 'adaptive coordination conditions', one could go further and look at the properties of the coordination enablers for the task, the attributes of distributed perception, goal generation and collaborative planning, and the modes of communication available and how they relate to external environmental factors or demands of the task. For example SMEs may be able to define some key types of external information that must be shared amongst team members performing the task in response to a particular environmental cue.

In our earlier example of a rock climber that has climbed out of sight of her belayer, team SA has been inhibited – impacting the awareness of task variables such as the amount of rope paid out and climber status. Might we predict or expect that if we generated these conditions in team training, these subjects should be expected to crop up in conversation? The content of these coordination enabling factors we would suggest is what a team of expert planners might generate in the planning process: discussion of the situation and its key features, the team aims, coordination and synchronisation requirements and so on. We would suggest these approaches are best built on top of a procedural task analysis approach, in that we first have to train a team task procedurally so that team members have a common point of departure for adaptive coordination. Discussions of the definition of instructional tasks as they relate to team training strategies (such as procedural training, cross-training or perturbation training) within TNA are to be found in Chapter 6 on the training overlay.

In the next two sections we discuss how concepts of procedural coordination and adaptive coordination may be integrated into team task analysis.

Procedural Coordination 'Crawl-Walk-Run'

Procedural Coordination: Sequential Activity

The activities that teams undertake frequently have timing and sequencing dependencies which are based on the state of the environment. This applies in a team task context just as it applies in an individual task context, and forms the basis of procedural coordination. For example; when making hard-boiled eggs we must boil our eggs before we attempt to peel their shells off. If we attempt to reverse the actions and peel the eggshell before boiling the egg we generate a different state of the environment. Component actions performed in a different sequence will generate different results or outcomes in the environment. This principle is illustrated in Figure 4.3 using HTA notation, looking at the task 'Fire Weapon at Target' which has been defined in terms of three constituent actions or activities.

Figure 4.3 Task 'Fire Weapon at Target'

Here following the plan of performing activity one, then activity two, then activity three in sequence (shorthand to 'plan 1 > 2 > 3') (Annett 2005) leads to satisfying our goal of firing the weapon at a target. However these operations performed in a different sequence will not generate this outcome, nor will omitting steps in the sequence, such as activity 2 'aim weapon'. This means that outcomes are dependent not just on actions, but also on the intermediate states of the environment that preceding actions generate. This means that activity 3 – 'pull trigger' – acting on the environmental state 'weapon loaded and aimed' takes us to a different place than the state 'weapon unloaded and/or not aimed' would, causing the target to be missed, or the weapon not to fire.

Our overarching task is to fire the weapon at the target (and our goal is to hit it). This task can be defined by its objective (that is, intended outcome generated which fulfils goal criteria). In this example, we fire the weapon at the target, we hit it and we achieve a certain score or measurable level of success. Where an individual is performing this activity on a personal weapon these steps are generally discrete as the individual cannot be both loading and aiming simultaneously. An alternative hierarchical notation for Figure 4.3 is shown in Figure 4.4. Here the information contained in the plan (1 > 2 > 3) has been expressed graphically with arrows linking boxes.

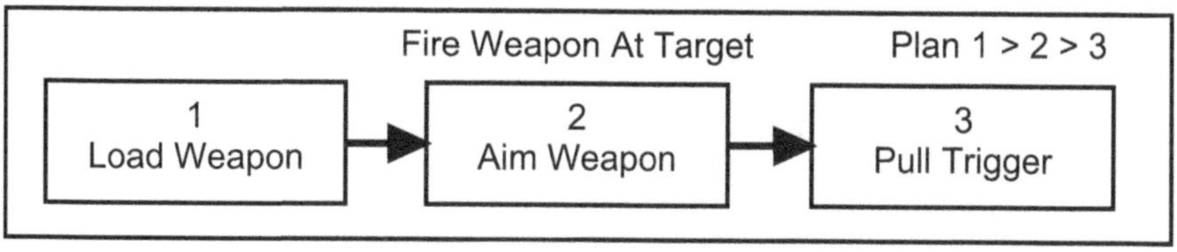

Figure 4.4 Alternative representation for the 'Fire Weapon at Target' task

This is still a hierarchical approach, as the three activities are all 'children' of the parent task 'Fire Weapon at Target'. Activities one, two and three as actions all sit within the overarching task of 'Fire Weapon at Target', just as in the prior hierarchical notation in Figure 4.3 compositional membership or inclusion is indicated by upwards linking lines. Multi-level hierarchies can be built using the approach illustrated in Figure 4.5 using successive levels of description at different system levels (Mesarovic et al. 1970: 42).

Adopting the view that tasks are transformations means one can start to bring the environment into the task. We can start linking elements of the environment to the activities that are being performed. Loading a weapon requires a

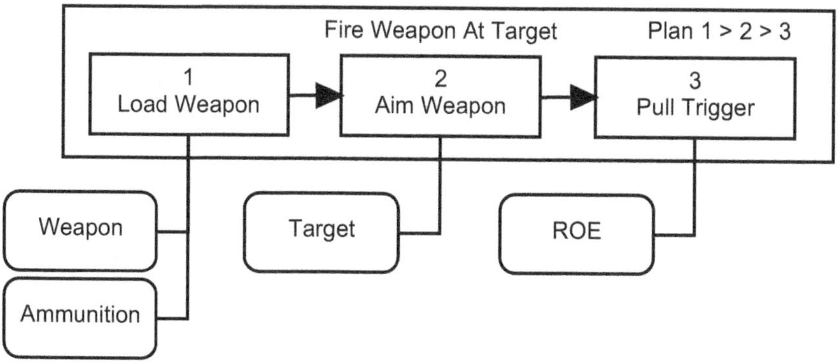

Figure 4.5 Task 'Fire Weapon at Target' with resources and information added

weapon and ammunition, aiming requires a target, and pulling the trigger will be performed within certain criteria or rules. These resource or information aspects are not actions or operations (that is, transformations), but are states of the world and are distinguished in the team task notation as rectangles with rounded corners.

The ROE box stands for the 'rules of engagement', for example in clay pigeon shooting this defines what constitutes a valid target; a clay disc moving within the bounds of the shooting field. This ROE information has been previously defined and briefed or given to the shooter, or is assumed through familiarity with the situation or the 'rules of the game'. For example, a clay pigeon can break on being launched meaning that it is not a valid target to shoot at; this fact might be called out verbally by the clay trap operator, to avoid a shot being wasted.

So far we have been discussing operations or actions; if desired one could add the initial environmental states and outcomes of the task to this diagram. As these new nodes are states of the environment they are shown as rectangles with rounded corners. This is expressed in Figure 4.6.

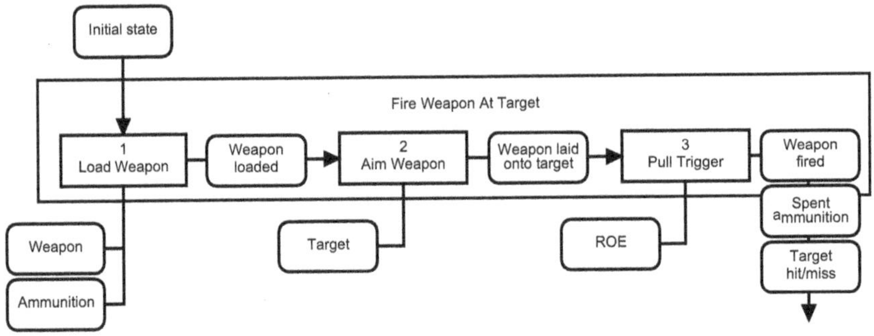

Figure 4.6 Task 'Fire Weapon at Target' with intermediate outcomes indicated

The final environmental state of the overarching task 'Target hit or miss' is the outcome generated from the task 'Fire Weapon at Target'. This notation also gives us the opportunity to define the initial state of the task – the weapon issued is hopefully unloaded at the start of the task and some form of command will be issued to indicate that the task can commence. This notation can also capture other less common outcomes, such as when the trigger is pulled nothing happens, either a misfire, some sort of weapon technical issue or we have simply run out of ammunition. Outcomes and intermediate outcomes and information generated from tasks may be judged by task product criteria, whereas the actions themselves can be addressed by process criteria. For example if the first activity 'load weapon' takes five minutes when it should only take one minute then our process needs addressing, whereas if the weapon is somehow mis-loaded that's a task product (intermediate outcome) not meeting goal criteria.

If necessary we can also start breaking the task down into sensing, doing and decision-making activities. For example the lines connecting weapon and ammunition resources to the action 'Load Weapon' are doing-based actions, the line connecting the target to action two 'Aim Weapon' could be characterised as a sensing action, whereas the line connecting ROE information to 'Pull Trigger' is related to internal decision making.

Procedural Coordination: Simultaneous Activity

Alongside sequential activity we have simultaneous activity. In a team context each team member is generally performing one or more simultaneous activities. Referring to Figure 4.7: (1) an environmental state may act as the trigger for multiple actions, (2) multiple actions may generate one state or (3) multiple states may invoke one activity.

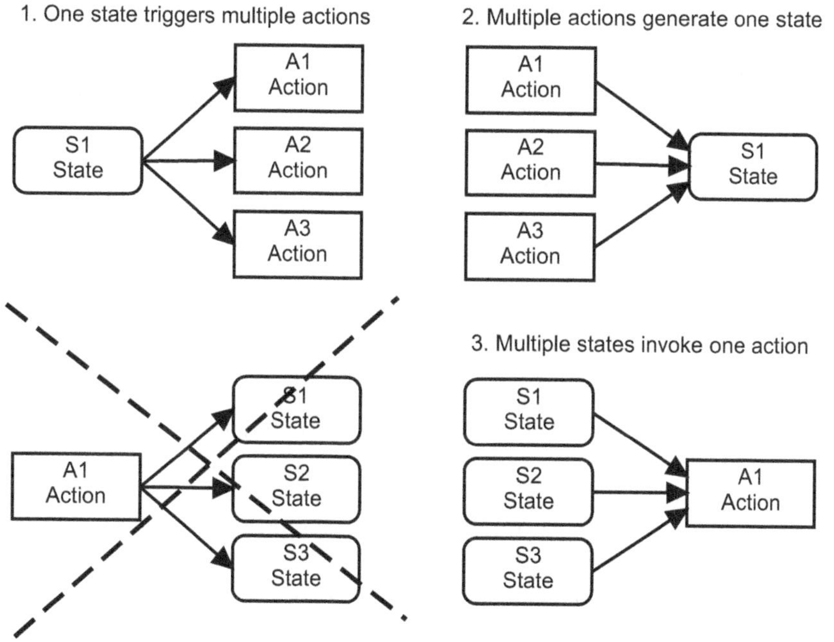

Figure 4.7 Simultaneous activity notations

We do not recommend using a notation where one action is generating multiple states, but instead have one action generating a single state with a more complex description. This keeps environmental state descriptions 'bundled up', and action transformations simpler to describe.

When one moves from looking at an individual task to a team task one of the things that becomes apparent is that a vast number of activities are occurring (and need to occur) simultaneously. This is why individuals have been assembled to form a team in the first place: the task is simply too large for an individual to undertake alone. For example, imagine a Road Traffic Accident (RTA) situation and the fire brigade are the first emergency service at the scene. All team members are responding to the same initial environmental situation and each team member is trying to generate his own intermediate outcomes. An example team task is defined in Figure 4.8.

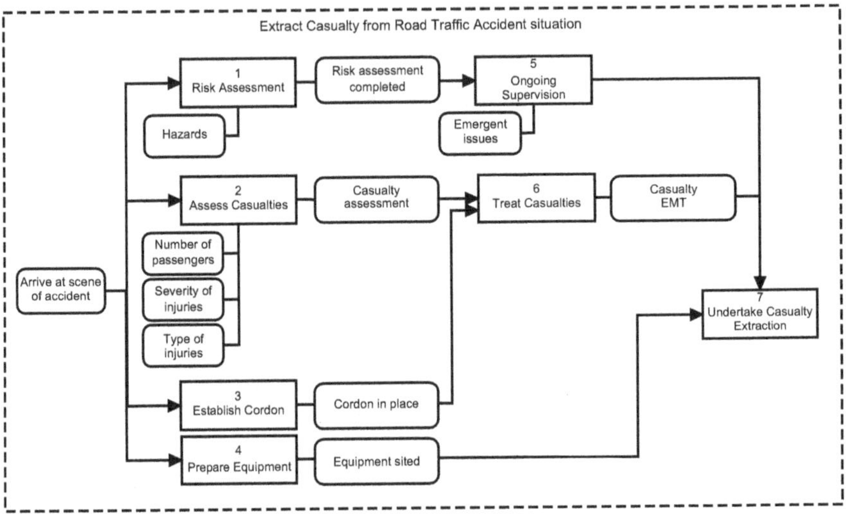

Figure 4.8 An illustrative example of a team task: extract casualty from a Road Traffic Accident situation

A number of features become apparent. Firstly we have multiple actors meaning that we have multiple parallel strands of activity occurring. We have situations where one environmental state such as 'Arrive at scene of accident' acts as a cue for multiple actors to perform activities. We also have points of convergence where for example one action 'Undertake Casualty Extraction' is predicated on three previous environmental states having being reached; 'Risk assessment completed', 'Casualty EMT (Emergency Medical Treatment)' and 'Equipment sited'. The environmental state 'Risk assessment completed' is information generated by one team member who is responsible for safety within the team. This information might have to be distributed within the team, and communicated externally. 'Casualty EMT' is an environmental state relating to the casualty or casualties and likely to be monitored as often as possible given the constraints of the situation. 'Equipment sited' is an environmental state relating to the location and preparation of a task resource that will be needed for the activity of 'Undertake Casualty Extraction'.

If we had four team members in our team we could see how one team member might be responsible for activity one, one for activity two and so on. These discrete tasks may proceed in parallel, with team members helping each other out when they have completed their first stage of the task. For example the team member performing activity three 'Establish Cordon' will start to assist in activity six 'Treat Casualties' once a cordon has been established. We may also have the situation where we have multiple team members assigned to a single task activity. For example if we had multiple emergency teams arriving at a major incident we might have six team members engaged in activity two 'Assess Casualties'. The commander on the scene may have the task of assigning or directing personnel to specific operations or actions according to the demands of the situation.

In a real RTA incident each situation will be different, with a variable number of injured people, with varying levels of severity of injury, the scene may be complex with a large cordon that needs to be put in place, getting equipment into place maybe very difficult and time consuming and the vehicle involved in the accident may be on fire and/or carrying a hazardous cargo. While an idealised team procedure may be outlined as in Figure 4.8, we cannot define with any certainty which of activity 1 through to activity 4 will be completed first. Both open and relatively closed (human-designed deterministic) environments may exhibit very high levels of environmental variability and exhibit complex behaviour which may challenge attempts at complete proceduralisation. In a similar way a complete procedure to win a game of chess has yet to be defined.

The impact of the environmental variability is that attempting to define procedures without due consideration of environmental conditions, as well as the preceding intermediate outcomes generated by the task, is impossible. For example activity 1 'Risk Assessment' may identify a risk that precludes specific types of cutting equipment from activity 7 'Undertake Casualty Extraction', however the severity of casualty injuries and their location within the vehicle may require their immediate removal from the vehicle by the fastest means. Here we can see the environment generating a set of potential challenges to task procedures and team coordination. In a similar vein it would be a poorly performing team that allowed the occupants of a vehicle to die because team members who could have been performing first aid were instead readying equipment for casualty extraction. Similarly, a commander who sacrificed the lives of the men under his command because a risk assessment had not been performed adequately would also be judged as a suboptimal performer.

Dynamic and complex situations involve team coordination and communication, and the supporting enablers of coordination. Understanding of the situation must be shared within the team, goals prioritised and plans made to enable coordinated action. Clearly proceduralisation of tasks is a necessary first step to define how things are generally planned to be dealt with in the future situations that the trainees will be reacting to.

Adaptive Coordination: 'Adjust-Adapt-Innovate'

Anyone who has ever experienced frustration with an automated telephone system whose predefined menu of options doesn't match the exact situation in hand, has experienced a procedural vs adaptive episode. A procedure has been designed that is supposed to cover most eventualities, however this does not cater for all eventualities. Procedures work to the extent that system states and predefined responses can be determined and predicted, everything else needs adaption of some type. Adaptive coordination may be based on a modification to a pre-established shared plan, or may involve complete improvisation and innovation. We may even have an exit condition built into the procedure – such as 'press star [button] or hash [button] to speak to an operator' in our automated (procedural) telephone system.

The team will generally have some sort of idea of the situation that they are in, the ultimate goals that they are trying to achieve and the opportunities and constraints of the situation. There may have been an original plan which has had to be modified in the light of new information. Adaptive coordination as compared to procedural coordination is not necessarily a binary state but could be considered to operate on a continuum of improvisation – it may involve minor modifications (adjustments) to established procedures or plans, or may involve major adaptations and novel sequences of actions, or may involve completely new solutions to problems (innovations). Plans may be dynamically generated or improvised in adaptive phases and as such cannot be predefined as a part of task analysis. If plans are conceived as routes describing activity that generates successive transformations in the task environment, while the specific route taken may not be reliably predicted, the intermediate outcomes as points on the potential task trajectory between current situation and final outcome may be. As an example, consider a situation which requires a change of plan – a group has gone on holiday and on arriving at the airport for their return journey find that all flights home that evening have been cancelled.

Here the existing plan is no longer valid due to environmental factors outside the team's control. The goal to get home remains the same and the start point of the task is the same (the airport) but circumstances force a complete change of plan. The plan for getting home in this example involves understanding the opportunities of the situation (other airlines, distance to other airports or train stations, amount of money available, acceptance of credit cards, chance of recovering money spent from the budget airline, taxi availability) and the constraints (price of next day travel, cost of overnight accommodation, cost of taxis). These factors contribute to the generation of a set of intermediate outcomes through a forward-chaining process. In forward-chaining, the next reachable intermediate outcome from the perceived situation is determined, in this example the main city airport or city train station. Back-chaining involves working backwards from our final outcome which satisfies goal criteria, in this example the closest major city to home served by rail and acceptably close alternative airports to home. Human factors research indicates that forward chaining is used more frequently than back chaining in expert decision making (Lipshitz et al. 2001). Forward-chaining is valuable because it immediately exploits opportunities that may not be available in the near future; as such one can commence the task and consider subsequent stages of re-planning *en route*.

Defining adaptive coordination as part of task analysis is challenging given the lack of a single correct response and the range of responses that could be adopted by the team. Many of these issues sit in detailed team training development, rather than in the TNA process. One approach is to:

- accept the necessity of adaptive coordination, even within procedural team tasks – this has implications for the training overlay and training environment design;
- define a standard example task procedure as a point of departure for adaption;
- specify the environmental conditions that are likely to necessitate adaptive coordination;
- use SMEs or personnel with relevant expertise to define acceptable responses to these adaptive coordination conditions;
- implement these measures in relation to the coordination enabling processes and anticipated communication that accompanies this.

If team task processes involving communication are mediated through technology then features of the information environment (such as Link16 picture, planning documents, orders issued) will also provide information on adaptive coordination responses.

Accepting adaptive coordination as an object for instruction has implications for training environments and the instructional functions contained within the training overlay. To support adaptive coordination in trainee teams, the instructional team must engage in adaptive coordination. In planning and execution tasks the trainees may modify plans – this is adaptive behaviour and providing the plan is valid, does not impact on safety, and satisfies the objectives of the exercise then this form of behaviour should be supported. One issue is that trainees changing the plan generates additional workload for the instructional team. Trainee adaptive coordination in an exercise will necessitate instructional staff adaptive coordination within the delivery of the training overlay. Ultimately changes in trainee planning should feed into events and behaviour within the training environment delivered by the training overlay. This may involve extensive instructional team technical support and instructional capacity to instantiate these changes in reaction to trainees doing the unexpected. Perhaps the worst thing both trainees and instructional staff can do is 'over-plan' the flow of events and the conduct of the exercise. This is not to say that training environments should lack detail, or that we should not have a master scenario event list or 'bone chart', but rather that we should not be over-prescriptive about the sequence and timing of events presented to allow trainees some flexibility in solving problems and engaging in adaptive coordination. If the instructional team has a very detailed plan there may be a temptation to keep trainees rigidly 'on track' and to corral them to the next situation – this could run contrary to the trainee team engaging in adaptive behaviour.

In this sense suitably developed environments, supplemented with skilled instructional staff who can reactively manage the environment, will offer the trainees more opportunities for adaption. Background environmental detail includes setting or world data that is appropriate to task and may include descriptions under the headings of Diplomatic, Information, Military and Economic (DIME) or Political, Military, Economic, Social, Infrastructure and Information (PMESII) systems. This data actually supports adaptive responses as all necessary detail has been fleshed out in advance, leaving the instructional staff to deal with the dynamics of the situation rather than having to generate content or lengthy answers to requests for information. For example if a unit may transport by rail, road or air from a staging point to a forward location relevant detail should be developed for each option, rather than assuming that trainees will adopt a particular route and only developing detail on that option. If we allow freedom in planning we should also allow freedom in execution. Plans that lack detail both provide capacity for adaptive coordination and necessitate adaption as coordination must be delivered somehow, if not through pre-planned means. This approach may carry some risks that require management, but equally over-planning generates risks because flexibility of task execution or ability to react to unforeseen events may be compromised.

Incorporating and expecting adaptive coordination has implications for exercise design and management including the skills required by members of the instructional team. Other considerations concern the ratio of instructional effort through the exercise design process, the objectives of team training (procedural or adaptive) and whether the trainee teams are at a skill level where adaptive coordination is a reasonable expectation.

Environmental Factors Necessitating Adaptive Coordination

Orasanu and Connolly (1993) identified the following as characterising the properties of naturalistic environments:

- high stakes
- time constraints
- competing goals
- multiple players
- dynamic environments
- shifting goals
- uncertain environments
- ill-structured problems.

Related factors labelled 'environmental stressors' (Cannon-Bowers and Salas, 1998) were identified as:

- threat
- performance pressure
- time pressure

- high workload/information load
- requirement for team coordination
- rapidly changing, evolving scenarios
- incomplete conflicting information
- multiple information sources
- adverse physical conditions
- auditory overload/interference.

We suggest that all of the above factors necessitate adaptive coordination in teams. These environmental factors, or adaptive coordination conditions characterise half of the adaptive coordination equation. Once an environmental factor presents itself the team must react and change behaviour, this being the other half of the equation. This may entail the team engaging in coordination supporting processes – distributed perception, team goal generation and collaborative planning – to support adaptive coordination of action, as outlined in Figure 3.4.

Teams are capable of generating adaptive coordination conditions through weaknesses in their own processes (NATO Joint Warfare Centre personal communication), as well as having these conditions imposed on them from an external source. For example, an organisation may possess conflicting goals because of poor management, or may be operating under time constraints because individuals or departments cannot generate the required outputs in a timely manner, environments may be uncertain because team SA is deficient or information collection priorities not defined. The fog and friction of war potentially exists within the team or collective processes as well as existing externally. Failures of team or individual performance or in the coordination enabling processes multiply and compound the complexities of the external environment.

In a team context we have *multiple players* and a requirement for *team coordination*. Adaptive coordination conditions may originate in the external environment as constraints that occur because of *dynamic environments* or *rapidly changing, evolving scenarios*; in the previous example the airline flight was cancelled necessitating a change of plan.

Figure 4.9 illustrates a simple plan, and how the dynamics of the environment may interpose causing a plan to become unworkable. The dynamics of the environment may prevent actions (for example an aircraft fault), change our current state (the flight may be overbooked or our ticket cancelled) or make the intermediate outcome we are trying to generate unreachable (our destination airport being closed due to bad weather). More seriously our outcome state might be unreachable, in this example we may discover that our house has been flooded or otherwise rendered uninhabitable. These possible situations are indicated in Figure 4.9 with an 'X'.

Dynamic environments and/or *rapidly changing, evolving scenarios* generate *time constraints* and *time pressure*, and a *high workload/information load*. Actions have a set of dependencies such as requiring team members, needing resources, information or time. Time constraints may prevent or limit actions including

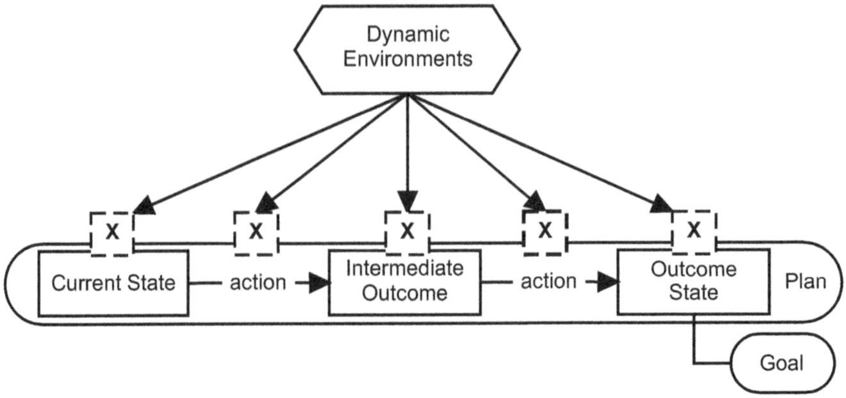

Figure 4.9 Dynamic environment impact on plans

communication, or may be related to issues of dynamic environments outlined above. These constraints may both disrupt the procedural flow of the task and/or cause errors to occur.

Any of the time or workload factors listed above along with *high stakes*, *threat*, *performance pressure* and *adverse physical conditions* (such as sleep deprivation, tiredness or physical fatigue) may cause mistakes or critical errors to occur, or tasks may be performed acceptably but 'other outcomes' may start occurring. Resources may be used sub-optimally, wastage may occur, information quality degraded, or team members may start falling behind with their activities and start cutting procedural elements. Slowly the team task may become degraded and latent errors start accumulating as the enacted task conditions drift away from the idealised task situation. Errors may snowball and generate critical consequences which are outcomes the team did not want to generate. These errors and consequences challenge coordination within the team and generate additional overheads as the team has to recover the situation while still performing the task. *Auditory overload/ interference* may be caused by *time pressure* or technical failures, impacting on team actions by degrading communication and thus disrupting coordination.

Information factors may also dislocate and uncoordinate team activity. These factors include *uncertain environments*, *multiple information sources* and *incomplete conflicting information*. These challenge team SA building by generating overheads for the process of distributed perception. These overheads may increase latency of team response, which may drive other concomitant factors such as *time pressure* and coordination challenges associated with *dynamic environments*.

Information factors and *dynamic environments* will also cause challenge to processes of goal generation, maintenance and priority. *Shifting goals* reflect these uncertainties and changes. Shifting goals may be self-generated by the team in response to environmental features exposed by team SA, or may be an endogenous variable originating from a higher echelon, such as a change in organisational priorities or command intent. *Competing goals* may also be generated by poor SA processes in the sense of 'the left-hand not knowing what the right-hand is doing', by non-integrated goal generation processes caused by poor organisational design, or a failure of deconfliction in planning.

Uncertain environments and *incomplete conflicting information* may also challenge the collaborative planning process. Internal factors self-generated by the team such as organisational barriers or information bottlenecks may compromise planning, causing flawed plans and incomplete plans generating task execution challenges, especially in *dynamic environments. Ill-structured problems* defy obvious simple procedural treatment and may include treatment of systems exhibiting complex behaviours with multiple competing feedback loops and time lags requiring complex analysis.

Scenario Specification

Scenario specification is where sample scenarios are defined in detail. Primary sources of information are SMEs and personnel familiar with the task environment or staff involved in the capability definition area. Scenarios instantiate the task and generate the detail for both training environment analysis and training overlay analysis. One needs to generate a range of scenarios (or range of features within an extended scenario) to deliver conditions to develop both procedural and adaptive coordination within team tasks.

Scenarios for team task procedures are relatively straightforward to define: we need to capture the range of conditions of the task according to the task analysis. For example, in maritime force protection where the crew of a ship are protecting the ship against close range threats, tasks may be conducted in open sea conditions or in harbour, during day or night, with a number of ROE profiles and team manning states. Measures of workload can be related to a team task analysis as a basis for further discussions with SMEs. In a maritime force protection context, workload factors might include: the number of suspect vessels in the surveillance area; how clear or cluttered the maritime surveillance picture is; and what level of complexity of maritime features such as territorial waters, traffic separation schemes and islands occur. For planning and execution tasks the task analysis captures the 'procedure for planning' and the 'procedure for execution'. The environment within which these occur can be defined in a number of ways, for example scenarios based around most likely, most dangerous or most demanding situations.

Potential axes for scenario specification include:

- *environmental challenge* – the problems generated by the environment which need to be solved, from few, simple, routine cases to numerous, highly complex, challenges.
- *environmental dynamics* – from static environment to very dynamic environments.

- *environmental uncertainty* – from situations for which we have good information to more ambiguous, less certain environments.
- *environmental complexity* – this reflects the signal-to-noise ratio, how much irrelevant information is available compared to what the team needs.
- *resource limitations* – the environment has high demands in relation to the team's capacity in terms of resources; shortages of fuel for example.
- *degradation of communication* – from clear to degraded channels of communication.

Team workload may be generated by any of these areas above.

Scenario specification may include the specification of the adaptive coordination conditions that cause procedural coordination to bend or blend into adaptive coordination. This may form a useful adjunct to the specification of plan-based procedural coordination inherent within team procedures. Related scenario specification approaches including the definition of team stressors can be seen within Event Based Approach to Training (EBAT) (Johnston et al. 1995), Stress Assessment Methodology (Hall et al. 1993) and Team Adaption and Coordination Training (TACT) (Serfaty et al. 1998) discussed in Chapter 6.

References

Annett, J. (2005) Hierarchical Task Analysis (HTA), in N. Stanton, A. Hedge, K. Brookhuis, E. Salas and H. Hendrick (eds) *Handbook of Human Factors and Ergonomics Methods*. Boca Raton, FL: CRC Press. 33-1 to 33-7.

Annett, J., Duncan, K.D., Stammers, R.B and Gray, M.J. (1971) *Task Analysis*. Training Information Paper No. 6. London: H.M.S.O.

Boyd, J.R. (1987) *Organic Design for Command and Control*. Available at http://www.ausairpower.net/APA-Boyd-Papers.html [accessed 20 February 2013].

Buzan, T. (1974) *Use Your Head*. London: British Broadcasting Corporation.

Cannon Bowers, J.A. and Salas, E. (1998) *Decision Making Under Stress: Implications for Individual and Team Training*. Washington, DC: American Psychological Association.

Chung, T.C. (1991) *Sun Tsu – The Art of War*. Singapore: Asiapac Books.

Cleary, T. (1988) *Sun Tzu – The Art of War*. Boston, MA: Shambhala Publications.

De Bono, E. (1982) *De Bono's Thinking Course*. Bath: BBC Worldwide.

Eisenhower, D.D. (1957) *Address at the Centennial Celebration Banquet of the National Education Association*, 4 April 1957.

Entin, E.E. and Serfaty, D. (1999) Adaptive Team Coordination, *Human Factors*, 41(2): 312–25.

Gorman, J.C., Cooke, N.J. and Amazeen, P.G. (2010) Training Adaptive Teams, *Human Factors*, 52(2): 295–307.

Grote, G., Kolbe, M., Zala-Mezo, E., Bienefeld-Seall, N. and Kunzle, B. (2010) Adaptive Coordination and Heedfulness Make Better Cockpit Crews, *Ergonomics*, 53(2): 211–28.

Hall, J.K., Dwyer, D.J., Cannon-Bowers, J.A., Salas, E. and Volpe, C.E. (1993) Toward Assessing Team Tactical Decision Making Under Stress: The Development of a Methodology for Structuring Team Training Scenarios. *Proceedings of the 15th Annual Interservice/Industry Training Systems Conference*. Washington, DC. 87–98.

Hughes, D.J. (ed.) (1993) *Moltke on the Art of War: Selected Writings*. New York: Presidio Press.

Johnston, J.H., Cannon-Bowers, J.A. and Smith-Jentsch, K.A.S. (1995) Event-based Performance Measurement System for Shipboard Command Teams. *Proceedings of the First International Symposium on Command and Control Research and Technology*. Washington DC: The Center for Advanced Command and Technology.

Kirwan, B. (1992) Introduction, in B. Kirwan and L.K. Ainsworth (eds) *A Guide to Task Analysis*. Boca Raton, FL: CRC Press. 1–15.

Klein, G. and Pierce, L. (2001) *Adaptive Teams*. Paper to the 6th International Command and Control Research and Technology Symposium. U.S. Naval Academy, Annapolis, MD, 19–21 June 2001.

Lipshitz, R., Klein, G., Orasanu, J. and Salas, E. (2001) Focus Article: Taking Stock of Naturalistic Decision Making, *Journal of Behavioural Decision Making*, 14: 331–52.

Macmillan, J., Entin, E.E. and Serfaty, D. (2004) Communication Overhead: The Hidden Cost of Team Cognition, in E. Salas and S.M. Fiore (eds) *Team Cognition: Understanding the Factors That Drive Process and Performance*. Washington, DC: American Psychological Association. 61–82.

Malcolm, D.G., Roseboom, J.H., Clark, C.E. and Fazar, W. (1959) Applications of a Technique for R and D Program Evaluation (PERT), *Operations Research*, 7(5): 646–69.

Mesarovic, M.D., Macko, D. and Takahara, Y. (1970) *Theory of Hierarchical Multilevel Systems*. New York: Academic Press.

MIL-STD-881C (2011) *Department of Defense Standard Practice: Work Breakdown Structures for Defense Materiel Items*. Available at https://acc.dau.mil/adl/en-US/482538/file/61223/MIL-STD/20881C/203/20Oct/2011.pdf [accessed 16 September 2014].

Nixon, R. (1962) *Six Crises*. Garden City, NY: Doubleday.

Orasanu, J. and Connolly, T. (1993) The Reinvention of Decision Making, in G.A. Klein, J. Orasanu, R. Calderwood and C.E. Zsambok (eds) *Decision Making in Action: Models and Methods*, Norwood, CT: Ablex. 3–20.

Poulton, E.C. (1957) On Prediction in Skilled Movement, *Psychological Bulletin*, 54: 467–78.

Reigeluth, C.M. (1999) The Elaboration Theory: Guidance for Scope and Sequence Decisions, in C.M. Reigeluth (ed.) *Instructional-Design Theories and Models*, Volume 2. Mahwah, NJ: Lawrence Erlbaum Associates. 425–53.

Romiszowski, A. (1999) The Development of Physical Skills: Instruction in the Psychomotor Domain, in C.M. Reigeluth (ed.) *Instructional-Design Theories and Models*, Volume 2. Mahwah, NJ: Lawrence Erlbaum Associates. 457–81.

Salas, E., Sims, D.E. and Burke, S. (2005) Is There a 'Big Five' in Teamwork?, *Small Group Research*, 36: 555–99.

Schaab, B.B and Moses, F.L. (2001) *Six Myths about Digital Skills Training*. Research report 1774. Alexandria, VA: US Army Research Institute for the Behavioural and Social Sciences.

Serfaty, D., Entin, E.E. and Johnston, J.H. (1998) Team Coordination Training, in J.A. Cannon-Bowers and E. Salas (eds) *Making Decisions under Stress: Implications for Training and Simulation*. Washington, DC: American Psychological Association. 221–45.

Shepherd, A. (2001) *Hierarchical Task Analysis*. London: Taylor & Francis.

Thordsen, M. (1998) Display Design for Navy Landing Signal Officers: Supporting Decision Making under Extreme Time Pressure, in E. Hoadley and I. Benbasat (eds) *Proceedings of the Fourth Americas Conference on Information Systems*. Madison, WI: Omnipress. 255–6.

US Army document TC 25-10 (1996) *Lane Training*. Available at armyrotc.msu.edu/resources/TC25-10LaneTraning.pdf [accessed 3 October 2014].

Van de Ven, A.H., Delbecq, A.L. and Koenig, R.J. (1976) Determinants of Coordination Modes within Organisations, *American Sociological Review*, 41: 322–38.

Vicente, K.J. (1999) *Cognitive Work Analysis*. Mahwah, NJ: Lawrence Erlbaum Associates.

Wellens, J. (1974) *Training in Physical Skills*. London: Business Books.

Chapter 5
The Training Environment

Introduction

An essential activity within team TNA is to produce specifications for training environment(s) and elements within them to enable team task practice and assessment. The specification of training environments is predicated on an understanding of the task – if we fail to analyse or define our tasks correctly it is likely our training environments will not support all of the necessary conditions or transformations that the task requires.

We can distinguish between *task environments* as places where tasks are performed in operational settings, and *training environments*, which are environmental substitutes for operational settings. The attributes that training environments must support are determined from analysis of operational tasks which are anticipated to be required to be performed by a team on the completion of training. Once these operational task environmental attributes have been determined, one can start to look at how specific training environments might deliver and support those tasks.

Task analysis is the lens through which we define which properties of the operational environment are relevant and must be replicated safely and cost effectively in training. Task analysis involves analysing the transformations of the environment that are enacted by team members, so in the course of task analysis we come to understand what is relevant from an environmental perspective. From a training environment perspective we require support for:

- A range of initial task conditions that constitute the potential starting points of the task.
- The team task actions that move us to our task outcome via a set of intermediate outcomes. Dependencies for actions include:
  - resources required
  - information required
  - other actors and agencies with whom the team is coordinating and communicating in the course of the task.
- Conditions that impact on task actions and team task processes. Where adaptive coordination is required we must support the adaptive coordination conditions that necessitate and allow adaptive coordination to occur, as discussed in Chapter 4.
- Our training environment, which must fully support the delivery of task outcomes and the (safe) generation of mistakes and 'other outcomes' in the course of task practice and assessment.

The elements of the task environment listed above form part of the description or specification of the task environment that our training environment must support. The specification of the task environment is couched in solution-neutral form, describing the environmental properties that must be supported without making assumptions about how these properties might be delivered by a training environment. Examples of training environments might include live training or simulation, or a mixture of the two.

Task environment specifications are descriptive – they describe characteristics and element attributes that should be provided in training environments. This is in contrast to approaches where numerical scores are applied to features or elements in the training system. Such numerical approaches, while allowing priorities or preferences to be identified, do not actually describe what is required to be designed and built, but only what is considered to be relatively important in relation to the task. Specifications for environmental characteristics within TCTNA are either mandatory (the training environment must support this attribute for this task), or optional (the training environment supporting this characteristic is of some benefit but is not essential). Descriptive task environment specifications (including essential fidelity information) enable organisations to procure training environments which are fit for purpose and support all aspects of the task required, rather than replicating all aspects of technical systems and possibly omitting non-technology centric aspects such as essential knowledge or skills required by role-players interacting with the team through the training environment.

Previous research in media selection has established the 'no significant difference' principle in relating learning media to learning effectiveness (Clark 1983) – provided a task is appropriately supported by the training environment (which includes instructional media), it does not matter exactly how the environment is delivered from a task perspective (for example book, printed PDF, online webpage) as long as the essential characteristics required by the task are supported. Training environment characteristics could be seen as a necessary feature, but not sufficient in themselves to deliver effective training (Pike and Huddlestone 2007). As an example having a fully equipped chemistry teaching laboratory is part of the training system that delivers chemistry teaching in a university but is not the whole solution on its own – we require lecturers, laboratory assistants, a teaching syllabus, learning activities and examination papers. If, however, our laboratory has been improperly specified and contains no materials to support task practice we can see that this would negatively impact our ability to conduct training practice and assessment. Going beyond the essential characteristics of training environments to support tasks, the means of training environment delivery (for example live vs synthetic) may differ in features such as offering additional benefits such as: support for a range of task conditions outside the core task conditions set required, wider integration of other concomitant tasks, higher capacity, greater safety, higher level of availability, the ability to leverage existing assets or lower costs (for example lower per student cost, lower per course cost, lower overall cost or a shift of costs between departments).

Fidelity is a term used in simulation that describes the extent to which the simulation replicates the actual environment (Alessi 1988; Gross et al. 1999, cited in Liu et al. 2008: 62). Within Training Environment Analysis we apply concepts of fidelity beyond simulation, to enable us to judge potential training environments, however they are delivered. Previous authors have identified a number of types of fidelity: these include physical, visual-audio, equipment, motion, psychological-cognitive and environmental fidelity (Liu et al. 2008: 62). By defining the major elements of the training environment one can place the types of fidelity identified by previous authors into a broader system which includes:

- the substitution of team members within the team, such as seen in team augmentation during training events
- the skills of role-played non-team members
- the interfaces that instructors and non-team members (such as role-players) use to control equipment such as sensors and effectors
- the content, quality and richness of the information environment with which trainees are interacting
- the use of real equipment in training, whether standalone or integrated into synthetic environments.

Considerations within these areas may have traditionally been treated separately from established areas of fidelity; however, we would consider these elements to be just as critical from a TNA standpoint. If we fail to make necessary considerations within these areas above, support for our task may be compromised, even within what may otherwise be very high fidelity environments.

Taking consideration of all task environment components in TNA is critical in delivering the task to the user in a training environment. For example, if a team task involves the real-time management of human crewed resources outside the team, the team members will need to communicate with non-team members and ask questions about their current status and requirements. Those non-team members will need to interpret their current situation, respond with appropriate terminology and make appropriate requests based on their tasking. The implication is that if those non-team members are role-played in training they will need to have a set of specialist knowledge, skills or experience. If this is not the case, adequate support materials and pre-training may be required so role-players can act as credible non-team resources to be managed by the core team. Furthermore, the interfaces that those role-played non-team members have access to, must present sufficient information (such as current fuel status or type of weapons being carried) to the role-player to enable them to engage in a realistic conversation with the team being trained. Since team training is focused at task coordination, communication is a critical process and must be adequately supported. Within the model of the task environment one can also place the major categories of simulation – live, virtual and constructive simulation – and identify hybrid categories of simulation that may be used as training environments.

Task Environments and Training Environments

In Chapter 2 the task environment was characterised as undergoing a transformation when tasks are performed, from an initial environmental state to an outcome state produced by the task or generated as an effect. As the task progresses intermediate states of the environment are generated, for example in the task of boiling an egg the intermediate state of 'soft-boiled egg' is reached before the state of 'hard-boiled egg' occurs. For any tasks in an operational setting (here the term 'operational' is being used to distinguish from a training situation) tasks will be performed under a range of conditions or initial environmental states that define the context of the task and the situation that the team has to deal with. In turn, these operational tasks will generate a range of task outcomes, including (hopefully) success – where the team's goals align to the outcomes they generate. The alternative is varying types of outcomes which do not satisfy our goals which may include critical errors or accidents occurring. The task environment is the totality of environmental states and dynamic conditions that are of relevance to that task. The task environment and associated concepts are illustrated in Figure 5.1.

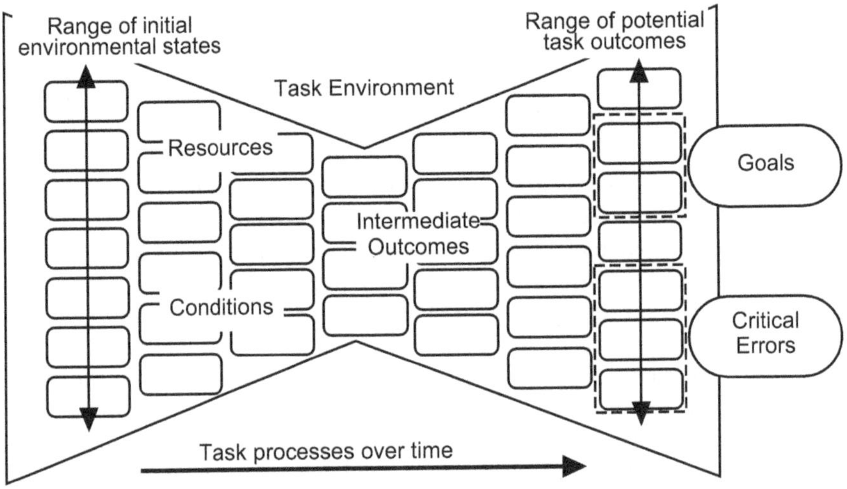

Figure 5.1 The task environment

It is within the task environment that task(s) which are the subject of the TNA at hand will be performed, after the relevant training has been conducted and completed satisfactorily. This task environment is the future operational environment for the task. The term 'future' is used as at the point the TNA is conducted, trainees are not yet trained and are therefore not yet capable of working independently in the operational task environment. The term 'future' operational task is also used to distinguish from training opportunities in the live peacetime environment which may support some, but not necessarily all of the characteristics of future wartime tasks. Wartime attributes of tasks might include extreme low level flying, weapons release and countermeasure deployment. In warfighting tasks the future operational environment is a real conflict situation. Not all future operational tasks, however, involve warfighting; vehicle maintenance for example, has to occur in peacetime and during times of conflict. Similarly for replenishment-at-sea tasks, the future operational environment for the task is performing the task 'live' at sea, whether in war or while deployed on routine operations. In the case of vehicle maintenance, the future operational environment is the mechanic working on real vehicles with real mechanical and servicing requirements as part of her 'day job', not under instructional supervision.

Generally, we do not want to let teams or personnel loose in the future operational environment without training because unnecessary waste, substandard performance, damage to expensive equipment, injury, death and/or mission failure may occur. From a training audit perspective organisations require evidence that individuals and teams have been adequately prepared for their future roles. The analysis and specification of the training system, as the mechanism for delivering this performance improvement and risk management process for operational tasks, has a significant role to play.

TNA makes statements about what training environments are required to prepare teams for employment in the future operational task environment. Clearly training environments as rehearsals for the 'real thing' need to be sufficiently similar in that realistic tasks are supported. However, they need to be different in the sense that risk, injury and expense are minimised. There is a compromise to be struck here; for example having soldiers saying 'bang' instead of firing their weapons is very safe (in the short term) but does not satisfactorily represent the weapon firing task. Teams and team members will hopefully be deemed competent, validated or certified in the tasks in question before they transition to the operational task environment. Figure 5.2 outlines the relationship of task environment and training environment.

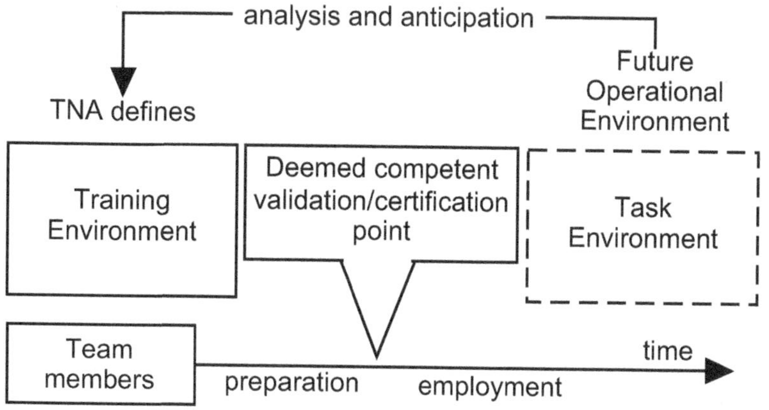

Figure 5.2 Training and task environments

As an example, the task environment for servicing a vehicle is:

- team members performing the task
- vehicles in need of servicing
- tools used in servicing such as spanners, test equipment, wrenches and voltage meters
- resources such as vehicle lifts, diagnostic equipment, spare parts, power and lighting, running water and containers for used engine oil
- task relevant information, which may be in the form of vehicle reference manuals, checklists, operating guidance, outputs from diagnostic equipment or online resources for referencing and ordering parts
- information systems which enable the information above to be displayed, retrieved, modified, stored, printed or emailed.

The vehicle servicing task might be performed in a garage, hanger or dock or might be performed at the side of a road depending on the exact situation. In any case we would expect the broad categories listed above to occur.

As a second example, a task environment for amphibious landing operations might be:

- team members performing the task
- a beach and landward location to secure
- resources and equipment – landing craft, helicopters, amphibious assault vessels, naval vessels, air support assets, command elements
- task relevant information, including the mission aim, expected enemy forces, current situation, weather reports, satellite imagery, signals intercepts
- own force information and communication systems.

In an amphibious landing we have an enemy who will react to the team's actions. Resultantly we also have to include enemy forces and their equipment:

- enemy forces
- enemy equipment, vehicles, command elements

The enemy uses information in their decision-making cycle and has a command intent and established information processes including signals communications occurring in and between the enemy command elements and force elements. So we must include:

- enemy information which includes the enemy's mode of operation and concepts and doctrine, preferred management style, national goals and factors such as whether they are seeking to escalate or de-escalate the situation
- enemy information systems that transform, store and communicate information the enemy is using.

Use of the Operational Task Environment for Training Purposes

An important question that TNA should address is could, or should one use a real operational task environment for training purposes? Use of the task environment for training purposes (that is, using the operational task environment as a training environment) hinges on positive answers to two questions: firstly, does the task environment exist at a time when training needs to occur? Secondly, can the task environment support training activity safely and cost-effectively, in alignment with the broader organisation and its ways of working?

Contrasting the example of vehicle servicing and an amphibious landing task, we can see that in the maintenance example we *might* be able to use a real work situation (task environment) to do training. This is because the operational task environment exists at a point in time when training needs to be conducted. Just because a task environment exists it does not mean that it can be utilised for training. There are a number of barriers to task environments being used for training. In the vehicle maintenance example these might be:

- management buy-in, ensuring alignment between training and operational functions within the organisation
- having to transport training team members to the location of the task environment
- having vehicles spare to be able to practise with (potentially for a longer period of time than routine task operations)
- operational risk factors in the workplace being accepted and managed (real equipment being out of action for longer, risks to equipment and trainees, higher resource implications)
- space in the workshop and other capacity management issues (spare tools, training activity conflicting with operational activity)
- spare tools and spare staff to conduct the training
- access to required information and information systems
- real life support aspects – accommodation, food, transportation.

In an amphibious landing example the operational task environment does not exist at the point training is conducted (there is no 'real-life' enemy, no civilians to protect/evacuate). As there is no operational task to piggyback onto we need to create our own dedicated training environment as part of a training or exercise event.

Some future operational environments are fairly predictable and cover routine, regular, scheduled activity – vehicle maintenance, for example, fits into this category. Other future operational environments such as future conflict are unpredictable and thankfully not routine. Some future operational environments can be predicted to occur because there are always tasks to be performed in these environments (such as maintenance tasks, or 'at sea' tasks part of routine naval deployments). Other future operational environments, such as those in relation to strategic nuclear tasks have thankfully never occurred. Where we move from routine, predictable tasks to more varied and unpredictable tasks one challenge is to make an assessment of what the future operational environment looks like so that training priorities may be determined, and suitable training environments created.

Ideal Properties for Training Environments

Training environments act as proxies for future operational environments which in a military setting include emergency or conflict situations of varying types and degrees of intensity. In an ideal world the training system would support a training environment (or set of training environments) that would allow the complete range of initial conditions for the operational task to be supported and would allow a wide variety of realistic task outcomes to be generated – this includes supporting the ability of things to go awry. This ideal training environment, while supporting all conditions and resources that the task required, would not consume resources or cause any damage or injury normally associated with the task. Furthermore the environment would be both low-cost, highly available, support all of the required training audience and would run with minimal instructional overheads. Unfortunately, such a training environment is yet to be invented – the job of Training Environment Analysis is to get as close to this ideal as possible by defining a specification for what is required. This specification is prepared in a solution-neutral form in that it discusses *what* is required rather than *how* this is to be delivered. This specification is based on the future operational task, and so is an anticipation of what might be required. Once this specification has been prepared one can then engage in the training option selection stage which is where candidate solution environments may be assessed and compared. External constraints such as personnel availability will also impact on training environment decision making.

In many tasks a dedicated training environment (such as an aircraft simulator) has to be used in place of an operational task environment. However, substitution of the operational task environments may not always be possible, leaving us either having to accept a level of risk in training, or having a training gap where practice of the full task is unsupported. This is because there are events that may occur in the operational task environment which we cannot allow to occur within training on the grounds of *risk*. Alternatively, *availability* or *cost* factors necessitate the use of a dedicated training environment.

- *Risk* – as an example, in freefall parachute training instructors do not deliberately fail to pack student parachutes properly just to cause student emergency drills to be performed, because errors would lead to fatalities. Since the freefall and under-canopy parachuting environment cannot be adequately replicated in a dedicated designed training environment which ensures safety, this leads to this particular emergency drill not being practised in full in training.

- *Availability* – some operational task environments do not exist or cannot be made available for training necessitating dedicated training environments; where training environments are not available a training gap occurs.
- *Cost* – cost relates to factors such as resources used in training in the operational task environment, which might include fuel, weapons expended or cost of servicing. These factors may make training prohibitively expensive. Training environments may be cheaper than operational task environments in certain situations.

Often all three factors occur together, especially in the context of warfare tasks. Warfare environments are high risk, low availability and very expensive. Where operational environments cannot be adequately replicated as training environments this may lead to a training gap as the task is not fully supported in a training context. We may have the situation where one training environment does not suffice for all aspects of the task, for example with stages of a freefall parachuting task one finds the operational task is substituted by a number of training environments, as outlined in Table 5.1.

Table 5.1 Example of training environment substitution in task practice

Operational task stage / conditions	Training environment for task practice
1. Walk to aircraft, board aircraft	Real aircraft on the ground used for familiarisation
2. On-board equipment checks	On the ground – equipment checks in hangar
3. Climb out	On the ground – real aircraft and door mock-up training aid
4. Exit	No substitution – operational task environment
5. Freefall	Wind tunnel, dirt-dive on ground for freefall patterns
6. Parachute deployment	No substitution – operational task environment
7. Canopy ride	No substitution – operational task environment
8. Approach and landing	No substitution – operational task environment

As an example the task stage of 'Freefall' may be handled by a vertical wind tunnel, but this wind tunnel will not support the task stages of aircraft exit, parachute opening or landing. Where a task stage is being supported by a dedicated training environment we may find that an environment used as a substitution may support some aspects of the task but not others. For example, a vertical wind tunnel does not (for reasons of limited tunnel diameter) support long distance lateral movement of parachutists as seen in tracking or back sliding, nor are large groups of individuals in freefall supported.

Training environments may also only cater for a sub-set of the range of conditions expected in the task environment – in many cases we don't know what the future operational environment looks like, so it's difficult to specify exactly what environmental properties we need to support in training. If we haven't defined or analysed the properties of the future operational environment (for example the potential enemy threat, types of missions) it unlikely these things will crop up incidentally in the environment used for training. The main means to mitigate this is to define or reference appropriate scenarios and then drive training environment selection from the scenarios of greatest concern – for example most likely, most dangerous, most demanding situations.

Training Environment Support for Environmental Conditions and Tasks

Training environments may support some aspects of a task but not others. This can be a positive factor if understood and incorporated into design as an opportunity that might otherwise be missed, or a negative factor if we are practising a task which is incomplete in comparison to the operational task environment and the potential risks to future performance this may cause have not been mitigated.

From a negative perspective training environments may only present a section of the required task environment, for instance:

- A training environment may present a restricted number of initial environmental states. For example, low level attack aircraft in groups of four may be required for a realistic task but only two slower aircraft may be available. The implication is that not all of the variety or complexity of the task can be trained. As another example, Norwegian training areas in winter are suitable for arctic warfare and not jungle warfare training, meaning that if the trainee task requires both potential environments then two training areas are required.
- A training environment may present a restricted set of intermediate outcomes. For example weapons may not be fired (in either direction) or targets do not exhibit characteristics that enable ROE criteria to be applied as part of decision making. As an example, a task may require a certain cue that has to originate from the environment, such as a suspect aircraft illuminating the trainees' vessel with a fire control radar. If the training environment did not support this type of radar emitter, the cue would be absent so any expected trainee reaction would be compromised. Here the implication is that not all decisions or actions can be supported possibly because environmental information is absent. In these situations one of the functions of the training overlay is to step in as a proxy and provide

substitute stimulus. This stimulus can be provided in the form of an instructor making a statement to the team as to the event that has occurred, or by the inject being provided directly into the shipboard combat systems.
- A partial or incomplete set of representative task outcomes: for example weapons effects in terms of accuracy or real impact may not be modelled. This is akin to a shooting range where the target does not register bullet strikes. This means that success or failure of actions cannot be determined and in turn means no accurate performance data is available on which to make assessment judgements. An equivalent critical issue is to have performance data potentially available, but have no standard statements of level regarding the expected performance defined as a task benchmark.

Training environments partially supporting tasks can be seen in training methods such as the Simplifying Conditions Method (SCM), tasks conducted under restricted conditions or Part-Task Training.

In SCM (Reigeluth 1999: 442) environmental characteristics that impact on actions are simplified, for example by removing a factor that generates a coordination requirement. From a positive perspective partial environmental support for tasks may make them easier to learn, either by deliberate incorporation of design features into training equipment, or through the removal of external complicating environmental conditions.

- Design features – as an example the stabilisers on a child's bicycle that remove the coordination requirement of the child pedalling and steering simultaneously, by the prevention of the outcome of 'bike falls over [due to lack of forward momentum]'. The removal of this potential outcome removes a required concomitant task 'maintain forward momentum through continual pedalling'. By simplifying the conditions we simplify the task by removing the number of simultaneous coordinated actions required. This allows the child to focus on steering, rather than the act of steering and pedalling together.
- The removal of complicating conditions is achieved through, for instance, only allowing training to occur in certain environmental conditions, and through monitoring. For example student parachutists are limited by the speed of wind they are allowed to jump in to avoid landing complications and student scuba divers are limited by factors such as depth and duration of dive.

Simplification or restriction of conditions also helps ensure safety by removing contributing factors to accident severity. Conditions that complicate task actions and make task actions more difficult, may also be those that may make accidents, when they occur, more serious. For example, in scuba diving nitrogen narcosis is caused by diving deep, and may make underwater tasks and decision making more difficult. Diving deep also reduces the time before decompression is required and increases the consumption rate of air. Deep diving can be avoided as a contributing accident factor by limiting depth in training exercises. Restricting conditions also helps ensure that certain 'other outcomes' cannot occur – for example, if a student scuba diver drops their weight belt while underwater and does an uncontrolled ascent to the surface, the depth and duration limits that have been established in the training environment prevent the risk of a decompression incident (though not the risk of a lung expansion injury).

Instructors supervising tasks may restrict conditions by removing risks of accidents or mishaps through active intervention. The instructor has the role of maintaining safety by preventing negative intermediate outcomes from occurring. This may include monitoring students undergoing training, counselling against bad decisions before they are enacted, or recovering emergency situations – for example a flying instructor recovering a stalled aircraft, or a jump instructor pulling the ripcord for a student parachutist. These are considered as part of a taxonomy of instructional tasks in Chapter 6.

Part-Task Training involves taking a section of a bigger task and training that in isolation. Part-Task Training has been used in individual and team training, for example enabling the training of aircraft navigation systems without the aircraft being in flight. In a team context, Part-Task Training may be conducted within sub-teams prior to the whole team being trained together. Training environments ideally need to be specified from a whole-task perspective, as coordination and communication processes are what team task training aims to develop and this requires a unified environment. This is not to say that everyone necessarily needs to be trained together at the same time, but rather training environment down-selection decisions should originate from a whole-task perspective.

Task Environment Components

The task environment embraces all things that change or can be changed by the task including: the starting conditions of the task in the future operational context, information relevant to the task, the resources used in the task and the products of the task both in terms of physical effects and information produced. The process of how to derive task environment features through task analysis is illustrated in Chapter 8.

The task environment is used to generate a specification for the training environment which forms part of the TCTNA output. The task environment specification breaks the environment down into a number of convenient elements and defines the types of interaction that occur between them.

The task environment consists of a set of elements, these being:

- *The Actors in the Environment* – these are either team members or non-team members that are capable of using information, making judgements and performing actions that may impact on and make changes to the task environment.

- *The Physical Environment* – the physical characteristics and features of the world, such as terrain, weather, runway layout, and/or geographic position of forces. The physical environment includes naturally occurring dynamic features such as weather, seasonal variations, tides and forest fires. Natural dynamic features account for transformations that occur in the task environment (for example: a heavy snowfall) which are not actor-driven nor generated but rather are determined by the forces of nature. The physical environment includes resources that may be used in the task.
- *The Information Environment* – this is the world of information generated stored and transmitted by mankind. Examples that may be relevant to tasks include:
  - internet and intranet webpages
  - email
  - internet chat
  - telephone calls in the telephone systems
  - satellite TV transmissions
  - CCTV that is streamed
  - books and manuals, aide memoires, quick reference cards
  - road signs
  - printed maps
  - satellite imagery
  - in a military context, the common operational picture, datalink information, readouts from sensors.

  The information environment has an associated physical aspect, for example map information may reside on a piece of paper, or within an LCD panel, or be projected on a screen. Similarly a road sign is printed onto a sheet of metal or sprayed on the asphalt surface of the road. The internet as an information environment is supported by physical hardware: computers, servers, switches and cabling. The physical and information environments are distinguished because information has a different set of characteristics from the physical environment that supports it.
- *Sensors and Effectors* – are the means by which changes are detected and generated in the environment. Sensors and effectors provide the link between actors and their environment. In complex systems these items may include dedicated sensors and effectors which are capable of generating effects at a distance. An example would be a radar system which generates radar pulses (effector) and then collects information back from the radar return (sensor). Sensors and effectors can also be taken to include elements of equipment which facilitate the sensing or doing actions of actors. A vehicle windscreen, for example, does not act as a sensor, but what it does do is facilitate driver and passenger vision in the vehicle. An equivalent on the effector side could be seen in the steering wheel of the car where driver effort in turning the wheel is directly transmitted by mechanical means to the wheels of the car. Other examples are devices generating mechanical

advantage: innovations such as the spear thrower, a wooden stick which acted as a multiplier for the innate spear-throwing force generated by hunters.

In most task situations there are a number of sensors and effectors and sensor and effector facilitators available to team members. For example in a car the sensors and sensor facilitators include: the windscreen and windows which facilitate vision by the operator, mirrors, parking sensors and the reversing camera system (if fitted). The effectors and effector facilitators are the controls of the car: the steering wheel, foot pedals, gear levers and dashboard switches and controls and the door handles and window controls. Sensors may present external stimulus indirectly such as car parking sensors producing audible signals to the user from reflected radar pulses sent externally. Sensor facilitators such as the car windscreen or car mirrors allow external stimulus to reach the driver. Effectors or effector facilitators make changes in the task environment under actor control – in our car example these would include: the car engine, the headlights, the doors, and position or rotation of the car wheels which ultimately move the car relative to the wider environment. An equivalent example of an interface into the information environment would be a camera with satellite uplink facility: video imagery is streamed or captured into an information system (sensor interface) which can then be stored and retransmitted via computer monitors (effector interface).

The relationship between the task environment components described is shown in Figure 5.3.

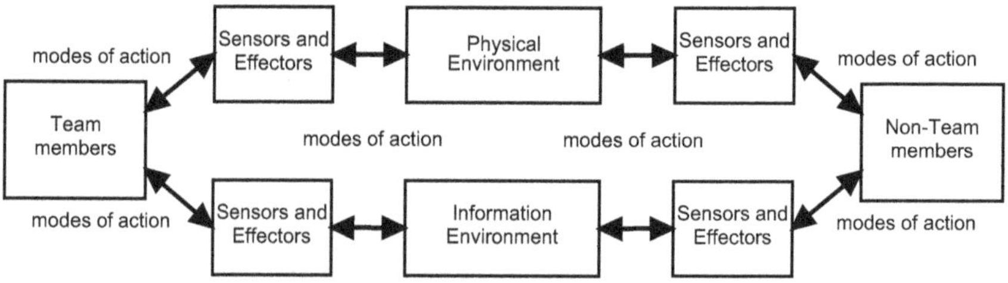

Figure 5.3 Task environment components

The arrows linking the task environment components in Figure 5.3 represent modes of action. Team members interact with the sensors and effectors available to them through system interfaces such as keyboards and display screens. Sensors and effectors have modes of action linking them to the task environment. A torch's mode of action generates light, whereas an electric kettle's generates heat. An example of a mode of action for a sensor is an astronomy telescope that may register electromagnetic radiation in particular frequencies or a sight system that may have a certain degree of sensitivity or magnification.

Team members interact with the world through equipment based sensors and effectors or sensor and effector facilitators, and directly through physical action and sensing. Team members make alterations to the physical and information environment through the tasks they perform. These changes may be detected and reacted to by non-team members with the sensors and effectors at their disposal. Figure 5.3 captures the *functional* connections between the elements: team members operate sensors, effectors and related technology to make changes in the physical and information environments. The separation of elements above is to generate labels of functional convenience for TNA purposes: team members (and sensors and effectors) obviously occupy the physical environment and constitute part of the state of the outside world. For example, if a team member is sat in a vehicle at a particular location, that team member and vehicle could be subject to enemy fire, just as much as the building that they might be parked next to. Similarly a surrendering enemy with their hands up in the air, while being a non-team actor, would also constitute an aspect of the physical environment, or task cue, which team members might have to react to directly.

We can also for convenience distinguish between internal and external environments. Where a team is occupying a shared location such as a vehicle or building, we can characterise the internal working space such as tables, chairs and the seating plan as the internal physical environment. Team members may also have access to local information such as printed documents, maps, identification charts and reference books which occupy the local environment. These items constitute the internal information environment, and may be interacted with directly such as by drawing on acetate overlays placed over printed maps. The wider environment beyond the bounds of the team can be described as the external environment made up of the physical environment and information environment.

The information environment and physical environment are both present in areas of conflict and military operations, the former (information environment) being characterised by propaganda, media operations, 'influence operations' or information such as media or social media that is contributing to the battle for hearts and minds. The physical environment is the area of kinetic warfare, traditional military operations, suicide bombs and Improvised Explosive Devices (IEDs). The two environments are linked: if a library is burnt down or a TV newsroom is taken out of action through kinetic effects then the corresponding information environment will be impacted. Information environments impact physical environments via two routes, the first could be characterised as cyber-warfare. For example, if the computerised control system for a power plant is compromised via a cyber-attack in the information environment, the plant may be damaged or shut down, then impacting the physical environment. The second means by which information environments impact the physical environment is through the decisions and actions people take in response to influences and messages from information environments, such as the internet or through face-to-face social gatherings. Effects may also be propagated kinetically *and* recorded as information. In this situation, both physical and information environments are addressed from the same event. Classic military activities such as camouflage, emissions control, deception or electronic countermeasures are seeking to either deny the enemy information relating to the physical environment or to generate spurious information in the enemy information environment. In either case the aim is to generate as big a difference as possible between the actual state of the physical environment and the enemy's information environment that relates to it.

Sensors and Effectors: Modes of Action

When team members operate equipment consisting of sensors or effectors or a combination of the two, there are two forms of interaction going on, the interaction between the team member and the interface of the equipment (or system) and the interaction between the equipment or system and the wider environment.

In the example of a motor car's hazard warning lights being activated, these two forms of interaction are:

- The team member interacts with the interface or controls of the equipment and gets feedback from the system. When a car hazard warning light button on a car dashboard is pressed, the button pressed stays recessed, illuminates and flashes and an audible intermittent signal is heard. This is feedback the user gets from the interface of the system.
- The sensor and/or effector system makes changes to the external environment according to its particular mode of action. The hazard warning lights of the vehicle are turned on and orange lights flash and are visible outside in the wider physical environment. The transformation, or mode of action being made in the physical environment is that orange coloured light is being emitted from a number of light sources which is capable of being spotted by other road users. The team member in the car will also be able to observe this mode of action by glancing out of the window, though they may not be able to ascertain whether all hazard lights are functioning without getting out of the vehicle and inspecting each light. This is feedback that the user receives from the wider environment, which in this case is direct to user.

In this example the mode of action has no particular associated hazards and is not expensive. However in a military context many modes of action are expensive to generate and potentially very hazardous. For example live firing is expensive because ammunition is expensive, and a shell landing outside a designated range area is an extremely hazardous event for those nearby. Hazards must be managed, meaning that training environments may be constrained to remote areas or the tasks or resources altered because of restrictions of space. In a team context we may be able to support the task, but not at the scale, complexity and intensity we might like. Other modes of action such as missile firings against manoeuvring targets present an additional level of challenge – not only is the mode of action dangerous and expensive, but the target being addressed will be destroyed if we are successful.

The target may be expensive and may be required to manoeuvre in a particular way – meaning that we must control our target remotely through technology. Enemy sensors and effectors will have modes of action as well, so we might be expected to detect radar in particular frequencies, or emitters to be associated with specific platforms; from a training perspective we need to represent these.

In summary our task environment can be described in terms of team and non-team actors, sensors, effectors and related items (sensor and effector facilitators), and the physical and information environment. Sensors and effectors have modes of action both to their operators (seen through the equipment interface to the user) and also modes of action in terms of the effects generated or data gathered from the environment. The task environment components and modes of action which exist between them outlined in Figure 5.3 all have a set of attributes, some of which we wish to preserve (for accuracy or realism) and others like cost or availability which we wish to alter – to make cheaper or more available.

Substitution of Task Environment Components

Training environments aim to replicate the key features of task environments while substituting other aspects. Any of the task environment components, interfaces or modes of action (interactions) in Figure 5.4 may be substituted in the training environment.

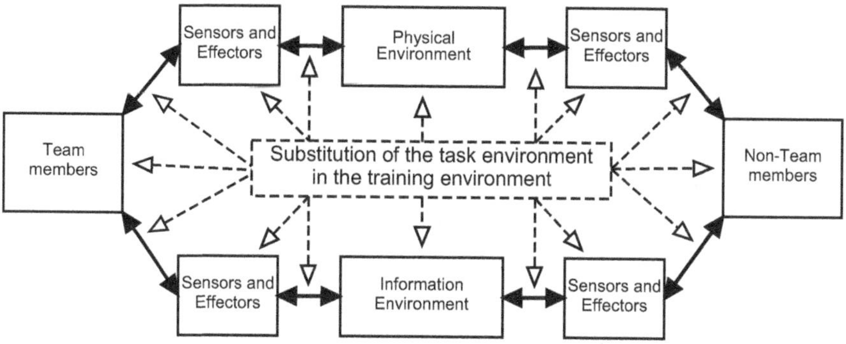

Figure 5.4 Training environment components

In training environments the following aspects of the task environment may be replaced or substituted:

- *Team Members* – certain team members or roles may be absent from the training audience and may need to be replaced by role-players, or instructional staff to fulfil vital tasks. Alternatively non-team personnel from within the organisation may step in to augment these roles. These augmentees will not necessarily deploy with the formed team on operations but constitute trained personnel within the organisation who can be called on if required.

- *Interfaces to Sensors and Effectors* – interfaces to specialist sensing and effector equipment which may involve specialist keyboards and other physical switches, dials and controls may be replaced by touch screen technology or similar. This may significantly reduce costs and increase flexibility as multiple equipment types may be reconfigured within a single set of instructional hardware
- *Team Member Sensors and Effectors* – sensors and effectors may be replaced by simulation or synthetic environment features, or substituted training resources that mimic the behaviour of real equipment. For example adding low energy laser emitters to conventional rifles substitutes live ammunition. Another example from the air domain is training practice munitions which may be live dropped and mimic the behaviour of real munitions in flight without the risk or cost of real equipment. Team member sensors and effectors also include communication and information access systems which the team uses to communicate through the shared information environment.
- *Modes of Action of Sensors and Effectors* – an example of substitution in this area is the use of blank ammunition in live fire training. Other forms of substitution include 'synthetic wrap' used on live fire training where lasers fitted to conventional rifles register hits on targets that have been fitted with laser sensitive sensors. Here a low energy laser substitutes for live ammunition in flight.
- *Physical Environment* – real task environments may be substituted by real physical replica buildings such as used in urban warfare training or replica shipboard environments used in firefighting and damage control training. Other examples in this category include using old airframes or other post-operational equipment for initial maintenance training, NASA's use of swimming pools to simulate zero-gravity for astronauts, or outdoor training being conducted in environments which are similar to anticipated future conflict environments. Some aspects of physical environments may be substituted by information environments – these considerations are discussed in the simulation section below.
- *Information Environment* – information environments may be substituted by proxy information systems, or training volumes created within operational military command and control systems which include access to dedicated training resources. The information environment also includes public information domains, the internet, cable TV and social media. Information environments may be visualised and presented to trainees through simulation systems acting as proxies for physical environments.
- *Non-team Member Sensors and Effectors* – these must always be substituted, unless equipment from potentially hostile nations can be sourced and made available. An example would be the Aggressor Squadrons flown by the US

armed forces which operate foreign aircraft types. Other examples include aerial target and decoy systems used by the US Navy to represent attacking aircraft and missiles.

- *Non-team Members* – it is unlikely that members of potentially hostile nations would be invited to 'play themselves' in an exercise, however foreign nations do observe each other's live training activities uninvited and surveillance ships such as Auxiliary General Intelligence vessels do frequently loiter and observe foreign nations' naval exercises. Not all non-team members in team and collective tasks are hostile or adversarial; we may have other government departments, non-governmental organisations (NGOs) and international organisations to deal with, so the possibility does exist to train with these 'white cell' [neutral non-team] actors. Most non-team members are substituted by instructional or support staff, or are played by another section of the training audience in a two-sided exercise. Critical attributes in non-team members include accurate reflection of enemy training and doctrine, including TTPs and suitable knowledge, skills and experience of instructors or role-players to play the potential adversary in a realistic manner.

Fidelity

A key issue for training environment specification is how closely a training environment must resemble a task environment. In simulation this degree of similarity is called simulator fidelity (Allen 1986, quoted in Liu et al. 2008: 64). Fidelity principles apply across all types of training environments, whether delivered through simulation or not. For example a live training exercise could be said to have fidelity in relation to the operational task it is representing. Fidelity is a term associated with simulation, but it is important to note that the principle of the training environment being simulative of the task environment applies across all elements of the training environment: team members, interfaces, sensors and effectors, modes of action, the physical environment and information environment, non-team member sensors and effectors and non-team members themselves. The implication is that high fidelity simulation can be compromised if, for example, our role-players who are communicating with the team are unable to play their part effectively and support the team task. Fidelity is a key issue in TNA, as fidelity is a factor in assessment of simulation quality, training transfer and cost effectiveness of simulation device design (Liu et al. 2008: 61). Fidelity related to communication as the mechanism supporting coordination and coordination enabling processes would seem to be a critical factor in supporting the team task environment. Research into the use of low-fidelity PC-based simulators in multi-seat Crew Resource Management (CRM) training has demonstrated positive training transfer (Nullmeyer et al. 2005). This might be attributed to communication and task coordination being adequately supported, even within an otherwise lower fidelity training environment.

In TCTNA we have adapted the simulation fidelity definition to cover all training environments. Simulation fidelity is 'an umbrella term defined as the extent to which the simulation replicates the actual environment' (Liu et al. 2008: 62). In TCTNA, training environment fidelity is a term that defines to what extent the training environment replicates the operational task environment. The term fidelity applies to the degree of correspondence or replication between the operational task environment and the training environment, in all elements of the task environment: team members, interfaces, sensors and effectors, modes of action, physical and information environments and non-team members and their corresponding means to sense, alter and communicate through the environment (non-team member sensors and effectors, interfaces and modes of action).

Fidelity measurement in simulation compares the 'extent to which a representation [for example a simulation] reproduces the attributes and behaviours of a referent [in our terminology a task environment]'. A 'referent is described as being an entity or collection of entities and/or conditions – together with their attributes and behaviours – present within a given operational domain' (Hughes and Rolek 2003, cited in Liu et al. 2008: 63). In the TCTNA environment model, the referent is the task environment and all of its elements and properties as described in Figure 5.1.

The definition of the task environment is that it encompasses all things (initial conditions, resources, information, intermediate outcomes, goal aligned outcomes, critical errors and consequences) that can be changed by or affect the performance of the task under consideration. The implication is that fidelity is not an absolute value but is always referenced to the task in question, and therefore the elements in the environment that relate to the task. For example a flight simulator may be very detailed and accurate with respect to flight characteristics, however there may be a particular task which involves dealing with smoke in the cockpit; if the simulator cannot deliver this input and allow the aircrew to respond meaningfully then the conditions of the task are unsupported and hence simulator fidelity in this particular task is very low. In this task example we would potentially look to use a smoke emitter in the cockpit, but might then find that the lack of a functional oxygen system in the simulator again impacted the task possibly with a safety implication as well.

Previous fidelity dimensions identified include:

- *Simulation fidelity* – 'Degree to which device can replicate actual environment' (Gross et al. 1999; Alessi 1988)
- *Physical fidelity* – 'Degree to which device looks, sounds and feels like actual environment' (Allen 1986)
- *Visual-audio fidelity* – 'Replication of visual and auditory stimulus' (Rinalducci 1996)
- *Equipment fidelity* – 'Replication of actual equipment hardware and software' (Zhang 1993)
- *Motion fidelity* – 'Replication of motion cues felt in actual environment' (Kaiser and Schroeder 2003)

- *Psychological-cognitive fidelity* – 'Degree to which device replicates psychological and cognitive factors' (Kaiser and Schroeder 2003)
- *Task fidelity* – 'Replication of tasks and manoeuvers executed by the user' (Zhang 1993; Roza 2000; Hughes and Rolek 2003)
- *Functional fidelity* – 'How device functions, works and provides actual stimuli as actual environment' (Allen 1986)

The relationship of these fidelity dimensions to the Training Environment Model is summarised in Figure 5.5.

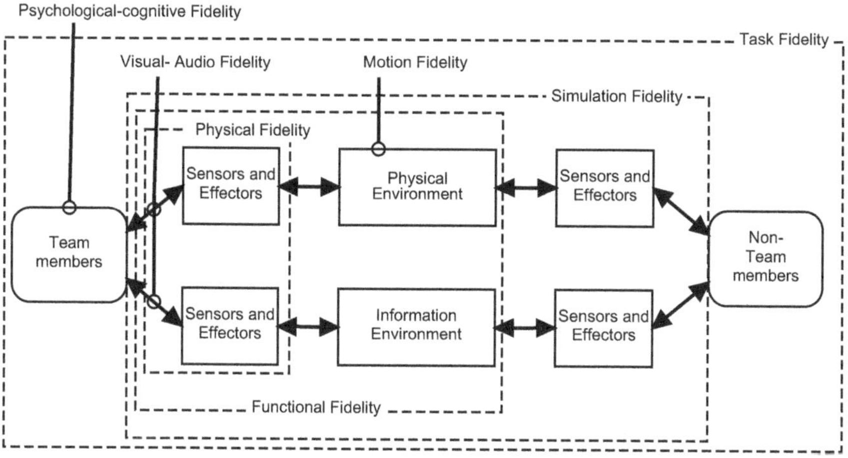

Figure 5.5 Fidelity dimensions mapped to the Training Environment Model

Whilst there are multiplicities of fidelity dimensions that have been identified in the literature, the majority can be considered to refer to instances of physical or functional fidelity, or in some cases both. For example, visual-audio fidelity and motion fidelity are aspects of physical fidelity. Equipment fidelity, on the other hand embraces both physical and functional fidelity. The replication of hardware speaks to its physical properties, whilst its software properties in terms of performance speak to its functional fidelity. That said, psychological and task fidelity are subtly different, as they are related to how all of the elements in a task environment act in concert to provide a credible task scenario which engenders appropriate aspects of task performance to be executed. Within TCTNA we consider physical, functional and task fidelity to be both a sufficient and complete set of fidelity dimensions appropriate to the specification of training environments and their management and control.

- *Physical fidelity* – the physical attributes of the element, such as look, feel, weight, size and sound that are experienced by human senses

- *Functional fidelity* – how the element behaves in terms of the responses that it produces to the inputs that it receives
- *Task fidelity* – the capability provided by the training environment to facilitate the set-up and control of the elements in the environment in order to deliver appropriate scenarios, which engender the performance of desired physical and cognitive task behaviours.

Where we are dealing with non-team actors, or augmentees within the team being trained, the terms 'appearance' (to reflect structural or physical characteristics) and 'knowledge, skills and experience' (to reflect task-relevant functional characteristics) are more meaningful labels than physical and functional fidelity.

Physical environment elements are specified in terms of physical and functional fidelity. Information environment elements do not have physical fidelity as such, as information may be expressed through a number of interfaces, so the description focuses on the properties of the communication systems and the properties of the data or content held in information systems. The properties of the information environment interfaces involve factors such as monitor size or specialised keyboards. In addition we need to describe local information environment elements such as printed map boards and reference documents. These tangible elements of information may have relevant physical properties (such as the printed size of the map) which also have to be identified, if they are of relevance to the task in question.

One observation is that while many measures of fidelity cover the user equipment-centric left-hand side of Figure 5.5, coverage of right-hand side factors (for example non-team member knowledge, skills, capabilities, doctrine) is rather tacit. Only task fidelity really covers these factors; 'Replication of tasks and manoeuvers executed by the user [in response to environmental conditions generated by the enemy …]'. If adaptive coordination and adaption is a required team capability, we will need intelligent, reactive and appropriately modelled non-team actors for the team to be adaptive with (or against). Scenario specification through the description of adaptive coordination conditions supports the consideration of non-team member fidelity factors.

Use of Real Equipment in Training Environments

One way to minimise the differences between the task environment and the training environment is to use real equipment in training. Real equipment as sensors and/or effectors cannot be said to have fidelity because it is not a facsimile nor copy of something but is actually the 'real thing'. This said, using real equipment is not a complete solution. As we have seen, sensors and effectors used by team members are only one aspect of the training environment, meaning that we must consider team members, the characteristics of the physical and information environments, and whether all modes of action of the equipment can be used in training, when judging how accurately our use of real equipment is supporting the replication of the operational task environment. The implication is that we can create substandard training environments by using real equipment without wider consideration of factors such as context. For example, firing blanks from a real rifle supports the interface aspects of the weapon system perfectly – the user can aim the weapon, and the trigger pull weight and balance of the weapon is realistic. The feedback that the user receives from pulling the trigger – the report of a shot fired and the recoil – is likewise 'the real thing'. When we look at the mode of action of the rifle, however, we see significant differences. We cannot use blanks in target practice, cannot 'walk' blank rounds onto a target if we are firing a machine gun at a surface target at certain engagement ranges, and we cannot fire blanks at live targets at close range for safety reasons.

When looking at the use of real equipment one approach is to combine real equipment in a live environment with simulation in what is termed 'live simulation', where real equipment is used, with realistic interface characteristics but simulated modes of action in the environment. For example we may use real equipment with realistic aiming and trigger pull characteristics, but substitute the rifle bullet for a laser pulse, or a non-lethal training round that marks its target. Alternatively real equipment may be combined with a synthetic environment such as seen in US Navy training where ships' operations rooms are connected to a common synthetic theatre of war via the Battle Force Tactical Training (BFTT) system.

Characteristics of the Physical and Information Environment

Extending beyond the interface and mode of action characteristics of sensors and effectors we have the contextual characteristics of the environmental situation that are represented in training events. These aspects form the inputs of the vignette or scenario into which our trainees are placed, allowing them to exercise and practise the tasks that are required to be trained. We must represent the necessary entities that are required by our scenarios, if we are representing a village as a feature in the physical environment relevant for our task we need houses, roads, vehicles, people and agricultural features – animals, crops, hedges, ditches and so on. Likewise in an operational level scenario we must represent the information environment, and so represent the press and media channels, the historical backgrounds of the protagonist countries and social media as the inputs to the tasks being performed by the team. The configuration and attributes of the entities in the scenario should also be realistic, for example the road layout in the village, the size of the fields, the height of the vegetation, the number of people on the streets and their clothing should be representative of what is anticipated in the operational environment.

Physical and information environment characteristics will feed into psychological-cognitive fidelity that 'reflects the psychological or cognitive experience that a person receives from being in the simulator [that is, training environment]' (Liu et al. 2008: 66). Team task processes and products such as team SA, goals and plans contribute to psychological fidelity. Some of this information is given to the team undergoing training in the form of background scenario information: command intent, higher-level plans and information such as 'past pattern of life' that informs situational awareness, goal generation and planning. This information is provided in briefings that the trainees receive prior to the actual training scenario being started, and contributes to psychological-cognitive fidelity without comprising part of the simulator system when viewed from an engineering standpoint. The implication is that one could jeopardise a technically excellent training exercise by failing to brief the scenario adequately, or provide necessary background information.

Adaptive coordination conditions within the physical and information environment form part of psychological-cognitive fidelity. Our training environments must be capable of generating:

- external challenges and problems of suitable complexity and workload
- dynamic situations – situations should be capable of developing rapidly, entities should react to team actions
- uncertainty – conflicting, out of date and ambiguous information
- complexity – including background detail that may mask task significant cues
- resource limitation – to force decision making and coordination
- degradation of communication.

Our information environment must support realistic external communication demands, as we would expect team tasks requiring communication *within* the team to be challenged by this form of external information workload. For this to occur our training environments must be capable of supporting both the requisite number of non-team actors, with appropriate knowledge and skills, and the requisite number of simultaneous communication channels.

Simulation

Simulate as a verb in the Oxford English Dictionary is defined as working to 'Imitate the appearance or character of' or to 'Produce a computer model of' (OED online 2014). A technical definition is: 'A simulation may be considered as comprising three parts. These are, a model of the system to be simulated, a device through which the model is implemented, and an application regime in which the first two elements are combined with a technique of usage to satisfy a particular objective' (Rolfe and Staples 1988: 4).

Simulation involves replacing some aspects of the task environment (the 'system' in the definition above) with a training environment (the 'device') with suitable properties to support tasks safely and cost-effectively, within a wider context or scenario which involves instructional design, management and control ('technique of usage' in the definition above). The combination of the simulation environment, the scenario and support for the instructional tasks from the training overlay, deliver the task conditions to the user and support practice and assessment. This means that consideration of the simulator properties alone is not sufficient in TNA, we must make statements about scenarios and instructional conditions of use. These areas include the scenarios that are generated as part of the simulation procurement and the skills of non-team actors who may be interacting with the team in training whether through communication or controlling elements within the synthetic environment.

A widely adopted classification of simulation is live, virtual and constructive (LVC). These are defined as follows:

- *Live simulation* – a simulation involving real people operating real systems. Military training events using real equipment are live simulations. They are considered simulations because they are not conducted against a live enemy (DoD 2011: 119).
- *Virtual simulation* – a simulation involving real people operating simulated systems. Virtual simulations inject human-in-the-loop in a central role by exercising motor control skills (that is, flying an airplane), decision skills (that is, committing fire control resources to action) or communication skills (that is, as members of a C4I team) (DoD 2011: 159).
- *Constructive simulation* – a simulation that includes simulated people operating simulated systems. Real people stimulate (make inputs to) such simulations, but are not involved in determining the outcomes. A constructive simulation is a computer program. For example, a military user may input data instructing a unit to move and to engage an enemy target. The constructive simulation determines the speed of movement, the effect of the engagement with the enemy and any battle damage that may occur (DoD 2011: 85).

In the definitions of the types of simulation above (live, virtual and constructive) the term 'systems' encompasses the elements of the task environment: sensors and effectors, physical environment and information environment, and the modes of action that exist between these areas. We must also include non-team actors and the effects that they generate in the environment through their own sensors and effectors. The definitions of LVC are partially overlapping, for example a pilot in a virtual simulator could be viewing simulated units (that is, constructive units) on a data link display within the cockpit. As another example both virtual and live simulation systems could be associated within the same synthetic environment.

Live Simulation

In live simulation we have team members operating real equipment (that is, sensors and effectors and their associated user interfaces). This real equipment may be placed in an operational physical environment, for example if we are training a ship's operation room the ship may be at sea, alternatively the ship may be tied up alongside the quay in harbour. These differences describe whether the real equipment is in the operational physical environment ('at sea') or not ('alongside' the quay). In either example above, the sensor and effectors within the ship and their user interfaces are real, what has been substituted is the modes of action into the physical environment; missiles will not be fired and sensors such as radar will be taking a synthetic feed from the simulation system. The information environment sensors and effectors (and their user interfaces) and modes of action in the information environment remain real – if someone sends a chat message or email it is a real email, typed at a real keyboard. The information environment used for training may be a dedicated, stand-alone environment populated with appropriate content or may be a dedicated area or volume within an operational information environment.

If we were to adopt a live simulation approach in an aviation context, with the real aircraft on the ground we could deliver the information environment in a similar way to the maritime example, but would have additional requirements to support the replication of the physical environment. This is because the aircrew when they look through the aircraft canopy normally have a direct and unmediated view of the external physical environment. If we were to generate a proxy of the physical environment for the aircrew we could project the view of the synthetic environment onto a screen that surrounded the aircraft – here we would be substituting the physical environment for a view of an information environment. In an infantry training example the rifle used as part of the live simulation may be real equipment but the physical environment within which the task is performed is a facsimile of the real thing. Dedicated training equipment has been added to the rifle which substitutes some of its modes of action: when the trigger is pulled a live round is not fired. In this example some modes of action performed are real and some are substituted: when a solder runs across the road it is a real physical action that changes the state of their world, however when the rifle is 'fired' the mode of action has been substituted. In a maritime context operating real equipment in a substituted physical environment ('alongside'), has a number of implications: when the captain orders 'hard to port' we do not actually move the rudder, but rather have to substitute that mode of action, and then reflect the state of the environment as it would be (the fact that the vessel has turned) back into the operations room. In contrast, operating real equipment in an operational task environment means that when 'hard to port' is ordered the ship actually changes heading, so this equipment is not required.

While the physical environment for some tasks may be very close to what might be expected in a real operational task, other characteristics of the physical environment may be more difficult to play live. As an example, displaced persons and refugees fleeing a conflict area, and blocking roads with vehicles, animals and people on foot would be difficult to replicate at scale in the real physical environment. So while the physical environment in live simulation can be a facsimile of a real environment it may yet lack characteristics that might be seen in future operational tasks.

The live simulation environment:

- may have the capacity to support large and mixed types of team (such as armoured vehicles, dismounted infantry, helicopters, fast jets, surface naval units and submarines) within a unitary environment enabling us to train more people in more tasks. For example maintainers, logistics, command and control and force elements may all be trained together 'joined-up'. As we previously identified coordination and communication as the main team task processes, live simulation environments enable very large teams to be trained in these processes;
- supports a wide range of task transformations and modes of action – including air assaults, amphibious landings, combat engineering, ground manoeuvres with logistics, infantry manoeuvres, weapons effects though simulation;
- acts as visible demonstration of training activity, political will and capability to potential adversaries.

Virtual Simulation

Virtual simulation is 'real people interacting with simulation systems'. In virtual simulation, such as seen in an aircraft simulator, we are substituting real equipment for a dedicated simulator and the physical environment outside the cockpit for an image of the outside world – an information environment on a computer display, or projected onto a screen. The substitution of the physical environment outside the aircraft cockpit for an information environment means that the expensive modes of action of flying in the live environment (cost of jet fuel to generate thrust and servicing for example) can be substituted to new, equivalent modes of action in the information environment, which are mathematical calculations instead of physical transformations. Virtual simulation environments may be cheaper in some situations than live flying, though one should ensure that in comparing costs one is comparing like-for-like in terms of number of tasks being trained across the organisation. Live flying generates potential training opportunities within the organisation beyond the aircrew involved in the mission.

The synthetic modes of action in the environment must be fed to the team members in the simulator via specialist sensors and effectors that constitute the simulator. In an aircraft simulator the training system components include a visual display system for representing the world outside the cockpit, the internal avionics and instrumentation that constitute the aircraft information environment and potentially a motion platform to reflect the changes experienced by aircrew as the aircraft changes speed and direction.

Cockpit information such as airspeed, altitude and engine thrust settings are presented to the aircrew via multifunction displays, dials and gauges and the aircraft systems may be queried through the flight management system. Information systems and associated data along with the communication systems and other elements such as quick reference cards, checklists, maps and printed binders of material constitute the information environment to which the team members have access in the operational task environment. Within a virtual simulation environment this information environment must be replicated and linked to the section of the information environment now acting as proxy for the physical environment. In the real task environment errors may also occur in these areas: in a real aircraft in flight it is possible for flight instruments to start giving readings which are at odds with what the aircrew are experiencing or expecting. This may occur because either an effector or sensor is faulty (for example a pitot tube is iced-up), or the linkage between the two is broken. As an example, the aircrew may push the throttles to 95 per cent power and expect a corresponding acceleration to be felt and a multi-function display to show a caption reading 'nine five' on a dial, however the dial may read 'seven zero' due to a fault of some sort. Within our simulation system we must be able to reflect these situations where physical environment and information environments available to the team do not fully align.

Depending on how we chose to define the term 'simulation systems' we can see an overlap between the definitions of live simulation and virtual simulation in that we can have real people interacting with real systems (real equipment, that is, live simulation), within simulation systems (synthetic environments), as in our previous maritime training example. In live-virtual simulation situations we must add supplementary equipment or systems attached to real equipment to convert real modes of action (such as picking up a sonar contact or firing a missile) to a substituted equivalent.

Constructive Simulation

Constructive simulation is the team undergoing training interacting with 'simulated people interacting with simulated systems; computer generated forces (CGF), fully autonomous or otherwise'. Typically constructive simulation falls into the 'computerised war-game' category, where units with relevant properties are manoeuvred and interact with each other in a synthetic battlespace which is presented to trainees through a set of computerised interfaces such as seen in modern command and control systems. Other constructive simulations may involve specific aspects of military operations such as logistics, planning or media. Constructive simulation is commonly used as a complement to CPXs, where lower-echelon controlled units (LOCON) or force elements are controlled by the exercise managers in response to orders that have been issued by the trainees. While the movement and interaction of forces (which may be referred to as semi-automatic forces or CGF) is played out in a synthetic theatre of war, there is still a requirement for the exercise managers to represent LOCON communication, that is, to generate reports and responses to requests for information.

In constructive simulation we have a potential overlap with either (or both) live and virtual simulation – where the team is interacting with CGF (simulated non-team members) through information environments and/or through the physical environment being substituted by a view of an information environment. In the future one could anticipate artificial intelligence (AI) acting as a substitute for non-team members and replacing human-driven computer generated forces with fully autonomous variants. Integrating AI into non-team member sensor and effector equipment offers instructional opportunities such as autonomous robotic targets, and AI generated information environments – such as computer generated social media, press reports and representations of enemy C2 processes.

Simulation Summary

Within the three types of simulation we have a different range of substitution of environmental elements and modes of action occurring within training environments.

With live simulation we are substituting some but not all of the modes of action such as weapons effects that occur between team equipment (sensors and effectors) and the environment, while still supporting the physical environment directly. In live simulation we may also need to substitute some of the modes of action of systems such as radar and radio, so we may generate direct synthetic radar feeds into the ship's combat system or substitute radio for Voice Over Internet Protocol (VOIP) as the mode of communication between ships that may be engaged in live simulation but are doing so while tied up alongside at the quay in a naval base. In live simulation we are not substituting the information environment in that the teams operating communications and command and control equipment will likely be using the real equipment, real networks without modification. These information systems are likely to have been specifically configured for the training exercise, just as a representative physical training area has been selected. The instructional team in a live simulation context may be using the opportunity to fit transponders and the like to vehicles so that they can keep track of the trainees in potentially large exercise areas. This instructionally-specific trainee-tracking functionality is associated with the instructional task rather than trainee tasks *per se*.

In constructive simulation we have a partial substitution occurring within the information environment – we may be using real equipment and systems with CGF representing the enemy, or we may be running a CPX where our own LOCON are also represented by CGF. An alternative form of constructive simulation would be an information system being replicated and represented by another possibly lower fidelity and less classified information system running on different hardware. Non-team members in constructive simulation may be played by role-players, or by means of pre-scripted algorithms constituting AI. The reason that the substitution within constructive simulation is partial is that there are still modes of action occurring in the information environment which are not substituted. For example, when an instructor sends an email report to the trainees, or sends a chat message or updates an intranet webpage, or phones the trainee headquarters pretending to be a member of the press, these events are generally not substituted in the same sense that CGF are.

Finally in virtual simulation we have the exclusive replacement of all elements within the environment including the team member's equipment. Figure 5.6 maps the potential areas of substitution within training environments.

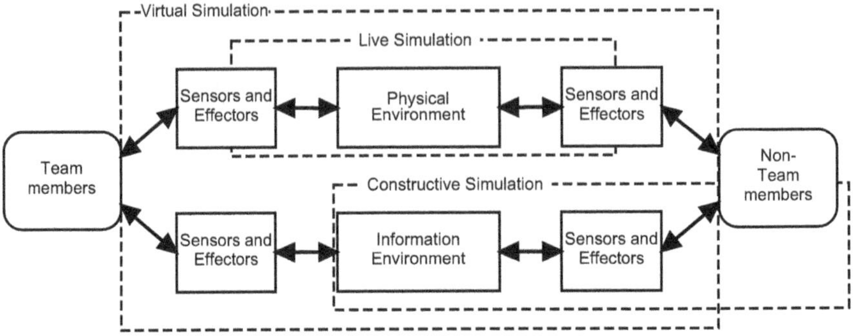

Figure 5.6 Types of simulation mapped to the Training Environment Model

Dashed boxes indicate the areas of simulation substituting elements of the task environment in the training environment. With live simulation and constructive simulation the area of overlap represents an area of non-exclusive substitution of task environment properties, so for instance we preserve some live modes of action (for example infantry man running) and substitute others (for example live rounds being fired). In the case of virtual simulation the area of overlap on Figure 5.6 represents exclusive substitution of task environment properties. In virtual simulation the environment and therefore the modes of action in the environment (for example flying) are completely substituted by synthetic means.

The Training Environment Model

In specifying the training environment, the aim is to describe the key features of the 'real world' task environment which must be present for the team to be able to carry out representative instances of their tasks. Once such a specification has been developed, it is then possible to examine how that training environment can be realised, making substitutions where necessary as described above.

In order to address the potential complexity of team and collective training environments, the Training Environment Model shown in Figure 5.7 illustrates categories of elements and their interactions which need to be captured during Training Environment Analysis, which is described in more detail in Chapter 11. The description approach for training environments is based on the task environment model outlined previously, with additional granularity to support guided analysis.

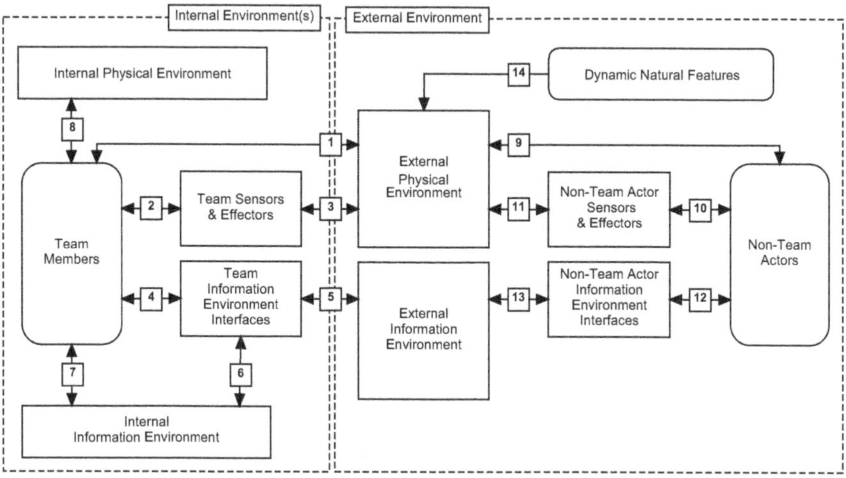

Figure 5.7 The Training Environment Model

The detail of the Training Environment Model is described below with references relating to numbered modes of action in Figure 5.7. The distinction between physical and information environments is drawn because they have different properties that need to be specified. The most obvious difference from the task environment model previously described is the distinction made between internal and external environments. Whilst some teams, such as dismounted infantry, operate directly in the physical environment, many operate from within a platform of some sort (be it vehicle, aircraft or ship) or a workspace such as a building or tent. In the Training Environment Model these dedicated workspaces are characterised as *internal environments*. Where multiple teams are involved, each team may have its own unique internal environment in which it operates. If we consider the example of maritime force protection (explored in full in Appendix A), the teams working on the upper deck, the bridge and in the operations room work closely together to defend the ship. Each of these separate but interconnected environments has to be replicated to support training. All other components are characterised as the *external environment*. The other change that has been made is the separate identification of dynamic natural features. By this we are referring to features of the external environment such as the tides and weather. They have the capability to interact with other elements in the environment (such as tides affecting the course of a ship) and as such, these interactions may need to be captured in the specification of the training environment. The attributes of the training environments that have to be specified are:

Internal Environment(s):

- *Team Members.* The team structure and representative manning states will have been determined as part of the task analysis process. The key data to be captured from an environment perspective is the distribution of the team across internal environments, ways in which each team member can interact directly with the external physical environment [1] and the range of sensors and effectors available to them.

- *Team Member Sensors and Effectors.* Each sensor (such as a radar or optical sight) and effector type (such as a weapon) needs to be specified in terms of the user interface which it provides and the interactions that user interface supports [2] and its performance attributes in terms of the modes of action on the physical environment that it supports [3] (such as the range and resolution of a radar).
- *Team Member Information Environment Interfaces.* Where team members interact with the information environment through an interface, as opposed to directly, the interfaces need to be specified. The specification should include the types of user interface interactions supported [4] and the modes of action with the information environment that are supported [5, 6].
- *Internal Physical Environment(s).* The pertinent features of the workspace within which the team or sub-team operates (such as workspace requirements, relative positioning of consoles and background noise levels) have to be captured. Where the team is split into different work areas or compartments, each of these local environments will need to be specified. There may be interactions with the local physical environment that need to be captured, such as adjusting lighting levels [8].
- *Internal Information Environment.* Internal information environment components such as intercom systems and information systems that are purely local to the internal environments in which the team/sub-teams are operating need to be identified. The internal information environment also includes elements which are accessed directly [7] rather than through a computer interface, such as printed maps, telephone directories and printouts from information sources.

The External Environment:

- *The External Physical Environment.* The external physical environment encompasses all of the physical characteristics and features of the land, sea and air domains within which the team operates and all of the elements of interest within them such as infrastructure and vehicles.
- *The External Information Environment.* This is composed of information and communications systems through which the team(s) communicates with other actors and agencies in the environment and send and receive information. There is a need to specify not only the interfaces to the information environment, but also the nature of the content with which it must be populated as this will have to be developed if it is not taken from a live source.

- *Non-Team Actors.* Non-team actors are all of the players outside the team that have to be represented. These can include friendly forces, enemy forces and neutral elements such as the local population. Non-team actors may appear in the physical environment and interact directly with it and the team, in which case their physical attributes (such as appearance and dress) need to be captured along with the ways in which they interact and behave [9]. The sensors, effectors and information environment interfaces available to each non-team actor also need to be identified. Since non-team actors are often role-played, it is also necessary to capture and describe the knowledge and skills (such as doctrine, tactics and voice procedures) required to perform the role.
- *Non-Team Actor Sensors and Effectors.* Non-team actors may use sensors (such as optical sights and radars) and effectors (such as weapons and countermeasures) to interact with the environment and the team (such as firing on them or moving their own assets). The user interface interactions [10] and the mode of action with the environment [11] and related system performance (such as radar range and vehicle speed) must be specified.
- *Non-Team Actor Information Environment Interfaces.* Where non-team actors interact with the information environment such as sending orders or publishing news stories in the media, the interfaces need to be specified, including the user interface interactions [12] and the modes of action with the information environment that are supported [13].
- *Dynamic Natural Features.* Natural features of the physical environment (such as the sea and weather, day/night) act to alter the physical environment, which can in turn impact on the actions being performed by the team. These dynamic features need to be identified and their modes of action [14] captured, for example an ocean current affecting vessel movement or fog impacting sensor range.

Team task environments can be enormously complex. The Training Environment Model provides a structured framework to aid the analysis of training environment requirements. By identifying the key components of the task environment that need to be represented in the training environment using this model, and defining their fidelity requirements, it is possible to produce a specification for the required training environment which can be used to inform Training Options Analysis. If a specification is produced in this way is has the advantage of being solution-agnostic, it is not skewed by constraints of any given implementation. Once such a specification has been developed, consideration can then be given to alternative implementations that could be realised (which may be live, synthetic or some mix of the two) as part of the generation of training options. In determining possible implementations, consideration can also be given as to how aspects of the training environment may be designed, controlled or selected to facilitate instruction.

References

Alessi, S.M. (1988) Fidelity in the Design of Instructional Simulations, *Journal of Computer Based Instruction*, 15(2): 40–47.

Allen, J.A. (1986) Maintenance Training Simulator Fidelity and Individual Difference in Transfer of Training, *Human Factors*, 28(5): 497–509.

Clark, R.E. (1983) Reconsidering Research on Learning from Media, *Review of Educational Research*, 53(4): 445–59.

DoD (2011) *Department of Defence Modelling and Simulation Glossary*. Modeling and Simulation Coordination Office, Alexandria, VA. Available at http://www.acqnotes.com/Attachments/DoD M&S Glossary 1 Oct 11.pdf [accessed 16 March 2011].

Gross, D.C., Pace, D., Harmoon, S. and Tucker, W. (1999) Why Fidelity? *Proceedings of the Spring 1999 Simulation Interoperability Workshop*. Orlando, FL.

Hughes, T. and Rolek, E. (2003) Fidelity and Validity: Issues of Human Behavioural Representation Requirements Development. *Proceedings of the 2003 Winter Simulation Conference*. New Orleans, LA.

Kaiser, M.K. and Schroeder, J.A. (2003) Flights of Fancy: The Art and Science of Flight Simulation, in M.A. Vidulich and P.S. Tsang (eds) *Principles and Practice of Aviation Psychology*. Mahwah, NJ: Lawrence Erlbaum Associates. 435–71.

Liu, D., Macchiarella, N.D. and Vincenzi, D.A. (2008) Simulation Fidelity, in D.A. Vincenzi, J.A. Wise, M. Mouloua and P.A. Hancock (eds) *Human Factors in Simulation and Training*. Boca Raton, FL: CRC Press. 61–73.

Nullmeyer, R.T., Stella, D., Montijo, G.A and Harden, S.W. (2005) Human Factors in Air Force Flight Mishaps: Implications for Change. *Proceedings of the 27th Interservice/Industry Training, Simulation and Education Conference (I/ITSEC)*. Orlando, FL, December 2005.

Pike, J. and Huddlestone, J.A. (2007) Instructional Environments: Characterising Training Requirements and Solutions to Maintain the Edge. *Proceedings of the 29th Interservice/Industry Training, Simulation, and Education Conference (I/ITSEC)*. Orlando, FL, December 2007.

Reigeluth, C.M. (1999) The Elaboration Theory: Guidance for Scope and Sequence Decisions, in C.M. Reigeluth (ed.) *Instructional-Design Theories and Models*, Volume 2. Mahwah, NJ: Lawrence Erlbaum Associates. 457–81.

Rinalducci, E. (1996) Characteristics of Visual Fidelity in the Virtual Environment, *Presence*, 5(3): 330–45.

Rolfe, J.M. and Staples, K.J. (1988) *An Introduction to Flight Simulation*. Cambridge: Cambridge University Press.

Roza, M. (2000) Fidelity Considerations for Civil Aviation Distributed Simulations. *Proceedings of the AIAA Modeling and Simulation Technologies Conference*. Denver, CO.

Zhang, B. (1993) How to Consider Simulation Fidelity and Validity for an Engineering Simulator, *Flight Simulation and Technologies*, August: 298–305.

Chapter 6
The Training Overlay

Introduction

The Team Performance Model, described in Chapter 3, provides a framework for analysing team tasks and identifying the training environment requirements to support team task execution in the training context. According to Reigeluth and Schwartz (1989) one of the key factors that affected the effectiveness of a simulation system used for training was the instructional overlay. They characterise this as embracing both the design of the simulation system itself, in terms of having appropriate levels of fidelity and control features, and the instructional design of the training that is carried out on it. In using the term 'training overlay', we extend the concept of the instructional overlay to embrace all the components of the training system that must be in place for effective training to be delivered. As described in Chapter 1 (Figure 1.2), the training overlay is composed of the training strategy and methods, along with the training staff and their tasks, supporting systems, and resources necessary to deliver training in accordance with the strategy.

The key components of the Training Overlay Model and the relationships between them can be elicited by considering two examples of team training, one on a small scale and the other on a large scale.

Small-scale Example: Pilot Type-conversion Training in Commercial Airlines

Pilot type-conversion training refers to the training of already qualified pilots to fly a new aircraft type, such as an Airbus A320 or a Boeing 777. The training strategy that is often adopted for this type of training is commonly referred to as 'zero flight time training' as the flying component is all taught in a flight simulator. The first time that the pilots will fly the actual aircraft is on a scheduled flight carrying fare-paying passengers. The operation of commercial aircraft is a team activity with different roles and responsibilities allocated to the two pilots. The Pilot Flying (PF) is responsible for flying the aircraft whilst the Pilot Non-Flying (PNF) monitors his actions, runs checklists and manages the aircraft systems. Effective coordination of their actions, maintenance of a shared understanding of their situation and communication between the pilots are critical to the safe conduct of the flight.

The choice of simulator-based training is predicated on a number of factors. The cost of simulator training is substantially lower than live flying (typically an hour of simulator time would cost in the order of £400–£500 per hour whereas the operating cost of an aircraft would be in the order of £10,000 per hour). Safety is another significant factor as emergency situations can be practised such as engine fires and system failures which cannot be fully replicated in the live environment. A wide variety of environmental conditions can also be presented on demand in the simulator such as adverse weather conditions, day and night flying, and a variety of airports from which the pilots have to depart and arrive. The zero flight time strategy is tenable for this type of training as the key entry condition is that the pilots hold an Airline Transport Pilot's Licence. As such, they already have extensive experience of live flying. Contemporary flight simulators which are certified for zero flight time training are of the highest fidelity and can support all of the necessary cues and responses required for this type of training.

A typical simulator training exercise on a course of this type starts with a briefing by the instructor about the simulator exercise content, and the key learning points and assessment criteria. In a typical full-flight simulator used for this type of training the flight deck is replicated in full and the trainees occupy the pilots' seats as they would in the real aircraft. Located immediately behind them is the instructor's console. This console provides the instructor with all of the necessary controls and displays to operate the simulator. The simulator configurations for each of the planned exercises are usually selectable from a menu system. This allows the instructor to quickly configure the simulator for the exercise with regards to the departure airport, weather and the systems state of the aircraft (such as fuel and passenger load). Once the exercise commences, the instructor's tasks include instructing the trainees and monitoring their performance, and noting key points for the post exercise debrief. Monitoring performance involves not only observing the trainees but also monitoring the environment. For example, during an approach to an airfield the instructor needs to monitor the speed of the aircraft, as well as its position in relationship to the required approach flight path. Plan and elevation displays of the flight path and key variables such as speed can be presented on the instructor station's displays to assist with this task. The instructor can also exploit the capabilities of the simulator for delivering instruction. For example, if the trainees are getting it horribly wrong the instructor can freeze the simulator so that incorrect actions can be discussed and instruction provided on the correct actions. The instructor can then unfreeze the simulator for the exercise to continue or even reset the simulator in time to an earlier point in the exercise. The events required for the exercise, such as the onset of poor weather or an engine fire, can also be generated from the instructor station. In a four-hour simulator detail each exercise will usually be run twice so that each trainee undertakes the exercise as PF and PNF.

Once the exercise is complete, the instructor debriefs the students on their performance. The latest flight simulators have debriefing systems which have the capability to record flight deck audio and video, avionics display data and selected aircraft parameters. Debriefing rooms have debriefing equipment installed which allows playback of the session recordings which the instructor can use to bring out key points from the exercise. This is a powerful tool as it presents the trainees with clear, observable evidence about their performance, including how they interacted as a crew. Trainee performance in each exercise is recorded in the trainee records, as part of the training management process.

The course culminates in two tests, the first is a test of instrument flying, conducted by a specialist Instrument Rating Examiner, giving the pilots an Instrument Rating which is required in order to fly in poor weather with limited visibility and in the airways. The second is a Licence Skill Test, which is the flying equivalent to a driving test. This is conducted by a specialist Type Rating Examiner, and gives the pilot a Type Rating which they must hold to fly that type of aircraft.

What becomes apparent from this description is that the instructor is carrying out two roles: firstly, as an instructor in carrying out the tasks of briefing, instructing, monitoring and evaluating performance, and debriefing; secondly, managing the training environment by setting up and controlling the simulator during the exercise. In order to carry out these two roles effectively the instructor will have required training both in flight instruction and in the operation of the simulator and its use as a training tool. There is also a requirement for specialist examiners, who carry out separate assessment roles.

Such a training course is a complex affair. In line with the Systems Approach to Training described in Chapter 1, its development will have necessitated extensive analysis to determine the training requirements, followed by a considerable amount of design activity to put together the simulator exercises and all of the supporting materials, such as exercise briefs and simulator configurations, often referred to as lesson plans. Such is the importance of training and the standards that must be achieved, regulatory bodies, such as the Federal Aviation Administration, issue regulations concerning the capability of the simulators that are to be used, the content of the Instrument Rating and Licence Skill Test, and the minimum number of hours of simulator training that must be undertaken. Furthermore, these courses have to be accredited by the regulator before they can be run. Aircraft manufacturers also issue guidance about training course content. In addition, the instructors are regularly checked and examined by the regulating authority to ensure that standards of instruction are maintained. Such training courses fall into the category of 'design once, deliver many times'. The day to day business of training is focused on training delivery and evaluation, with analysis and design being essentially a one-off, up-front activity.

The key points related to the training overlay from this example are:

- The simulator has to be capable of presenting the required training environment.
- The simulator must provide the instructor with the capability to configure the simulator and control the environment during simulator exercises.
- A capability to record and provide feedback on student performance is required.
- Instructors have instructional and environmental management roles, both of which have a training and competence prerequisite.
- Multiple categories of instructors with different skills may be required.

Large-scale Example: Exercise Joint Warrior

Exercise Joint Warrior is a large-scale UK military exercise run by the Joint Training Exercise Planning Staff (JTEPS) based in the Joint Headquarters at Northwood in northwest Greater London. It is run every six months and provides an opportunity for maritime, land and air force components to exercise together, hence its 'Joint' title. To give an idea of what 'large-scale' means in this context, the range of participants in a typical Joint Warrior exercise may include:

- two naval task groups, each comprising an aircraft carrier and supporting ships with its command staff on board the carrier
- multiple submarines
- an infantry or marine battle group, with its command staff
- an air component of many tens of aircraft, including airborne command and control assets
- a Joint Force Headquarters staff.

In terms of training strategy, it is a live exercise which is conducted in maritime, land and air training areas located in and around the north of Scotland. It lasts two weeks and is split into two phases, each of which last a week. This structure is predicated by maritime training requirements. For the maritime participants, the first week is spent undertaking Combat Enhancement Training (which includes such tasks as gunnery and resupply at sea) conducted by individual ships, and Fleet Integration Training, in which the ships in a task group practise operating together as a task group (which includes such tasks as anti-air warfare, anti-submarine warfare, and fleet protection against a Fast-Inshore Attack Craft threat). The second phase is a free-flow exercise which is in essence a two-sided exercise driven by a scenario based on increasing tensions between two countries, ultimately leading to kinetic warfare. Land participation tends to be threaded into the second phase, with air participation occurring across both phases. Another significant aspect of the training strategy is that the exercise is opened up to participants from other nations. This provides significant benefit as a much richer and larger set of assets can be brought together to train. For example, the UK could not generate two maritime task groups for training at the same time. In addition, since the demise of the Nimrod maritime patrol aircraft, the UK does not have a long-range, airborne maritime patrol capability so training for the integration of such a capability with naval task groups cannot be trained in the live domain unless other nations contribute that capability.

Such a training strategy facilitates the delivery of very rich training experiences for all the participants, including the opportunity to train alongside and integrate with forces from other nations. However, it is not without its complications. These include:

- The training requirements are the sum of the requirements of each of the participating elements.

- The elements participating in the exercise can change significantly from exercise to exercise and there can also be significant late-notice changes in declared participation months or even weeks before the start of the exercise.
- The start state of each participating element can vary considerably depending on their previous employment and staff turnover.
- The training equipment available is that owned by the participating elements.
- As the training is live, the training environments available are limited to the maritime, land and air training areas available in the UK.
- In live training weather may have a significant impact on which serials or evolutions can be conducted – it may cause naval vessels to require relocation during exercise play and air serials to be cancelled.
- The number of instructional staff required to carry out the tasks of briefing, monitoring, evaluating and debriefing is determined by the number and nature of the participants in the exercise.

The most significant consequence of these factors, and the main reason for using this particular example, is that a 'one size fits all' exercise design simply could not be developed; consequently, each exercise is in essence a bespoke event. This means that the JTEPS team has to engage in both analysis and design activities before training delivery can be undertaken. In the case of Exercise Joint Warrior, this planning process begins five months before the exercise is scheduled to run, when an initial planning conference is held. This is attended by representatives of all of the participating elements, with the purpose of discussing their training requirements and identifying the likely start state of each participating element. Based on an analysis of these requirements, the JTEPS team can then develop an outline plan and scenario for the exercise. This is presented back to the representatives of the participating elements at a subsequent Main Planning Conference. Inevitably, with a wide variety of participants and their associated training requirements, it is unlikely that all of the requirements can be accommodated in an exercise of limited duration and as such compromises have to be made. These are discussed at the Main Planning Conference. The JTEPS team then go on to develop the exercise in detail. The detailed design is presented to the representatives of the participating elements at a Final Planning Conference approximately a month before the exercise runs. Notwithstanding the bespoke nature of each exercise, there is scope for the reuse of overarching scenarios and detailed scenario components. Furthermore, knowledge of which elements worked well and which were less successful and why can inform the development of future exercises.

At the time when the authors were observing Exercise Joint Warrior, the JTEPS staff numbered 25, the majority being warfare specialists from the maritime and air domains. In addition, there were logistics and communications specialists and a small administrative support team. Land warfare specialists were also assigned to assist with the planning of the land components of the exercise. Whilst this number of staff was sufficient for exercise planning purposes, considerably greater numbers were required to manage the exercise, with the staff being augmented to a total of 125 for the duration of the exercise. It should be noted that this team of 125 was not responsible for the detailed briefing, monitoring, coaching, assessing and debriefing of all of the participating units. One of the conditions of participation in the exercise is that the owning organisations of the participating units are responsible for the provision of training staff to undertake these functions.

The need to undertake analysis and design activities to develop bespoke exercises is common, if not universal, for training conducted at the collective (team of teams) level. At the small team level there is a greater likelihood that more generic training may be suitable but this cannot be assumed to be true for all cases.

The key points for the training overlay from this example are:

- Team and collective training often requires bespoke training events to be developed.
- The training team responsible for training delivery are likely to have to undertake training analysis and design activities prior to training delivery and evaluation taking place.
- Exercise delivery may well require many more training staff than are required for exercise design.

Overview of the Training Overlay Model and the Team Training Model

The Training Overlay Model provides a framework for the analysis of the training overlay. When the Training Overlay Model is linked to the Team Performance Model the result is the Team Training Model which provides the complete framework upon which the TCTNA methodology is based.

The High-level Training Overlay Model

Based on the key points identified in the two examples discussed above, the high-level Training Overlay Model can be constructed as shown in Figure 6.1. The training strategy has arrows linking into training delivery and evaluation and into training analysis and design, as it is the strategy that drives the tasks, resources and supporting systems for each of these components. The process sequence follows the SAT cycle in that analysis and design feed into training delivery and evaluation. Training delivery and evaluation is split into two parts. Instruction and exercise management is concerned with the instructional tasks and training environment management tasks, each with their associated resources and supporting systems to reflect these two distinct but related roles. The two-way links between these components capture the nature of the relationship between them in that monitoring of the environment informs the evaluation of trainee performance and the environment can be controlled in response to this evaluation.

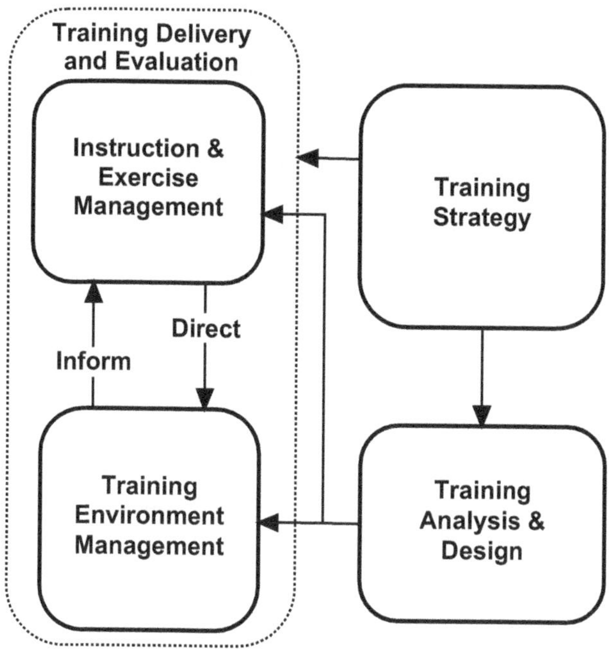

Figure 6.1 High-level Training Overlay Model

The High-level Team Training Model

The Team Training Model is formed by the linkage of the Team Performance Model and the Training Overlay Model, as shown in Figure 6.2. The links from the instruction and exercise management component of the Training Overlay Model and the Team Performance Model reflect the interactions between the instructor and the trainees. The links between the Instruction and Exercise Management component of the Training Overlay Model and the Training Environment Model reflect the interactions between those responsible for managing the training environment and the training environment itself.

Training Strategy

The Oxford Dictionary (2014) defines a strategy as 'a plan of action designed to achieve a long-term or overall aim'. A training strategy can therefore be considered to be the plan of action for delivering training to meet a given training requirement. Before the areas that a training strategy should address are discussed, it is worth considering the factors that influence its development.

Factors that can influence the formulation of the training strategy are shown in Figure 6.3.

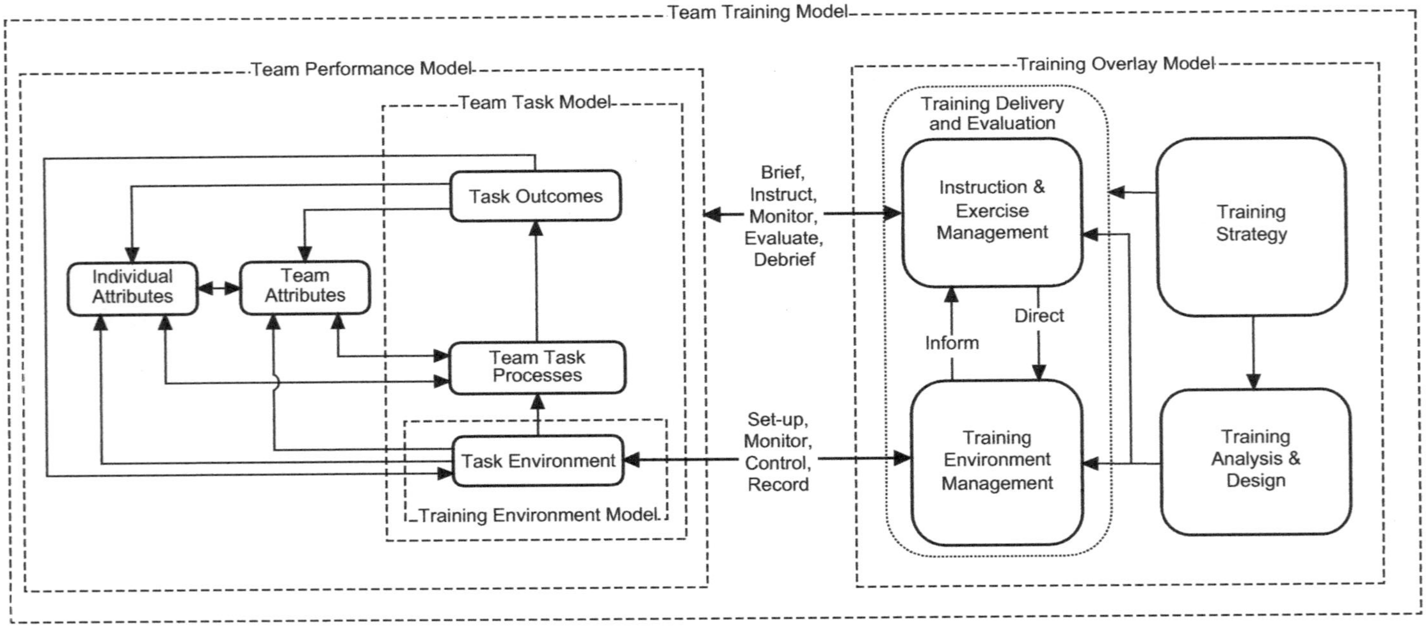

Figure 6.2 The high-level Team Training Model

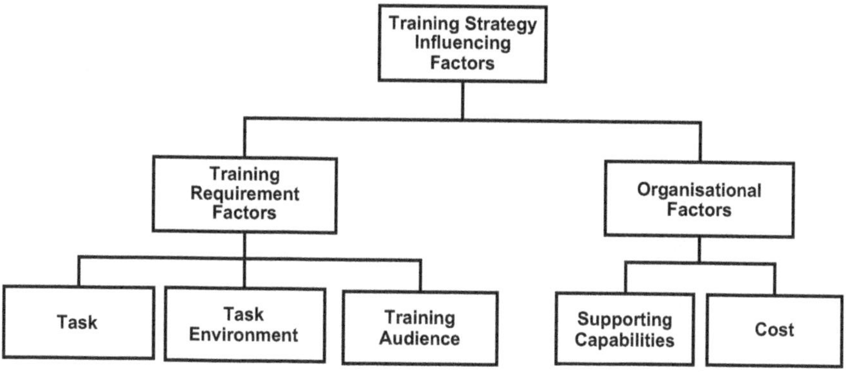

Figure 6.3 Factors that influence the training strategy

These primarily serve as constraints, as for a training strategy to be considered viable it must: satisfy the training requirements; be feasible to implement from an organisational perspective in terms of cost and the capability of the organisation to support the solution (for example in terms of manpower and infrastructure); and make sense in terms of the broader training context of related training activities (such as individual training and higher order collective training). They may also afford opportunities. In the remainder of this section each of the factors that influence training strategy are considered in more detail.

Training Requirement Factors

The task

The precise nature of the team task will have a significant impact on the choice of training methods. For example, in Chapter 1 it was noted that Lt Col Collins was particularly concerned that his troops should be proficient at anti-ambush drills. Survival when ambushed requires instantaneous team responses, there is no time to engage in lengthy analysis of potential courses of action. Consequently, training will need to focus on repeated rehearsal of well established, proven procedures to ensure that the troops can take the correct actions instantaneously. Tasks which require the team to adapt to novel situations will require a different training approach. Task duration also affects training design. For example, command headquarters typically operate a planning and execution cycle that lasts over a number of days. An exercise to provide the opportunity to work through this cycle more than once would need to last the better part of a week. By comparison, ambush drills can be completed in a matter of minutes, so many repeated rehearsals could be achieved in a matter of hours.

The task environment

Training environment requirements can have a significant impact on training strategy. For example, training in arctic and mountain warfare requires access to arctic conditions and mountainous terrain.

Training audience characteristics

There are a number of characteristics of the training audience that are of significance from a training strategy perspective. These include:

- the size of the team/collective organisation
- the number of teams that will require training over what timescale
- the availability of the team for training
- the stability of the team (rate of turnover of staff)
- the geographical dispersion of the teams to be trained
- the starting state of the team in terms of knowledge and previous experience.

A critical consideration in developing a training strategy is the volume of training that needs to be delivered and the profile of that volume requirement in terms of peaks and troughs. As an example, the authors were asked to apply the TCTNA approach to the training of two man crews for a new class of emergency escape craft for oil platforms in a hostile environment. The crews were to be drawn from the manning of the oil platforms. Four crews were required for each platform at any one time in order to provide coverage of two crews for the two craft on the platform over the day and night shifts. They worked a two week on, two week off profile. This meant that eight crews per platform had to be trained during the initial surge when the escape craft were to be introduced into service. Crew training could not be allowed to interfere with oil production, so training could only be conducted in days before the crews went back onto the platform (delaying them getting home after two weeks on the platform was not considered viable). This meant that initial training could only be delivered at a training centre once a month. It was estimated that staff turnover would be in the order of 25 per cent per annum, based on previous figures. Therefore, in the steady state condition, two new crews per platform would have to be trained each year. Based on the number of platforms, the multiple simulators were considered necessary in order to ensure that the course duration was kept acceptably short, despite the fact that the simulators would be unused for more than 50 per cent of the time.

Organisational Factors

Supporting capabilities

It is common within large enterprises for different aspects organisational capability to be managed as parallel but interdependent strands, of which training is but one. Table 6.1 shows the dimensions of capability that are employed for capability management in the UK MOD, the US DOD and the Australian

Defence Organisation. It can be seen that they are broadly equivalent, although slightly different labels are used (for example, Command and Management in the Australian Defence Organisation encompasses doctrine).

The development, delivery and maintenance of a training system requires interactions with these other areas of capability, often posing constraints on the nature of the training system that can be developed. These constraints may take the form of such factors as resource limitations and policies. For the remainder of this book we use the UK MOD Defence Lines of Development dimensions, but any other model could be substituted.

Table 6.1 UK, US and Australian capability dimensions

UK MOD (Defence Lines of Development)	US DOD	Australian Defence Organisation (Fundamental Inputs to Capability)
Training	Training leader development	Collective training
Equipment	Matériel	Major systems
Personnel	Personnel	Personnel
Information	–	–
Doctrine	Doctrine	Command and management
Organisation	Organizations	Organisation
Infrastructure	Facilities	Facilities
Logistics	–	Support, supplies
Interoperability	–	–
–	Policy	–

The significance of each of these areas of capability includes:

- *Training* – Any training solution that is developed will have to operate in the context of the extant training capability of the organisation, and be compliant with its policies, processes and systems. It should also exploit the opportunities afforded by existing training capability and take account of related training development activity. One of the more challenging issues to deal with in developing a training strategy is the training context, by which we mean how the component of training fits into the continuum of individual, through team to collective training. Arguably, the worst outcome will occur if the training strategy for a team is developed without reference to the training solution for the organisation that it fits into at the next level of aggregation or the provision of individual training below it. Such a 'stove pipe' approach to training system provision can lead to the duplication of training systems, at significant cost to the organisation. For example, if a new weapons platform is being procured training systems will be required for individual training for each of the operator roles involved in operating the platform. However, as they operate in a team, their individual training will need to take account of the team context in which individual tasks are conducted. This means that, in all probability, the individual training systems will need to have the capability to support individuals operating in all roles so that the interactions between team members can be practised. A simulator supporting all of the roles in the team is a likely candidate training device. Formed teams in operational units operating that type of platform will also need access to a suitable training system. It is likely that a simulator which looks incredibly like that required in the individual training system would be a good solution. Equally, when multiple teams operating such a platform need to train together, multiple simulators of a similar type may well be required. Uncoordinated acquisition of exactly the same type of simulator to satisfy each of the levels of training is unlikely to be an efficient approach financially, and is likely to lead to the duplication of capability and redundant capacity.

From a different perspective, if teamwork concepts have not been taught in an organisation before, but it is deemed that such an approach should be introduced, there is likely to be a surge in training requirement, where staff at all levels in the organisation require training. When that surge has passed, and a steady state is achieved, consideration will have to be given to how best to sustain the impetus of the application of those principles. Repetition of the initial training would become dull, so a different approach would have to be developed. Furthermore, as the CRM literature suggests, training in teamwork skills alone is unlikely to be successful, there has to be a programme put in place to embed the application of teamwork skills into everyday practice (Helmreich and Wilhelm 1989).

- *Equipment* – If the training requirement is associated with the introduction into service of new equipment, the TCTNA will have to be closely coupled to the acquisition process both to glean information about the equipment and to input information such as cost estimates into the acquisition process. Any requirement for live equipment to be used in the training system will impact on the amount that has to be procured. Training equipment such as simulation systems may also need to be procured. Equally, budgetary constraints may limit the numbers of systems that the organisation can afford which may limit the availability of systems to be used in training.
- *Personnel* – The personnel area will need to be consulted with reference to the training audience characteristics (numbers and so on, as above) and the training staff requirements. Constraints on the potential availability

of suitably experienced personnel to act as training staff need to be fully understood.

- *Information* – The training may include the requirement to use and exploit extant information systems or new systems that are being developed. The training system itself may have to integrate with extant training related information systems such as training management systems and exercise management systems.
- *Doctrine* – The concepts and doctrine area will need to be consulted to inform the TCTNA about doctrine applicable to the task to be trained and also in relation to training that doctrine.
- *Organisation* – The development of a new training system may well have an organisational impact, such as structural reorganisation if additional capability is added with extra staff.
- *Infrastructure* – It is likely that the development of a new training system will have associated infrastructure requirements, such as buildings to house simulators and debrief facilities. This could involve a new build or the modification of an existing infrastructure.
- *Logistics* – A new training system will almost certainly have logistics requirements. For example, training armoured battle groups in Canada required the supply of spare parts to maintain a battle group of armoured vehicles, and the transport of thousands to troops from their home bases to Canada and back. Whilst in Canada the soldiers had to be fed, watered and accommodated. There was also a requirement for live and blank ammunition for the armoured vehicles and personal weapons.
- *Interoperability* – The requirement for interoperability between services and with other partner nations may have an impact on the training solution that is chosen. The requirement for the interoperability of synthetic training systems with extant systems will also typically be a key requirement.

Cost

Team and collective training systems represent a significant challenge from a cost perspective on account of their sheer scale and complexity. Significant amounts of resource have to be brought together for events that are relatively short in duration and may be infrequent. Costs associated with personnel, equipment, infrastructure and logistics may be substantial. For a training solution to be viable it has to be achievable within the available budget both in terms of the cost of acquisition of any required training systems and equipment, and the ongoing costs of delivering the training.

Training Strategy Components

Areas that a training strategy should address are shown in Figure 6.4 and amplified below.

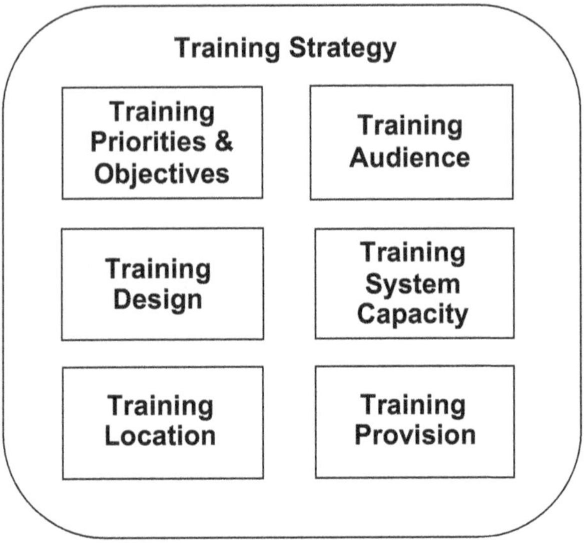

Figure 6.4 Training strategy

Training priorities and objectives

Analysis of the task should yield a complete description of the task that has to be trained and the full range of environmental conditions within which it could potentially have to be executed. The organisation may choose to prioritise both the aspects of the task that are to be trained and the range of conditions under which it is to be executed. For example, training under arctic conditions may be assigned a low priority, and so may be considered out of scope for the majority of units to be trained, with perhaps a specialist unit being trained in those conditions to retain organisational competence in that area.

Training audience

This can be a simple statement of who the training audience is. However, where multiple teams are being trained together the issue of training audience primacy can arise, whereby the training requirements of one particular component of the overall training audience are given greater priority. For example, NATO runs an exercise in which the Joint Force Headquarters and its subordinate Land, Air and Maritime Headquarters components, nominated to lead the NATO Response Force for a period of time, train together. However, the primary purpose of the exercise is to validate formally the Joint Headquarters as being ready to lead the NATO Response Force. As such, the Joint Headquarters has primacy, as its training requirements have priority over those of the other participating components.

Training design
As described in Chapter 1, TNA embraces both analysis and high-level training design. The first stage of design activity occurs within the development of the training strategy. Training design sits at the core of training strategy development and embraces the structure and sequence of training, the selection of training and assessment methods, the determination of training duration and the selection of suitable training environments. These are discussed in full in the next section.

Training system capacity
The determination of the necessary capacity for the training system is of critical importance as it will have a fundamental impact on the level of resources that are required which in turn impacts on cost. Not only does it have to take account of the duration of training and the annual throughput required, it also has to factor in initial surge training requirements, and training audience availability (which may generate peak loadings).

Training location
The choice of training location can be influenced by a number of factors including:

- the geographical location of the units to be trained
- the location of suitable training environments (such as training areas for live training and extant synthetic training environments)
- infrastructure requirements to accommodate new training facilities
- the location of related training.

For example, in the case of Maritime Force Protection training discussed in Appendix A, this training has to be integrated into Basic Operational Sea Training, which is conducted at Her Majesty's Naval Base (HMNB) Devonport, near Plymouth in the southwest of England. If a synthetic training solution were to be adopted, there would have to be a very compelling case to locate it anywhere other than at HMNB Devonport, given that the nearest alternative location is at Portsmouth dockyard, some 170 miles away. If the training system were to be located in Portsmouth, the training audience would probably spend longer travelling on the 340-mile round trip than they would in undertaking the training.

Training provision
A key decision relating to training delivery is who is to conduct the training, and whether or not the training system or some part of it is to be outsourced or provided entirely from internal organisational resources. In the military context, a common model is for simulation facilities to be owned, operated and part-staffed by external providers. This model can reduce the up-front cost of acquisition of a training system and can confer advantages of continuity of staffing in parts of the instructional team.

Training Design in Training Strategy Development

Within TCTNA we suggest that it addresses:

- training structure and sequencing
- training and assessment methods
- training duration
- training environment selection.

Training structure and sequencing
The training strategy needs to define how training is to be structured. In the simplest case, the whole team or collective organisation would conduct all of its training together in one environment in the same training event. However, depending on the scale and nature of the task, the scale and nature of the team and all of the constraints that may apply may be neither achievable nor desirable. Therefore, consideration needs to be given to whether or not it is necessary or appropriate to decompose the overall training requirement into sub-components, with training the whole organisation together being one of those sub-components.

One approach to determining the structure is to consider how the task breaks into sub-tasks which can be trained separately before whole task training is conducted. Candidate sub-tasks for separate training would include:

- *Sub-tasks which do not require the whole team.* If there are sub-tasks that only require a sub-team it may make sense to train those tasks separately as a precursor to whole team practice.
- *Sub-tasks which have different environmental requirements.* If there are sub-task components which require only a sub-set of the environmental elements it may make sense to train these tasks separately from the perspective of efficient training resource utilisation. Such sub-tasks can be identified from the input and output fields of the Task Description Tables.
- *Sub-tasks which are prone to skill fade and require additional practice.* Sub-tasks which are considered to be particularly prone to skill fade may well benefit from additional practice if sufficient practice cannot be achieved during practice of the whole task.
- *Sub-tasks that are separated in time.* Where sub-tasks happen in sequence there may be a case for training them separately. An example would be the separation of the planning and execution phases of an operation for a Headquarters staff. Planning could be conducted some time before the execution phase is scheduled. This has the advantage of providing the opportunity for exercise planning staff to tailor the events in the scenario to match the team plan and test it more effectively.
- *Sub-tasks that are conducted in parallel which would benefit from separate practice.* If the team has to undertake tasks in parallel that are not tightly interlinked, there may be a case for practising these tasks

> separately before practising their parallel execution, particularly if each sub-task is demanding in its own right. An example of a pair of parallel tasks of this type occurs in the training of air-to-air refuelling tanker crews. When an air-to-air refuelling tanker is supporting a group of fast-jet aircraft on a long transit flight, such as from the UK to the US, the fast-jets are refuelled at repeated intervals during the journey. The detailed scheduling of these refuelling intervals is carefully planned in advance, ensuring that all of the aircraft will have sufficient fuel to reach a diversion destination at all times during the journey. However, if there is an equipment malfunction on the fast-jets (such as an underwing fuel tank valve sticking closed) or the tanker (such as a problem with one of the refuelling systems), the refuelling plan will have to be modified in-flight. The fast jet with an underwing tank problem would need refuelling more often since it would in effect have a smaller overall fuel capacity. Modifying the refuelling plan can be a complex task. The worst time for this to occur would be when the crew are busy refuelling fast-jets. As the re-planning task is essentially a paperwork exercise, which is unrelated to flying the aircraft and conducting refuelling, it can be practised effectively in the classroom. Once the crew become proficient in this process, they can then move on to practising it in the more challenging situation of having to simultaneously conduct the refuelling sub-task.

The end result is a form of part-task/whole-task training but with the potential for different parts of the organisation to train independently but simultaneously. For example, Command Element training and Force Element training can be conducted separately, in different environments, before the Command Element and Force Elements are brought together in Whole Force Training as illustrated in Figure 6.5. It is common for command elements to train in a constructive simulation environment without the need for their subordinate force elements to be deployed on the ground.

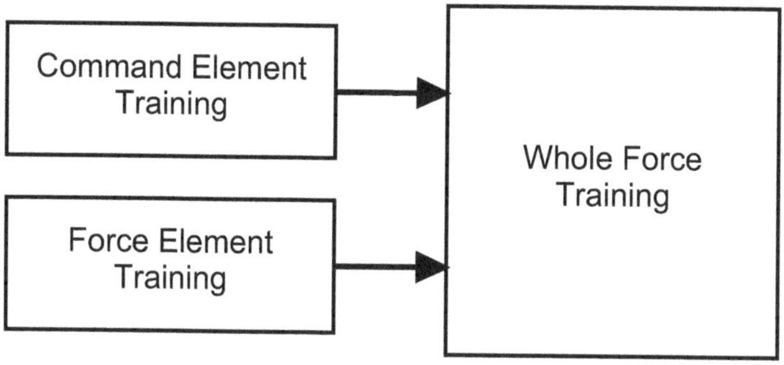

Figure 6.5 Part-task/whole-task training in the collective context

For large-scale, complex organisations which may conduct a range of different tasks, this principle can be extended into a consideration of what we shall term 'task domains'. A task domain is a sub-division of the overall task set which is conducted by a cohesive sub-group (that is, a sub-group that has tight connections/ interactions within its sub-components) of the overall organisation. This is illustrated in detail in the Carrier Enabled Power Projection case study at the end of the book (Appendix B).

Training and assessment methods

During the last 20 years there have been significant developments in team training and assessment methods. These developments have arisen primarily through research and development activity in the military and aviation domains. A significant contribution to the military research has been made by the TADMUS project, instigated in response to the Vincennes incident as described in Chapter 1. This has been focused mainly at the small team level. There has also been significant work done over the last 10 years in the US and the UK at the larger, collective level, in the context of the use of distributed simulation. In the aviation domain the central development within team training has been the emergence of the Crew Resource Management construct. Whilst there is much that is new, there are also well established methods that should not be disregarded. These include procedural training and the simplifying conditions methodology. In this section we review the developments in team training and assessment methods and discuss the implications for the TNA process.

1. Procedural Training

Procedural training focuses on the repetition of task procedures, with the associated assessment criterion of correct adherence to the procedure. The use of this approach is essential for such tasks as reacting to an ambush as described above in the task section. It is commonplace for procedures to be developed for handling a wide variety of routine tasks, as they provide a way of capturing organisational wisdom on the best way to carry out a task. In a military context they are often referred to as SOPs and TTPs.

Klein and Pierce (2001) observe that the pursuit of a procedural approach to training to achieve mastery can be appealing to commanders for a number of reasons:

- it is an easy approach that takes less work than more complex alternatives;
- it is easy to gauge a level of progress;
- it provides a feeling of mastery;
- it satisfies a leader to know that the team can perform its routines flawlessly.

However, they caution that pursuing such an approach to mastery can be counterproductive as it can lead to a false sense of expertise (which they refer to as 'experiosclerosis' and defined as the opposite to flexibility). They suggest that, once a team has achieved an acceptable grasp of the procedures, they should be exposed to situations where they have to adapt their approach if they are to be enduringly or consistently successful.

2. The Simplifying Conditions Methodology (SCM)

SCM, developed by Reigeluth (1999), provides guidance for the analysis, selection and sequencing of training content. The underlying principle is that the complexity and diversity of task conditions should be gradually increased, in other words progressing from simple to complex. Reigeluth (1999) acknowledges that this approach has long been used by practitioners intuitively. Whilst this approach is aimed at providing guidance for detailed design of training events, the point of significance for TNA is that consideration has to be given to the number of iterations that are required to move from simple to complex as it impacts on exercise duration. Klein and Pierce (2001) caution that over-emphasis on such a graduated progression may be counter-productive for the development of adaptive skills, as teams may be reluctant to abandon strategies learnt in such a progression. The implication is that earlier exposure to challenging conditions and the opportunity to learn from failure may be more valuable in this regard.

3. Event Based Approach to Training

Event Based Approach to Training (EBAT) is a practice-based method in which exercise scenario events are developed to target each of the training objectives for the team and specific performance measures are developed for each of the events. In EBAT the process for developing an exercise has five steps:

- identify, through task analysis, individual and team training objectives;
- translate training objectives into representative scenario events;
- identify and establish performance criteria that represent the achievement of targeted objectives for each event;
- translate performance criteria into event-based measures of individual and team performance processes and outcomes to evaluate training effectiveness;
- develop a framework for performance feedback that enables an instructor to provide details to teams regarding processes and outcomes, and performance goals for remediation.

Johnston et al. (1995) provide a detailed exposition of this process as first applied to the training of naval Anti-Air Warfare (AAW) teams from Aegis Cruisers (the type of team that was involved in the USS Vincennes incident). A notable feature of the approach is the identification of the stressors that impact on the performance of each training objective which are then used to inform the design of scenario events of differing levels of stress/difficulty. The process used to achieve this is called the Stress Assessment Methodology and is described in detail by Hall et al. (1993). For the example of a team performance objective of correctly identifying a commercial aircraft within 60 seconds of its track being displayed on the radar screen of an Aegis Cruiser, high workload and ambiguity of information were identified as the key stressors. SMEs identified conditions that would generate a high level of environmental stress (20 aircraft in the area with incomplete information showing on the track profiles) and then suggested an event which would require the team to carry out the performance required by the objectives. In this case the event was an unknown, slow-flying aircraft appearing as a track at a range of 40 nm.

An exercise is built up as a sequence of such events. Once the events have been developed and scripted, a detailed process is followed to design data capture formats for the assessment of team performance, which include both task process (including team interactions) and outcome measures. When the exercise is run, the delivery of the events in the scenario on the planned timeline ensures that: the training audience have the opportunity to exercise the required skills, be they procedural or adaptive; and the trainers know when specific behaviours related to the training objectives should be observed. Adherence to the timeline for the introduction of events ensures that the planned conditions for performance are delivered and standardisation of the scenario is achieved, which is particularly important if the exercise is to be run for different teams.

Whilst the concept of exercises run to a scenario with events occurring to test team performance is not new, the linkage of events to training objectives, the considered specification of conditions for each event, and the identification of measures of both team process and products has the potential to offer a robustness to exercise design which may not otherwise be achieved. The sparse or low uptake of employment of such a detailed exercise development process could be due to the level of effort required in the planning stage. However, the counter-argument would be that if exercise events are not carefully-planned, taking into account the impact of the prevailing environmental conditions, and if there is not a clearly articulated plan for measuring team performance when the event occurs, what training benefit will the team get from participating in the exercise?

4. Targeted Acceptable Responses to Generated Events or Tasks (TARGETS)

TARGETS is an approach to the evaluation of team performance that was developed in parallel with the EBAT construct. Dwyer et al. (1997) provide a detailed account of the method's development and testing. The TARGETS approach involves the development of detailed assessment forms for each of the events that occur in an exercise scenario. SMEs develop lists of the actions that the team should take in response to an event, including the interactions that they believe should occur between team members. These are then turned into checklists for use by the raters observing the team during the exercise. Each item on the checklist is a binary assessment, that is, it records if the action or interaction was present or absent. The checklist forms used also have space for comments. The team's score is determined by counting the number of correct items recorded.

The method was first trialled with helicopter crews of two pilots undertaking aircrew coordination training (a variant of CRM training) in a high fidelity simulator, where it was relatively easy to control the scenario and observe the interactions between the team members. The evaluations were conducted by two independent observers by analysing videotapes of the crews undertaking the exercise. In this initial trial it was found that there was high inter-rater reliability (they gave similar scores), and that the measures were sensitive in that a spectrum of scores was produced that enabled the performances of the participating crews to be differentiated effectively. It was then trialled in a distributed simulation environment for a Close Air Support task with a team of 19 players responsible between them for planning and executing Close Air Support in a number of scenarios each of 2–3 hours in duration. Dwyer et al. (1997) identify a number of lessons that were learnt from this trial:

- In order to evaluate the team interactions, assessors needed access to all of the communications nets that the team were using. However, they had insufficient access to these nets which compromised the assessment. This underlines the importance of ensuring that a training environment has the capability to support the observation, recording and playback of team interactions to inform assessment and facilitate the provision of feedback to the team.
- Assessors in different locations had different vantage points on the team's performance. Discussion between assessors during the preparation for providing feedback to the team enabled a full picture of how the team performed to be developed from the combination of ratings from each assessor.
- Converting data from the checklists and from multiple assessors to provide feedback proved to be very time consuming.
- The assessment checklists could not be used to capture performance associated with unplanned events that the team introduced into the exercise, which could have been very creative. In other words, the team could have generated a solution which was both novel and sound, yet would not be in alignment with the 'correct answer' captured in the assessment checklists.
- The checklists appeared to be manageable by the assessors but for larger-scale exercises the number of items could become unmanageable.
- Dichotomous scores could capture that a required behaviour had occurred, but not how well it had been executed.

The development of assessment measures that capture team processes as well as outputs, and are clearly linked to the training objectives of the exercise, is clearly desirable, if meaningful feedback on performance is to be given to the team. However, the merit of using a detailed checklist of dichotomous items is a matter for conjecture. The development of such a checklist is predicated on the presumption that there is a 'right answer' and that all the team interactions can be predicted, which would seem at odds with the nature of adaption.

5. Cross-Training

Cross-training involves providing team members with training about other team members' roles. This approach was motivated by the observation that experienced team members appeared to read each other's minds as they were aware of what they were doing, could anticipate when they needed information or assistance, and as such they were exhibiting implicit coordination that is, coordinating without communication (Blickensderfer et al. 1998). The underpinning idea is that training regarding other team members' roles develops the individual's mental model of how the team performs and that this knowledge should in turn facilitate implicit coordination, thus improving team performance. Blickensderfer et al. (1998) proposed three types of cross-training:

- *Positional Clarification* – This is a form of awareness training aimed at providing knowledge of the team structure and the tasks and responsibilities of each of the team members within it. Training methods include discussion, lecture and demonstration.
- *Positional Modelling* – Positional modelling involves direct observation of other team members carrying out their tasks in a simulated situation, and discussions about their tasks. It gives greater insight into team dynamics and how one team member's actions relate to and affect other team members in the execution of their tasks.
- *Positional Rotation* – This approach involves giving team members hands-on practice in other team members' roles so that they gain a better understanding of each task and of the interaction requirements between team members.

They further suggest that the choice of cross-training method should be based on an analysis of the level of interdependence within a team. Positional clarification is recommended for low-interdependence teams (such as quality control circles and advisory groups), which are characterised as operating with minimal amounts of communication and coordination, and with little or no mutual monitoring or feedback. Positional modelling is recommended for medium-interdependence teams (such as flight attendant teams and research teams), which are characterised as having constant internal exchanges of information and resources amongst team members whose roles are functionally related, where some communication and coordination is required and monitoring and feedback can be important. Positional rotation is recommended for high-interdependence teams (such as cockpit crews and surgical teams) where there is a critical need for cooperation and coordination, functional positions are unique and there is commonly a requirement for direct verbal communication and interaction. Blickensderfer et al. (1998) tested the efficacy of positional rotation in small team contexts where it was feasible for team members to interchange roles (such as a team of three aircraft controllers working in a military combat information centre) and demonstrated positive results. However, they also acknowledge that where differing roles have high complexity, full cross-training may be neither desirable nor achievable, citing surgical teams as an example, where positional rotation would be inconceivable as a training approach.

A simple but effective instance of cross-training was observed by the authors during a recent European Union battlegroup exercise. Two companies of Royal Marines from 42 Commando were to combine with a Lithuanian company and a Latvian company, along with Swedish and Dutch supporting elements to form the battlegroup. Before field training commenced, the commander got each of the units to put all of its equipment on display. Each unit then took turns to explain what its equipment was capable of and how they operated with it to each of the other units. Although we were located in a deployed headquarters which was operating at a level above the battlegroup, there was much discussion amongst the staff about what had been informally reported back 'through the grapevine' about unusual equipment that had been seen and the capabilities of the multinational partners. Cross-training effects had spread beyond the boundaries of the participating units, even to us as visiting researchers!

6. Team Adaption and Coordination Training

Team Adaptation and Coordination Training (TACT) is an approach that was developed to enhance team coordination skills and develop teams' abilities to exploit implicit coordination when operating under the effect of team stressors as discussed in Chapter 3 (Serfaty et al. 1998). It is predicated on the adaptation model put forward by Entin et al. (1994) which suggests that successful teams adapt to high levels of stress by shifting from explicit coordination to implicit coordination and engage more in mutual performance monitoring and support, and on the observation by Orasanu (1990) that they use periods of low stress to consider alternative strategies and courses of action for periods of higher stress. Training in adaptive coordination strategies is provided. Serfaty et al. (1998) selected five strategies: preplanning, the use of idle periods, information transmission, information anticipation and workload redistribution (these are consistent with the processes that support coordination identified in Chapter 3). This is followed by practice opportunities in conditions of low stress followed by conditions with higher levels of stress. Serfaty et al. (1998) trialled this methodology with five-man AAW teams in a simulator. The teams who received the training described above outperformed the teams that had not received training. They also trialled a modified strategy, referred to as TACT+ in which additional training was provided to team leaders on how to give brief, periodic situation-assessment updates to the rest of the team in order to enable them to update their mental models of the task situation (team SA, as discussed in Chapter 3). Team members were taught how to interpret this information and use it effectively. Teams which undertook TACT+ training outperformed teams that had undertaken TACT training.

7. Generic Team Training

Generic team training, as the name suggests, is based on the provision of training about teamwork that is not task specific. Ellis et al. (2005) and Rapp and Mathieu (2007) carried out controlled trials in which experimental group participants received generic training about teamwork skills before being allocated to teams to participate in practical exercises. Notably, this training was individualised. Ellis et al. (2005) provided 30-minute lectures based on case studies about planning/task coordination, collaborative problem solving and communication. Participants received these lectures in a one to one setting. Rapp and Mathieu (2007) provided their participants with a commercially available, self-administered, CD-based programme which covered the following areas:

- *transition* (five modules: developing team charters, selecting team members, determining member roles and responsibilities, planning for action, and goal setting)
- *action* (four modules: managing team performance, giving and receiving feedback, making decisions, and communication basics) and
- *interpersonal* (four modules: facilitating team interactions, building a collaborative environment, managing team conflict, and team problem solving).

Participants reported spending 30–45 minutes on each of the three modules.

In the Ellis et al. study, the 64 four-person teams of undergraduate students participating spent an hour undergoing training on the operation of a computer game. The game took the form of a command and control simulation in which the team had to monitor activity in a geographical region and defend it against unfriendly ground or air tracks. Once they had completed the training, they then spent an hour in the experimental session. The experimental group showed significantly greater performance scores than the control group. In the Rapp and Mathieu study, 54 MBA students were allocated to 3–4 person teams to participate in a business strategy game where they had to manage an athletic footwear manufacturing business. After two practice rounds, the game was run in eight rounds over a period of eight weeks. The experimental group that had received teamwork training achieved significantly higher results than the control group. Furthermore, after a very high score in the first week and a dip in the second week their results improved markedly week on week. They were observed to be exhibiting better teamwork skills than the control group. By comparison, the control group scores improved from week one to week two but then steadily declined.

These studies are significant as they suggest there is potential for the use of generic teamwork training which could be delivered using an individualised training strategy, as a precursor to team training exercises.

8. After Action Review (AAR)

AAR is a construct for providing performance feedback to the participants in training exercises. It was first developed by the US Army over 30 years ago and has become widely used across all armed services in many nations. The US Army characterise an AAR as:

> a professional discussion of an event, focussed on performance standards, that enables soldiers to discover for themselves what happened, why it happened, and how to sustain strengths and improve on weaknesses. (US Army Combined Arms Centre 1993: 1, cited in Morrison and Meliza 1999)

The description of AARs presented here is based upon the work of Morrison and Meliza (1999) who provide a detailed account of the foundations of the AAR process, and more recent research into AAR best practice by McKeown and Huddlestone (2010) who investigated AAR best practice in the US, Canadian and British Armies.

An AAR takes the form of a meeting held shortly after an exercise or exercise phase has been completed, so that events are still fresh in participants' minds, and can be characterised by the following:

- *Attendance* – Attendance depends on the size of the team or collective organisation that has been participating in the exercise. For a small team, typically the whole team would attend. For a larger organisation, such as an infantry company, the command chain would attend and then debrief the rest of the company afterwards. Although, at the British Army Combined Arms Tactical Trainer (CATT) in the UK it is possible for a complete squadron to attend the AAR. For a very large scale organisation, such as a battlegroup, there may be multiple levels of AAR. For example, at the British Army Training Unit at Suffield in Canada AARs are held at company/squadron level first, and then a battlegroup level AAR is held, attended by the battlegroup headquarters staff, the commanders of the companies/squadrons, and any attached units, such as engineers and signals. Representatives of the opposing force and the training staff who observed the exercise also attend.
- *Format and Content* – A typical AAR is a facilitated discussion based around the following questions:
  - What was supposed to happen? This section provides a review of the mission and objectives for both the training force and the opposing force.

  - What actually happened? The facts of pertinent events that occurred during the mission are presented and the training audience are encouraged to discuss what happened.
  - Why did it happen? Having reviewed the facts presented, the participants are then encouraged to discuss the causes of their actions and the outcomes.
  - What can you learn from this experience? The participants are required to determine what could have been done differently when things went wrong and what they did well, so that areas where success was apparent can be sustained.

  The aim is to create an open environment that encourages participants to discuss their perspectives openly and honestly such that the team can develop a better understanding of its strengths and weaknesses. A key component in achieving this is the presentation of factual evidence of what occurred in the exercise. The facility to replay events from the exercise by such means of tracking data from vehicles in live exercises or replays from the simulator recordings reduces any ambiguity as to who was where, doing what at any given time. An example of the power that such capabilities can afford in the hands of a skilled facilitator occurred at the CATT in the UK. A corporal gunner in a tank had given an incorrect target location which resulted in the fire from his squadron falling 1 km short. On first inspection, it looked as if he had made a basic error and stood to look quite stupid. However, the instructor was able to bring up the image from his guns sight which had the range indication on it and demonstrate that the error was due to the laser range finder clipping a ridge which hid dead ground between the tank and the target. A fraction of difference in alignment of the rangefinder would have yielded the correct range. The change in the atmosphere and body language in the room suggested that quite a few people in the audience realised that they could easily have made the same error.
- *Facilitation* – There are a number of important issues concerning facilitation. Firstly, facilitators need to be trained so that they can guide the discussion in a constructive and open way such that the audience identifies the areas of strength and weakness in their performance and develops their own solutions to the issues that are identified. Secondly, they need to be trained to exploit the debrief facilities that are provided (discussed below). Finally, they need to be perceived to be credible to comment on team performance and lead the discussion. McKeown and Huddlestone (2010) found that the expectation of commanders being debriefed was that the facilitator should be of at least the same rank, with command experience at the same level. Issues can arise where the facilitator is not perceived to be credible.
- *Timing and Duration* – To be effective, AARs should be held as soon as possible after an exercise or exercise phase (for longer exercises) has been completed, typically within four hours. The US Army recommends that

AARs at company level should be no longer than one hour, whilst those at battlegroup level should last no longer than two hours. The argument for longer AARs is based upon the complexity of the exercise and the range of issues that are likely to come up. A counterpoint to this is if too many points are discussed in too much detail then few may be remembered.

- *Frequency* – In an exercise lasting multiple days AARs may be held every day or every few days, depending on the nature of the exercise scenario and where the natural breakpoints in activity fall. The US Army typically stops exercise play for half a day after the AAR so that participants have time to debrief their teams and put fixes in place for the weaknesses that were identified before they reengage with the exercise.
- *Preparation* – Adequate preparation by the facilitator is critical for the successful conduct of the AAR. In a large-scale exercise, such as at battlegroup level, there will typically be a whole team of training staff who have been engaged in observing the performance of the training audience. To provide accurate feedback and highlight the key points from the exercise their different observations need to be collated so that a coherent picture of events is developed. Materials to support the AAR then need to be produced.
- *Location* – The location for an AAR needs to be accessible to the training audience, and have the necessary facilities to conduct and support the AAR. For an infantry section AAR this could be somewhere sheltered to sit on the ground, and the facilitator armed with his notebook. If the exercise has been conducted in a simulation system, there are typically debrief facilities in the form of a suitably sized room equipped with a debriefing system, which can replay events recorded in the simulation system during the exercise. Large-scale live exercises present more of a challenge for AAR conduct as the training audience can be very widely spread across a training area of some size. The solution provided at the US Army training facility in Fort Irwin takes the form of debriefing theatres built in large trucks which drive out to locations near the training audience. They have communications links back to the exercise control centre which are used to download pre-prepared materials to be displayed on debrief systems installed in the trucks. This saves the training audience from having to travel many miles to get to the AAR. Video teleconferencing is another approach that is used where the training audience is split across distributed sites (during live or synthetic training exercises).

Variants of the AAR construct have been developed, notably Team Dimensional Training (Smith-Jentsch et al. 1998), which adopts the same format but focuses on specific 'dimensions' of teamwork. The dimensions which they selected (information exchange, communication, supporting behaviour and initiative/leadership) are explained in the exercise pre-brief and are the focus of the debrief, which is conducted in accordance with the AAR format described above.

The AAR construct is significant not only due to its use in providing a mechanism for guided feedback, but also from the perspective of its potential to develop the ability of teams to critique their own performance without an external facilitator. The Canadian Army focuses on this aspect by encouraging a gradual transition from external facilitation to facilitation by the commander, over the course of an exercise. By the time the last AAR is conducted, the commander is responsible for both the preparation and delivery of the AAR (although assistance is provided in the preparation of materials). An example of this process having been adopted and exploited by a team in training was seen by the authors on board an AWACS aircraft during a UK joint exercise. During a gap in exercise play, the controllers sat together at the back of the aircraft to review their performance and determine what improvements they needed to make during the remainder of the exercise, and how they would change their approach to achieve these improvements.

9. Distributed Simulation Training

Extensive research has been conducted into the use of distributed simulation systems for collective training in the air domain over the last 15–20 years. This has included international collaboration between the US, Canada and the UK in the Coalition Training Research Programme, and national level developments in the form of the US and Canadian Distributed Missions Operations initiatives and the UK Mission Training through Distributed Simulation (MTDS) programme (Smith et al. 2006). Such has been the success of this work that multinational distributed exercises, in the form of Exercise Virtual Flag, are conducted annually in preparation for Exercise Red Flag, which is a live exercise conducted in the US. In the maritime domain, multinational synthetic training exercises occur, with simulators in the UK, France and Germany being connected to the US systems during the Fleet Synthetic Training-Joint series of exercises run by the US Navy Tactical Training Group Atlantic located in Norfolk, Virginia.

The research findings from the UK MTDS programme reported by McIntyre et al. (2013) are of particular interest from a training strategy and TNA perspective. Both Combined Air Operations (COMAO) exercises and Air Land Integration (ALI) exercises were undertaken within this programme. Not only did they receive very positive feedback from participants about the value of the exercises, but performance improvements were found to occur that matched the performance improvements that occurred in live training exercises, and these were seen to transfer into subsequent live training McIntyre et al. identify five key tenets which they deem critical to the successful use of distributed simulation:

- *Adopt a user-centric design approach* – During this programme extensive work was carried out to determine the precise needs of the training audience such that credible exercise scenarios could be constructed.

- *Create a total training environment* – This refers to the inclusion of all mission phases from planning through briefing, execution and then debriefing. Importantly, McIntyre et al. assert that the technology used must effectively support all phases, including the capability to run synchronised debriefs across all sites.
- *Do not underestimate the benefits of colocation* – Participants reported that one of the significant benefits of colocation was the opportunity to work together with members of other teams and develop a better understanding of each other's roles. In both the COMAO and ALI exercises, the opportunity to ask questions and clarify terminology was cited as a significant benefit by the exercise participants (effectively, colocation supported cross-training).
- *Provide a flexible and dynamic training environment* – The MTDS simulation system was sufficiently flexible such that the training staff (referred to as the White Force), having seen the planning undertaken by exercise participants, were able to adjust the scenario in order to fully test the plans produced. They could also adjust the scenario during exercise play to increase or decrease the level of difficulty in response to training audience performance.
- *Use an expert White Force* – Exercise management was conducted by an appropriately qualified and experienced military team with current operational experience. This level of experience was considered essential in order for the correct tactical points to be brought out during the debriefs and for the correct lessons to be taken away.

10. Crew Resource Management Training

Crew Resource Management (CRM) training, which is the term the aviation industry use to describe human factors training which includes training in teamwork principles, came into being as a consequence of safety concerns. Kanki et al. (2010) provide a comprehensive account of CRM training and its development since the inception of the concept, and the key points in this section are derived from that source. Whilst the introduction of reliable turbojet engines in the 1950s resulted in a marked reduction of aircraft accidents due to mechanical failures, there were still an unsatisfactorily high number of aircraft accidents occurring. Flightcrew actions were found to be causal in over 70 per cent of accidents causing aircraft damage beyond economical repair in the period from 1959 to 1989. A typical example cited was a crew, distracted by the failure of a landing gear indicator light, failing to notice that the autopilot was disengaged and allowing the aircraft to descend into a swamp.

Typical CRM training programmes include information and demonstration based components which include coverage of teamwork principles and Line Oriented Flying Training (LOFT). LOFT training comprises simulator training exercises which have scenarios based around operational sectors which run from take-off through to landing at the destination airfield. Events are threaded into the scenario to provide credible challenges to the crew which test their teamwork skills.

Kanki et al. (2010) summarise research findings concerning the implementation and impact of CRM training under the following headings:

- *Crewmembers find CRM and LOFT to be highly effective training.* This result is based on survey data of more than 20,000 flight crew members from both civilian and military organisations regarding CRM training and 8,000 regarding the LOFT component.
- *There are measurable, positive changes in attitudes and behaviour following the introduction of CRM and LOFT.* These effects have been measured through the use of questionnaires to determine attitude change and audits of flightdeck behaviour for behavioural change.
- *Management plays a critical role in the effectiveness of CRM training.* Greater acceptance of CRM training has been found in organisations where the senior management have demonstrated a real commitment to the concepts or CRM by implementing intensive and recurrent training.
- *Without reinforcement, the impact of CRM training decays.* Longitudinal studies have shown that, even where intensive initial CRM training was provided, if management support was weak attitudinal changes were not sustained and sometimes dropped to baseline levels noted before initial training had been delivered.
- *A small but significant percentage of participants 'boomerang' or reject CRM training.* Helmreich and Wilhelm (1989) found that there are multiple factors that seem to determine if someone will reject CRM training. Crew members who are lacking in traits associated with interpersonal skills, achievement and motivation are initially more prone to reject CRM concepts. Furthermore, group dynamics during training can have an effect. For example, if a charismatic individual openly rejects the training the level of acceptance by other participants can be reduced.

CRM principles and associated training content have been adopted in a range of other high-risk domains, including medicine and firefighting. The results from its implementation in the medical domain are of particular interest as they align with the findings above. Following a particularly unfortunate incident in the obstetrics department of Beth Israel Deaconess Medical Centre in 2000, a lady admitted for an induced birth lost her baby and had to have a hysterectomy. Poor teamwork, poor communication and poor coordination of care were considered to be contributory factors (Pratt et al. 2007). Following this incident a CRM programme was put in place. This included a four-hour introductory course followed by an implementation programme, which had a timeline for the introduction of each of the CRM concepts, coaches assigned to each shift, the development of communications tools (such as a pre-operative briefing template), and an information campaign to keep staff aware of each step of the process. The timeline called for one CRM concept to be implemented every two weeks, with the overall implementation to be completed within 6–12 months. The outcomes of this process were highly significant. Pratt et al. (2007) report that in the three year period following implementation nearly 300 fewer women experienced an adverse event than in the equivalent period before implementation. Furthermore, law suits, claims and observation cases (monies placed in reserves for potential claims) were reduced by 62 per cent. As a consequence, the physicians who participated in teamwork training were offered a 10 per cent reduction in their malpractice insurance premiums.

Selection of training and assessment methods

Unlike the case for individual training where there is an almost bewildering range of methods and associated media to choose from, there are relatively few training methods to choose from for delivering team and collective training. EBAT would appear to be the cornerstone from which to work, as it provides a meaningful context into which such techniques as procedural training, SCM, cross-training and team coordination training can be applied. The development of suitable measures for the key elements of teamwork would appear to be a particularly challenging area, and the implantation of an effective AAR strategy would seem to be essential. The potential for the development of generic teamwork training is an interesting avenue which merits further exploration, and raises some questions about how best to introduce the training of teamwork skills across an organisation.

Whilst the design of appropriate exercises is a matter for detailed design which occurs after TNA has been completed, the choice of methods informs:

- *training duration* – for example how many iterations of drills will be sufficient;
- *training environment requirements* – if there is a requirement to develop team adaptability, it must be possible to generate sufficiently diverse and challenging scenarios which necessitate adaptations in team coordination strategies, task processes and organisational structures;
- *training staff experience requirements* – the development of challenging, credible scenarios to provoke adaptive responses requires broader experience that would be required for the generation of scenarios to exercise procedural responses.

Training methods need to be matched to the nature of the task and the teamwork requirements, and the capabilities of the training audience. For example, within the British Army the planning process undertaken by headquarters staff follows a procedure known as the Combat Estimate. This takes the form of a series of questions which are addressed in sequence. The application of the Combat Estimate has both procedural and adaptive elements to it, in that it has a procedural flow but the planning problems that may have to be solved can be both diverse and range from simple to complex. Given that there is a constant turnover of headquarters staff, with individuals spending typically to two three years in post, when a formation changes role and begins its training for that role, the team is likely to be composed of members with a wide range of experience levels. Those with previous experience may well have been promoted and so have new roles within the organisation. Consequently, when a formation is transferred to a new role, such as the 1st Royal Irish Brigade transferring from Northern Ireland to undertake the air assault role (as described in Chapter 1), the Headquarters team can be considered a new team to the process. Therefore, training in the application of the Combat Estimate includes initial components of instruction in which teams develop an understanding of how the process works, the relationships between the stages, and the dependencies between the outputs of the sub-teams within the headquarters structure at each stage of the process. Once the team has become competent in the procedural application of the approach, they can then move onto implementing it in a more challenging exercise environment where they have to produce a plan and then oversee its execution. The combination of EBAT and SCM would appear to be the method of choice for this component of training.

By contrast, at a tactical level, the reactions to an ambush constitute a set-piece drill which has to be repeatedly practised so that the correct actions are taken immediately, with little instruction, as there is no time to plan how to address the problem. Procedural training in progressively more demanding situations would be a good training method for this task.

Training environment selection

Training environment selection may be influenced by policy considerations (such as a requirement to exploit synthetic training wherever possible) and security (for example certain aspects of combat training may only be conducted in a simulated environment to avoid revealing full system capability to potential observers of live training). Constraints such as limited availability or accessibility of suitable live training environments may also drive training towards synthetic training environments. Cost may also be a factor, as live training can be very expensive compared with synthetic training. Safety may also have to be taken into account.

Training Analysis and Design

From a TNA perspective, we are concerened with identifying the tasks that will need to be carried out in order to determine the staff, resources and supporting systems that will be requried to execute them. Staff training requirements also have to be identified. This breakdown is reflected in the expansion of the Training Analysis and Design component of the Training Overlay Model, as shown in Figure 6.6

The requirement for analysis and design activity prior to each instance of training delivery is largely determined by the degree to which the training has to be adapted to the training audience for each training event. In the aircrew training

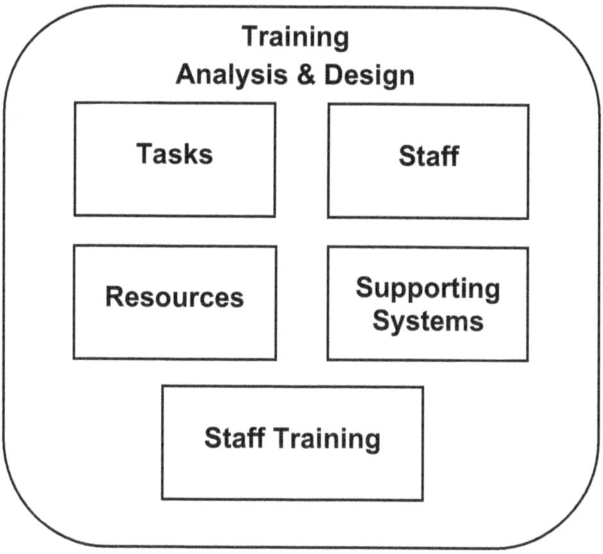

Figure 6.6 Training analysis and design components

example given at the start of the chapter the requirement is minimal, whereas for Exercise Joint Warrior the requirement is significant. If the requirement for adaptation is considered to be minimal, the up-front training design activity will have to be resourced.

Analysis Tasks

Typical training analysis tasks include:

- *Training task requirements capture* – Determination of the specific training requirements for an exercise will often involve liaison with the command staff of the training audience to identify priorities for training. These are likely to be influenced by developments in current operations or anticipated future operations.
- *Determination of team start state* – The currency of a team, affected by such factors as its recent employment and staff turnover, will influence training design in terms of the level of difficulty of the tasks presented at the start of training and the anticipated rate of progression. This would again be established by liaising with command representatives from the participating organisations.
- *Development of exercise objectives* – Once the training requirements and the start state of the audience have been determined, it is possible to construct objectives for the exercise which will inform the subsequent exercise design.

- *Identification of key scenario requirements* – Given the close coupling between tasks and the environmental conditions in which they are conducted, there are likely to be specific scenario requirements linked to the training requirements that are identified.
- *Identification of critical training enablers* – Critical training enablers are resources that are required to deliver the exercise and are external to the participating organisations. Examples in the live environment include support helicopters for the movement of land forces and contracted assets providing electronic warfare threats in the maritime environment.

Design Tasks

Typical design tasks include:

- *Training environment selection and booking* – Training environment selection and booking may be as simple as scheduling dedicated resources such as training areas or simulation facilities, though it can be more complex. For example, NATO runs regular exercises for Joint Force Headquarters to train with their subordinate Maritime, Land and Air Component Command Headquarters. These exercises involve the use of constructive simulation to provide the force elements for each component. The NATO Joint Warfare Centre in Stavanger, which runs the exercise, has the facilities to accommodate all of the participants in a dedicated exercise suite. When the authors observed one of these exercises, each of the participating Headquarters deployed into the field and were connected into the simulation system from their respective locations. The mix of maritime, land and air training areas used for Exercise Joint Warrior varies from exercise to exercise, depending on the numbers and nature of the participating forces.
- *Instructional strategy and methods selection* – Whilst the overall training strategy and choice of methods would not usually be revisited wholesale for each iteration of an exercise or training event, there may be a requirement for some adaptations to be made. For example, in a European Battle Group which the authors observed, the British Force Headquarters component was to be supplemented by staff from a number of participating nations. The British have a particular, question-based approach to the planning process referred to as the Combat Estimate, which the staff from other nations would not have been familiar with. Therefore, training in the conduct of the Combat Estimate was provided for these additional staff, and the Headquarters staff as a whole then participated in a command post exercise to practise as a team.
- *Scenario development, scripting, validation and rehearsal* – The development of the scenario and the associated events list sits at the core of exercise development. Validation of the scenario can take a number of forms, including review by the whole of the exercise staff by external

advisors, to ensure that the scenario events are likely to precipitate the required performance from the training audience in line with the exercise objectives. Representatives from participating organisations may also be asked for comment. Rehearsals may take the form of walkthroughs of the scenario by the staff to check that there are no issues with the sequencing and timing. For simulation based exercises, pre-programmed events may be run through to check they are in alignment with the plan.

- *Information product development* – Information products required in the exercise have to be developed/sourced. These may take many forms including maps, imagery, narrative reports and webpages.
- *Development of exercise materials* – Development of materials supporting the conduct of the exercise can be a significant undertaking. These would typically include: materials for the training audience, such as the orders which outline the tasks they are required to undertake, and form the basis for their planning activity; and materials for the training staff to guide the delivery of the exercise events and the assessment of training audience performance.
- *Information and communications systems planning* – Information and communications systems planning can become quite complex, particularly in a multinational context where units from participating nations may not have common communications systems that can easily be inter-connected, or connectivity is constrained by security considerations. In the case of the NATO exercise where Headquarters units deployed into the field, communications units had to deploy with each one to provide communications connectivity to the NATO Joint Warfare Centre and to the other deployed units.
- *Supporting systems planning* – There may be a need to plan how systems supporting the delivery of instruction, exercise management and training environment management are to be delivered. For example, on one of the Joint Warrior exercises observed by the authors, a GPS-based solution was developed for producing a ground picture to display the position of the ground forces during the exercise, which could be displayed on a large screen in the operations room.
- *Logistics support planning* – Logistics planning can also be a substantial undertaking, particularly for large-scale live exercises. For example, for Exercise Joint Warrior this includes planning for: the provisioning, fuelling and supply of engineering spares to the naval task groups; accommodation and base support for visiting aircraft; and contracting helicopter support for transporting training staff and visitors to and from maritime units. In addition, the team (augmented to its exercise delivery strength of 125) has to deploy to the location in Scotland where it runs the exercise, which is a logistical undertaking in its own right, requiring accommodation and transport bookings.
- *Development of role-player briefs* – Where role-players are used, briefs need to be produced to provide them with pertinent information for them to carry out their role successfully.

- *Augmentee training design* – If augmentees are to be used in instructional and exercise management roles there may be a training requirement to enable them to operate supporting systems or to conduct other aspects of their role, such as debriefing participants. Such training will need to be designed.

Staff

The delivery of the analysis and design tasks requires the following categories of staff:

- *Training staff* – Staff with appropriate backgrounds and experience are required to conduct the necessary analysis and design of the exercise. As described above, the JTEPS team that develops the Joint Warrior exercises is composed of warfare specialists from the air and maritime domains, with additional support provided by land warfare specialists.
- *Technical support staff* – There may also be a requirement for specialist technical support in such areas as logistics and information, and communications systems. If the training is to be simulation-based, then technical staff will be required to support the development of the scenario and programming of exercise events.

The TNA process should identify not only the categories of staff required but also the numbers of staff in each category that are necessary to successfully complete all of the required tasks.

Resources

The resource requirements for analysis and design tend to be modest, typically including office space and meeting rooms.

Supporting Systems

Information and communications systems are critical to the successful conduct of the analysis and design tasks. In addition, access to Training Management and Information Systems and Exercise Management Systems (particularly for simulation systems) will also be required if they are in use in the organisation.

Staff Training

The analysis and design of team and collective training is a complex task and would not normally fall within the experience of operational specialists who are typically required to undertake the role. Consequently, there is likely to be a training requirement for these staff. Whilst this requirement may seem obvious, the authors have noted that all too often it is conspicuous by its absence.

Training Delivery and Evaluation

The tasks that will need to be carried out during training delivery and evaluation have to be established in order to determine the staff, resources and supporting systems that will be requried to execute them. Staff training requirements also have to be identified. This breakdown is reflected in the expansion of the Training Analysis and Design component of the Training Overlay Model, as shown in Figure 6.7.

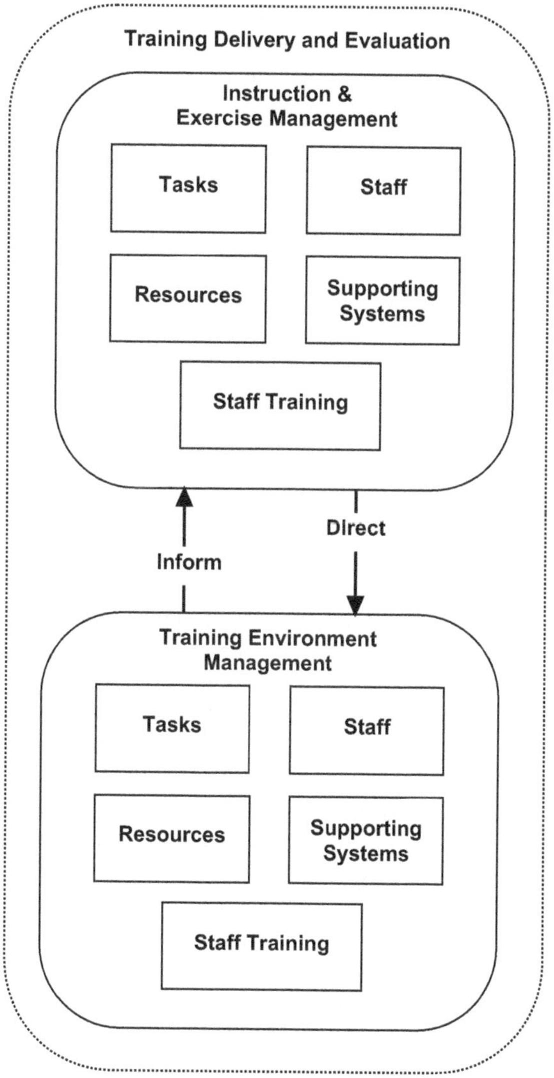

Figure 6.7 Training delivery and evaluation

Tasks

Typical training delivery and evaluation tasks include:

- *Instruction* – The tasks in the instruction category are all of those which encompass interaction with the training audience and include:
  - *Briefing* – Commonly at the start of an exercise or training event, or in a longer exercise at the start of each exercise phase, the training audience will be given a briefing about the purpose of the exercise, key focus areas and how their performance will be evaluated.
  - *Instruction* – Instruction can take a number of forms including initial instruction, such as that provided on the Combat Estimate to overseas staff joining the Force Headquarters of the European Battle Group, as described in the section on strategy and training methods in analysis and design above. Most commonly, it will take the form of coaching and mentoring during the conduct of an exercise. In large-scale US and NATO exercises witnessed by the authors, mentors have been assigned to senior commanders to provide personal coaching for the duration of the exercise. In the case of the US exercises for troops about to deploy to operational theatres, the mentors were commanders of equivalent rank who had just returned from theatre. The commanders that the authors spoke to found this form of coaching to be invaluable.
  - *Monitoring* – Monitoring of training audience performance during exercises is a critical function which supports the evaluation of performance and the provision of feedback. It should be focused on the interactions between team members and between teams as much as on overall task execution. A key element of monitoring is also the recording of performance to provide material to support debriefing.
  - *Evaluation* – Evaluation of the performance of the training audience is essential to the provision of informed feedback and to determination of the degree to which they are achieving the training objectives.
  - *Feedback* – The provision of feedback to the training audience about their performance is the mechanism that facilitates learning. It can be provided during the conduct of the exercise and at formal AARs at the end of exercise phases.
- *Exercise management* – The exercise management task is concerned with ensuring that the exercise progresses in accordance with the scenario and planned sequence of events, taking into account training audience performance. If the training audience is struggling with their task it may be necessary to simplify, delay or even cancel the planned events. In the case of live training, exercise management also needs to respond to factors such as weather constraints or equipment unserviceability, and adapt the

scenario accordingly. Achievement of this task requires feedback from the instructional team about the performance of the training audience and information about the state of the training environment.

- *Training environment management* – The training environment management task involves the delivery of the following functions:
  - *Set up* – Set up is concerned with the initial configuration of the training environment for the start of the exercise.
  - *Control* – Control is concerned with the delivery of the planned sequence of events and any adaptations of these events as directed by the exercise management team.
  - *Monitoring* – Control of the environment can only be achieved if the progress of the exercise is closely monitored. This also provides a crucial source of information for the exercise management team.
  - *Recording* – Recording of key events in the exercise provides invaluable factual evidence to support evaluation of the training audience performance and the AAR process.

Staff

The following categories of staff are typically required to conduct training delivery and evaluation:

- *Training staff* – Successful training delivery and evaluation is highly dependent on having sufficient training staff with appropriate expertise in the relevant task domains to be able to carry out the instruction, exercise management and training environment management tasks.
- *Role-players* – Where role-players are used it is critical that they have sufficient training and domain knowledge to play their role credibly when they interact with the training audience.
- *Technical support staff* – The nature of the technical staff required to support training delivery and evaluation is dependent in part on the nature of the training environment. Logistics and information and communications systems staff can play an important role in supporting live training exercises. Simulation systems will require specialists to maintain and operate them.

Resources

A wide variety of resources can be required to support training delivery and evaluation. Examples include: accommodation for the exercise management team, the training environment management team and their supporting systems; transport for instructional staff (such as helicopter transfers to ships and four-wheel drive vehicles to follow armoured vehicles across land training areas); and briefing and AAR facilities.

Supporting Systems

The systems required to support training delivery and evaluation merit particular attention in the TNA process:

- *Instruction* – Given that a key focus within team and collective training is the interaction within and between teams, the instructional team must have the facility to monitor these interactions. Consequently, they must have access to all of the communications channels which the team(s) have available to them (such as intercom, radio, email, chat rooms, email and web pages). Communication between members of the instructional team must also be supported if they are to coordinate their own actions and share their situational awareness of how the exercise is progressing and how the training audience are performing.
- *Training environment management* – Appropriate systems must be provided to facilitate the set-up, control, monitoring and recording of the training environment. If the training environment is simulated, either wholly or partially (such as for weapons effects in a live environment), the implication for TNA is that the requirements for the initial configuration, data capture and control during the exercise of each simulated element have to be specified. In the live environment, good communications with all the non-team actors are essential so that events can be controlled and any changes that are required can be communicated to the people on the ground (such as those acting as opposing forces) in a timely fashion. Tracking the movements and actions of the training audience also needs to be facilitated.
- *Exercise management* – If an exercise is to be managed effectively, the exercise management team needs to have ready access to an overview picture of what is happening in the exercise. They must also have effective communications with all of the training staff so that they can coordinate and control their efforts.

Staff Training

The training requirements for staff in instruction, exercise management and training environment management need to be considered within the TNA process. Areas which are often overlooked are exercise management and the exploitation of simulation systems to support training (as opposed to simply how to operate a simulation system).

The Detailed Training Overlay and Team Training Models

The key factors that have to be addressed in each component of the training overlay are shown in the detailed version of the Training Overlay Model in Figure 6.8

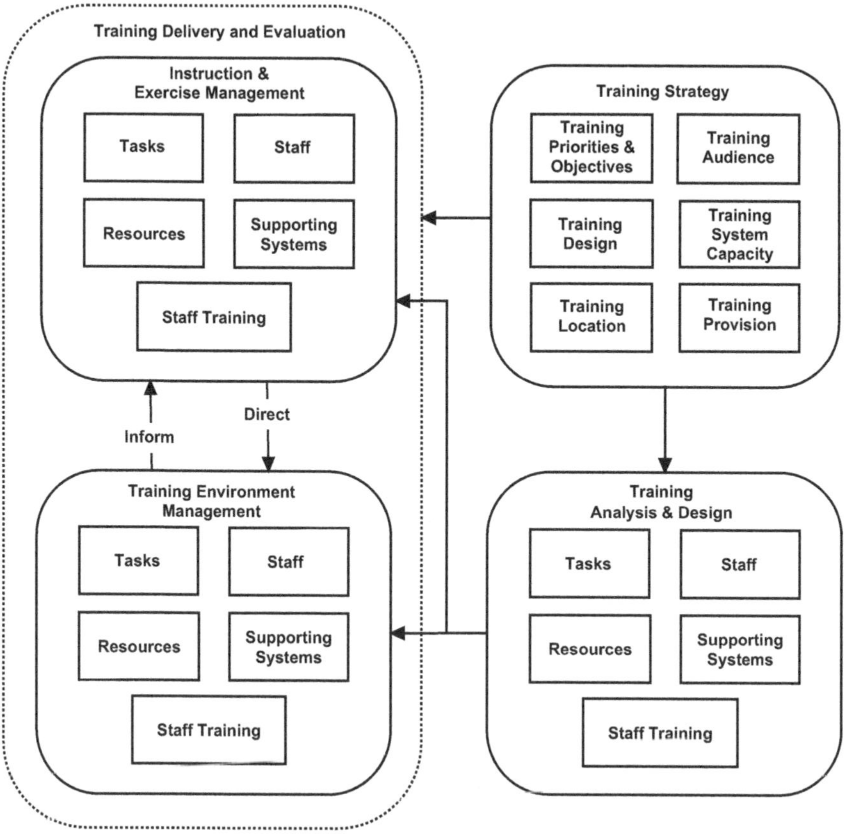

Figure 6.8 The detailed Training Overlay Model

Combining this model with the detailed version of the Team Performance Model forms the detailed version of the Team Training Model (Figure 6.9), which underpins the TCTNA methodology.

Summary

The analysis of the task to be trained, the environment within which it has to be executed, and the nature of the team that has to undertake it yields the detailed description of the training problem that has to be solved. The training overlay provides the means to deliver the required training effect. The analysis of the training overlay is concerned with the development of a training strategy and the specification of the means to deliver it in terms of the capability to undertake training analysis and design, followed by training delivery and evaluation. The Training Overlay Model provides a framework to guide this analysis. The Team

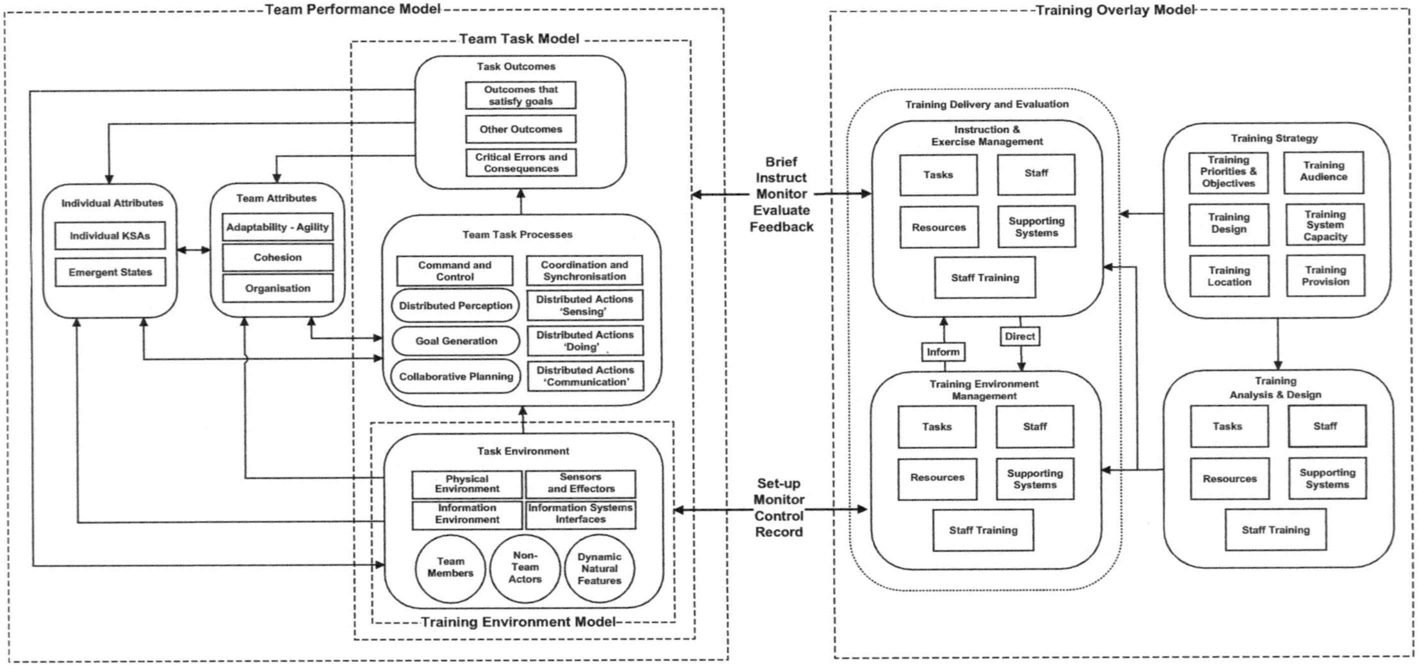

Figure 6.9 The Team Training Model

Training Model is, in essence, the conceptual map that guides the whole of the TCTNA methodology.

References

Blickensderfer, E.J, Cannon-Bowers, J.A. and Salas, E. (1998) Cross Training and Team Performance, in Cannon-Bowers, J.A. and Salas, E. (eds) *Making Decisions Under Stress*. Washington, DC: American Psychological Association. 299–311.

Dwyer, D.J, Fowlkes, J.E., Oser, R.L. and Lane, N.E. (1997) Team Performance Measurement in Distributed Environments, in Brannick, M.T., Salas, E. and Prince, C. (eds) *Team Performance Assessment and Measurement*. New York: Psychology Press. 137–54.

Ellis, A.P.J., Bell, B.S., Ployhart, R.E., Hollenbeck, J.R. and Ilgen D.R. (2005) An Evaluation of Generic Teamwork Skills Training with Action Teams: Effects on Cognitive and Skill-Based Outcomes, *Personnel Psychology*, 58: 641–72.

Entin, E.E., Serfaty, D. and Deckert, J.C. (1994) *Team Adaptation and Co-ordination Training*, TR-648–1. Burlington, MA: ALPHATECH.

Hall, J.K., Dwyer, D.J., Cannon-Bowers, J.A., Salas, E. and Volpe, C.E. (1993) Toward Assessing Team Tactical Decision Making Under Stress: The Development of a Methodology for Structuring Team Training Scenarios. *Proceedings of the 15th Annual Interservice/Industry Training Systems and Education Conference*. Washington, DC: National Security Industrial Association. 87–98.

Helmreich, R.L. and Wilhelm, J.A. (1989) When Training Boomerangs: Negative Outcomes Associated with Cockpit Resource Management Training. *Proceedings of the Sixth International Symposium on Aviation Psychology*, Ohio State University, Columbus, OH. 92–7.

Johnston, J.H., Cannon-Bowers, J.A. and Jentsch, K.A.S. (1995) Event-based Performance Measurement System for Shipboard Command Teams. *Proceedings of the First International Symposium on Command and Control Research and Technology*. Washington, DC: Centre for Advanced Command and Control Technology.

Kanki, B., Helmreich, R.L. and Anca, J. (2010) *Crew Resource Management*, 2nd edn. San Diego, CA: Academic Press.

Klein, G. and Pierce, L. (2001) Adaptive Teams. *Proceedings of the 6th International Command and Control Research and Technology Symposium*, June 2001.

McIntyre, H.M., Smith, E. and Goode, M. (2013) United Kingdom Mission Training through Distributed Simulation, *Military Psychology*, 25(3) 280–93.

McKeown, R. and Huddlestone, J.A. (2010) *A Comparative Review of After Action Review Practices for Collective Training across the British Army, the US Army and the Canadian Army*. HFI DTC Report No. HFIDTCPIII_T11_01.

Morrison, J.E. and Meliza, L. (1999) *Foundations of the After Action Review Process*, Special Report 42. Alexandria, VA: United States Army Research Institute for the Behavioral and Social Sciences.

Orasanu, J.M. (1990) *Shared Mental Models and Crew Decision Making*, CSL Report 46. Princeton, NJ: Cognitive Sciences Laboratory, Princeton University.

Oxford Dictionary (2014) http://www.oxforddictionaries.com/definition/english/strategy [accessed 22 September 2014].

Pratt, S.D., Mann, S., Salisbury, M., Greenberg, P., Marcus, R., Stabile, B., McNamee, P., Nielsen, P. and Sachs, B.P. (2007) Impact of CRM-Based Team Training on Obstetric Outcomes and Clinicians' Patient Safety Attitudes, *Joint Commission Journal on Quality and Patient Safety*, 33(12): 720–25.

Rapp, T.L. and Mathieu, J.E. (2007) Evaluating an Individually Self-Administered Generic Teamwork Skills Training Program across Time and Levels, *Small Group Research*, 38(4): 532–55.

Reigeluth, C.M. (1999) The Elaboration Theory, in Reigeluth, C.M. (ed.) *Instructional-Design Theories and Models*, volume 2. Mahwah, NJ: Lawrence Erlbaum Associates. 425–53.

Reigeluth, C.M. and Schwartz, E. (1989) An Instructional Theory for the Design of Computer-Based Simulations, *Journal of Computer-Based Instruction*, 16(1): 1–10.

Serfaty, D., Entin, E.E. and Johnston, J.A. (1998) Team Coordination Training, in Cannon-Bowers, J.A. and Salas, E. (eds) *Making Decisions Under Stress*. Washington, DC: American Psychological Association. 221–45.

Smith, E., McIntyre, H., Gehr, S.E., Schurig, M., Symons, S., Schrieber, B. and Bennett, W. (2006) *Evaluating the Impacts of Mission Training via Distributed Simulation on Live Exercise Performance: Results from the US/UK 'Red Skies' Study*, AFRL-HE-AZ-TR-2006-0004. Mesa, AZ: Air Force Research Laboratory, Warfighter Readiness Research Division.

Smith-Jentsch, K.A., Zeisig, R.L., Acton, B. and McPherson, J.A. (1998) Team Dimensional Training, in Cannon-Bowers, J.A. and Salas, E. (eds) *Making Decisions Under Stress*. Washington, DC: American Psychological Association. 271–97.

US Army Combined Arms Center (1993) *A Leader's Guide to After-Action Reviews*, Training Circular 25–20. Fort Leavenworth, KS.

PART II
The TCTNA Methodology

Chapter 7
TCTNA Overview

Introduction

The TCTNA methodology provides an integrated framework for conducting TNA for team and collective training, underpinned by the Team Training Model developed in Chapters 2–6. It is made up of the following six components:

- Project Initiation
- Team/Collective Task Analysis
- CARO Analysis
- Training Environment Analysis
- Training Overlay Analysis
- Training Options Analysis

The TCTNA process model in Figure 7.1 shows the connections between these six components and the nature of the information that is passed between them.

In the following sections the TCTNA processes and their linkages are outlined and the application of the TCTNA methodology in the context of acquisition is discussed.

TCTNA Processes

Project Initiation

A TCTNA represents a significant organisational undertaking, reflecting the complexity of team and collective tasks. With the possible exception of determining training options for small teams, it will typically require a team of analysts to execute the analysis. In all cases it will require extensive access to SMEs and a wide variety of information from different parts of the organisation (such as those responsible for training, personnel, infrastructure, logistics and equipment acquisition). Consequently, it is critical that it is fully supported by the organisation and that effective project management of the TCTNA process is undertaken. The Project Initiation phase is concerned with putting the necessary project management and organisational support infrastructure in place and planning the conduct of the TCTNA. It directly feeds the Team/Collective Task Analysis process by providing the initial definition of the scope of the analysis in terms of training audience and training task. CARO known at the initiation stage, such as policy constraints, are also identified and recorded.

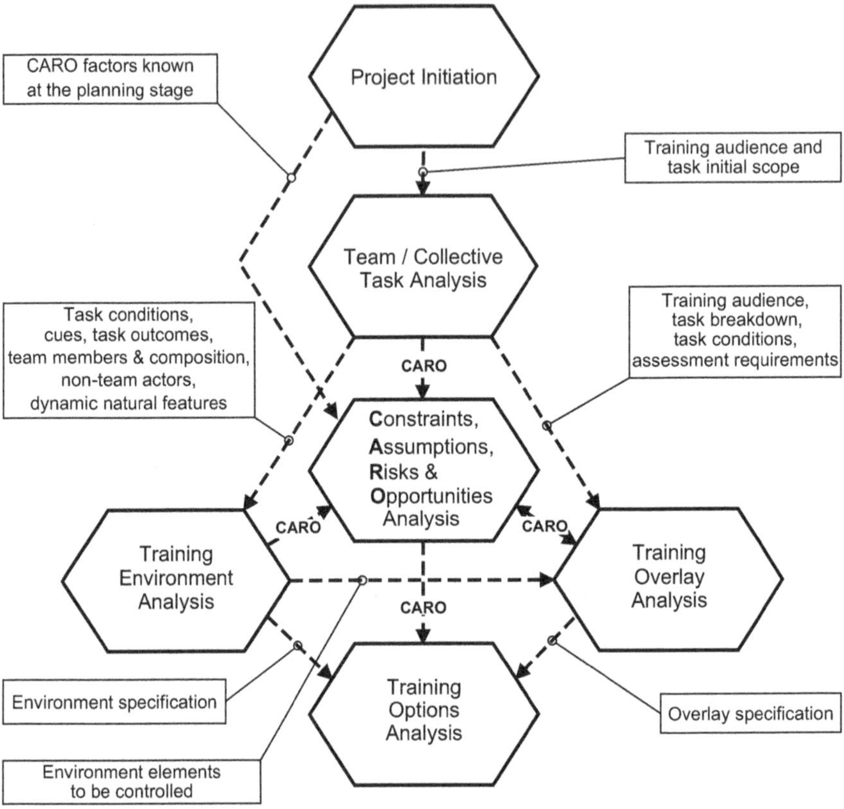

Figure 7.1 The TCTNA process model

Team/Collective Task Analysis

Team/Collective Task Analysis is the first of the analytical processes that is carried out and forms the foundation upon which the rest of the TCTNA is built. Given that a task can only be understood meaningfully in the context of the environment in which it takes place, and that inter- and intra-team interactions are critical components of team and collective task execution, all of these aspects have to be addressed. The analysis also identifies precisely who is involved in each of the components of task execution. The products of the analysis serve as the starting points for the subsequent Training Environment Analysis and Training Overlay Analysis processes.

CARO Analysis

CARO Analysis is concerned with the identification of all the constraints, assumptions, risks and opportunities that impact on the development of the training strategy and the identification of viable training options to deliver that strategy. It is a critical component of the methodology because if constraints, assumptions and risks are not captured effectively there is the distinct possibility that nugatory effort may be expended in exploring training options that are not viable, or worse still, a strategy and a related training option may be selected that subsequently cannot be implemented effectively. Equally, opportunities to exploit existing training capabilities, or to harmonise new training solutions at different individual, team and collective training levels may be missed. It is notable that that this process component is connected to all of the other processes in the methodology. Experience suggests that CARO are likely to come to light during Project Initiation and the subsequent analyses of the team and collective task, the training environment and the training overlay. Consequently it is a process that runs in parallel with each of the other processes.

Training Environment Analysis

The Training Environment Analysis follows on from the Team/Collective Task Analysis and provides a detailed specification of the training environment requirements that must be satisfied if all aspects of the team/collective task are to be trained. This encompasses: the physical and information environments in which the team operate, the means that the team use to interact with these environments (such as sensors, effectors and information systems interfaces), the non-team actors that must be represented and the means that they use to interact with both the environment and the team. The specification details both the elements that must be present and their respective fidelity requirements. It serves as an input to both Training Overlay Analysis and Training Options Analysis.

Training Overlay Analysis

Training Overlay Analysis is concerned with the initial development of the training strategy, informed by the outputs from the Team/Collective Task Analysis and the CARO that have been identified. It is also concerned with the development of a specification of the capability that must be provided to manage the training environment, including set-up, control, monitoring and data capture and display functionality. The development of this specification is informed both by the Training Environment Specification, which details all of the elements of the environment that must be managed, and the Team/Collective Task Analysis, which identifies the nature of the scenarios which it should be possible to construct and the information that is required in order to assess the performance of the training audience and to provide appropriate feedback.

Training Options Analysis

Training Options Analysis is the final process in TCTNA. Training options are identified based on the outputs of the Training Overlay Analysis and the Training Environment Analysis. The CARO that have been identified are used to determine whether potential options are viable and worth exploring in detail. A recommended option is determined based on a set of evaluation criteria that are developed, which include the degree to which each viable option meets the specifications that have been developed in the preceding stages of analysis.

Applying TCTNA in the Acquisition Cycle

A common context for the application of TNA is during the acquisition cycle for new systems. The acquisition of a major platform such as the Queen Elizabeth Class (QEC) aircraft carrier generates the requirement for a multiplicity of TNAs to be carried out across the full range of individual training for all personnel categories required to both man and maintain the platform, and at multiple levels of team and collective training. From a team and collective training perspective, TCTNAs would be required for each team involved in operating the platform (such as the operations room team, firefighting teams, the flight deck team and the Air Management Organisation), and for the collective training of all of these teams operating together on the platform. In addition, the integration of the platform into a task group has to be addressed, which can include consideration of training for new capabilities that the platform affords. In the case of QEC, this has included the development of training concepts for Carrier Enabled Power Projection (CEPP) (this is explored in the CEPP case study in Appendix B).

The nature of acquisition cycles poses a number of challenges from the TNA perspective. Figure 7.2 shows a representative acquisition cycle: the UK Ministry of Defence Concept, Assessment, Demonstration, Manufacture, In-Service, Disposal (CADMID) cycle.

The CADMID cycle is invoked when a capability gap has been identified and a decision has been made to acquire a system to address that capability gap. During the concept phase a range of concepts may be explored to address the capability gap and inform the development of the User Requirements Document which communicates the capability requirements without unduly constraining the scope of possible solutions. The Initial Gate (IG) is a key point at which the

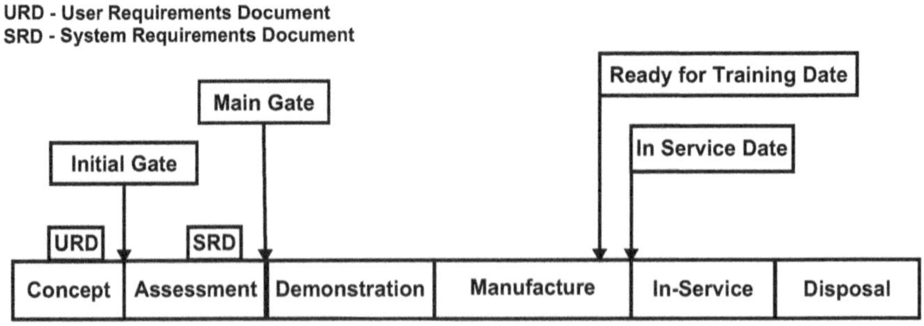

Figure 7.2 The CADMID cycle

decision is made as to whether or not to proceed further down the acquisition route and is informed by the IG Business Case. During the assessment phase a number of disparate system options may be explored which inform the development of one or more detailed System Requirements Documents (SRDs). At the end of the assessment phase the Main Gate (MG) decision point occurs where the decision is made as to whether or not to proceed into the demonstration phase, and if so which option is to be pursued. This is informed by the MG Business Case. Competing systems are then developed and demonstrated during the demonstration phase which leads to a final decision being made as to which system is to be manufactured.

The challenges from the TNA perspective are that, on the one hand detailed information about the nature of the system to be procured will only become available during the demonstration and manufacturing phases, but on the other hand the cost of training has to be considered in the IG and MG Business Cases. Training constitutes a significant component of the through life cost of a system and cannot be overlooked. Consequently, an approach is needed to inform the calculation of Rough Order of Magnitude (ROM) costs for training to inform the IG and MG Business Cases. Given that the costs of training are associated with the nature of the training audience and their tasks, the training environment, and the training overlay, these have to be considered in determining ROM costs. The approach that we suggest for addressing this issue is to apply the TCTNA approach iteratively, as illustrated in Figure 7.3.

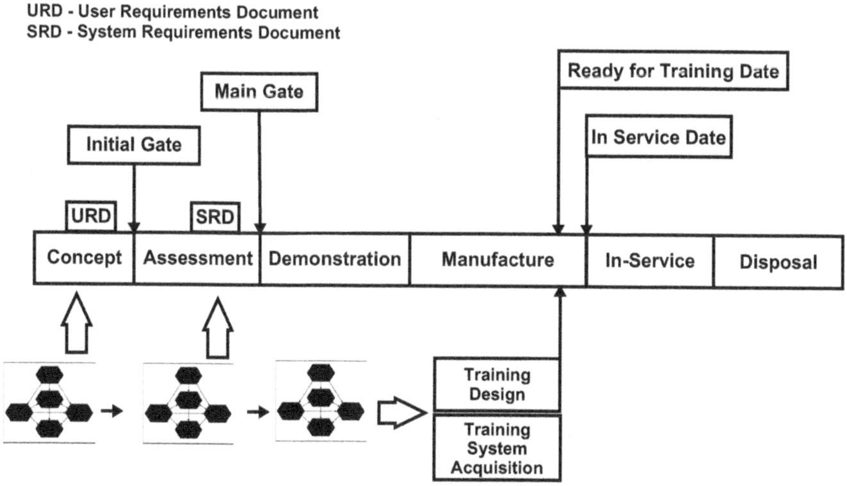

Figure 7.3 Iterative application of TCTNA in acquisition

The logic behind the iterative approach is that, whilst detailed information will not be available, reasoned assertions can be made about the nature of possible training solutions. As an example, consider the requirement for a long-range surveillance capability. At the concept stage a diverse range of concepts might be considered including a satellite-based system, a conventional manned aircraft system, an unmanned air vehicle system and a combat flying saucer. Given that it is a surveillance system, it will be equipped with sensors of some sort which will have to be managed and the outputs interpreted. The operation of satellite systems, conventional aircraft and unmanned air vehicles are well understood. They all require operators to control the system and sensor operators to deal with the sensors, and training systems for training teams to operate them exist at the moment. Therefore, ROM costs associated with potential training solutions could be based on those of extant systems.

Whilst the combat flying saucer is more problematic as an option, as there is no extant system on which to base estimates, some reasoned assumptions could be made based on what is known about the concept compliment. Imagine that it was envisaged as having an extreme duration of many days, and was to be supported by a ground station akin to an unmanned air vehicle ground station whose operators could control the system to assist or provide respite to the on-board crew (pilots and sensor operators). Then some parallels could be drawn with a synthesis of manned and unmanned air vehicle systems. One could envisage a synthetic training environment supporting the platform and its associated ground station, in which collective training of the ground and air teams was conducted, and dedicated ground stations and flying saucers employed for live training. TCTNA could be applied at a high level for each of these components to inform the development of ROM costs, during the concept stage, accepting that the accuracy and confidence levels of the estimates would have quite a wide error margin in the first instance, with many assumptions being made. As more details became available during the assessment phase, another pass through the TCTNA process could be taken for each option, to produce refined estimates with a higher confidence level and a smaller margin of error. Further refinement would be achieved through the demonstration and early manufacturing stages to yield a final TCTNA output with a recommended training option for the selected system. This would inform training design and training system acquisition in order to put a training system in place by the Ready for Training date, ahead of the In-Service date.

Structure of the TCTNA Guidance

The guidance chapters that follow each have a similar structure. They start with an overview of the particular stage of analysis and its relationship to the other analytical stages. This is followed by a summary of the process, its products and the information sources that may be relevant. The remainder of each chapter provides the detailed guidance for the conduct of the analysis at that stage. Each of the analysis components is illustrated with an example. The majority of the examples have been drawn from the Maritime Force Protection case study, provided in full at Appendix A. Other examples have been used where particular points could not be drawn from this case study. We have endeavoured to explain not only what to do at each stage, but why we suggest doing it.

When attempting to apply the guidance provided a number of points should be borne in mind. If the guidance presented is treated as a set of rigid rules it is likely to prove unsatisfactory. As Douglas Bader is attributed to have said, 'rules are for the obedience of fools and the guidance of wise men' (Brickhill 1954: 44). Given that the spectrum of possible contexts for the application of TCTNA is vast, ranging from established small team tasks to large-scale collective tasks involving the application of new capabilities, it is inevitable that you will at some point need to adjust, adapt or innovate when conducting a TCTNA. We hope that our guidance, with its supporting models, provides a sound platform from which to do just that.

References

Brickhill, P. (1954) *Reach for the Sky: The Story of Douglas Bader DSO, DFC*. London: Odhams Press.

Chapter 8
Project Initiation

Overview

Project Initiation is the first process in the TCTNA process sequence, as shown in Figure 8.1. The aim of the TCTNA Project Initiation phase is to develop the TCTNA Project Initiation Document which, when endorsed, would form the Terms of Reference for the analysis team.

Process

The Project Initiation process constitutes standard project management and planning activity of defining the requirement and deliverables and developing a plan to meet the requirement.

Products

The product of this phase is the Project Initiation Document which details:

- aim of the TCTNA
- context of the TCTNA
- links to other training analyses
- required outputs
- key project data
- management structure and process
- data sources and points of contact
- resources
- methodology
- plan and timescales
- CARO.

Information Sources

The key information sources for developing the TCTNA Project Initiation Document are likely to be the sponsor for the TCTNA and the acquisition team, if it is associated with a procurement project.

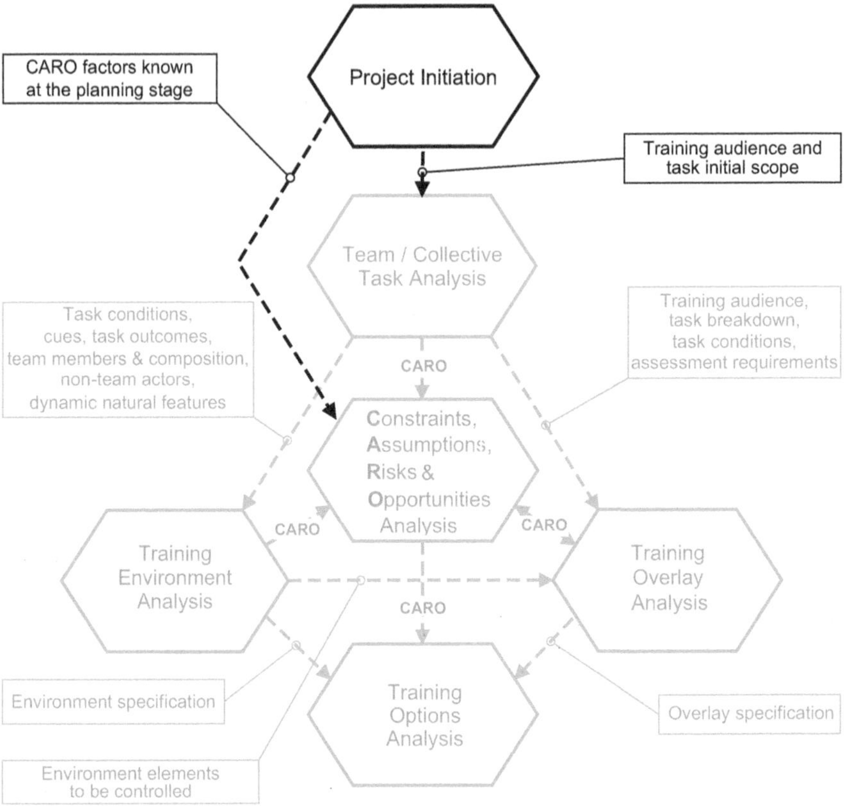

Figure 8.1 TCTNA process sequence – Project Initiation

Project Initiation Document Content

Aim of the TCTNA

This provides a short statement defining the aim of the TCTNA in terms of the training problem which it intends to address.

Context of the TCTNA

The context description provides a summary of the reasons for performing the TCTNA, identifying what is new or what has changed. Contextual factors may include:

- a requirement for the development of new capability, based around new or existing systems/platforms or some mix of both;
- new equipment to deliver an existing capability;

- new tasks or a change in task context caused, for example, by changes in doctrine, changes in team organisation, new Tactics, Techniques and Procedures (TTPs) or alterations in defence planning assumptions;
- alteration to the training environment such as a new training opportunity, or the loss of training area;
- the outsourcing of elements of the training system previously delivered in house;
- alterations to the instructional organisation or its remit;
- the definition of new team performance requirements;
- an alteration in team input standards;
- an increase in organisational turnover placing increased demands on the training system;
- assessment of team performance indicating or giving grounds for suspecting that the training system is inadequately preparing teams for the operational environment.

If the TCTNA is being conducted within the context of an extant training strategy, this training strategy should be articulated or referenced.

Scope of the TCTNA

The scoping statement sets the boundaries for the TCTNA task. For example, the introduction of a new system or platform may well have an impact through multiple levels in the organisation, and as such impact on multiple levels of team and collective activity. This may require multiple analyses to determine how training related to the system/platform and its use should be addressed. A clear definition of the scope of the TCTNA is required to establish a boundary to the subsequent analytical steps. This should identify the training audience and the task(s) which they are to carry out. This description provides the starting point for the subsequent analysis.

Links to other Training Analyses

If more than one TCTNA is being conducted, or if related individual TNAs are also being conducted, these should be identified and links established to ensure coherence in the development of training solutions.

Required Outputs

This section should define the required outputs for the TCTNA and the purpose for which they are being produced, that is, what subsequent activity are they going to inform.

The outputs to be generated by the TCTNA will be determined by the aim and scope. For example, if it is being conducted in the context of the acquisition of a new system or platform, then it may be conducted iteratively to inform successive stages of the acquisition process (such as progressively refined business cases), and then in detail to support the subsequent acquisition of a training solution. In such a case, the outputs from each of the components of the TCTNA will be required in increasing levels of detail at each stage, reflecting the growing quantity, maturity and granularity of available data. If on the other hand the TCTNA is being conducted to determine what enhancements may be required to an existing training solution to cater for an enhancement in capability of a system or platform, then the emphasis may fall on the questions relating to the training environment and training overlay.

Key Project Data

Key project data related to timings of decision points and training system delivery that inform the conduct of the TCTNA need to be identified. In the acquisition context, examples of such data include:

- sponsor
- the organisation responsible for the acquisition
- user (usually Front Line Command)
- project title
- key dates in the acquisition process such as the dates for submission of business cases, and the dates when the system/platform is to be brought into service and by when its full capability has to be delivered
- the date when training has to start being delivered
- the prime contractor for the system/platform
- planned mid-life upgrade.

Projects that are not acquisition-based will still have key dates that are significant to the conduct of the TCTNA.

Management Structure and Processes

The management structure and key stakeholders for the TCTNA should be identified. It is likely that the key stakeholders will form the Steering Group that will endorse the output of the TCTNA and have a key role in facilitating its conduct. The management process should be articulated, including such elements as progress reporting and review and risk management. It is common practice in military organisations to form a steering group of key stakeholders to oversee the process.

The following are suggested as potential stakeholders for the TNA and may form the Training Steering Group that would endorse the analysis products:

- *Chair* – It is suggested that the Chair would be a representative of the sponsor of the TCTNA.

- *User representative* – To provide the organisational perspective of the capability requirements.
- *Training delivery organisation representative* – To provide the training delivery organisation perspective on extant training capability and the implications for adopting alternative training solutions, which may include infrastructure requirements and logistics.
- *Human resources representative* – To advise on personnel issues.
- *Quality assurance personnel* – Ensures coherence with applicable Policy and Guidance.
- *Industry* – At the Chair's discretion if contracts have been let. Could be prime contractor and / or training solution / analysis contractor.
- *Subject Matter Experts (SMEs)* – As required. It should be noted that many SMEs may not have experience of the new capability, but will have experience of similar legacy capabilities.
- *Acquisition organisation human factors representative* – To advise on the integration and coordination of TCTNA activity with broader human factors activity associated with system/platform acquisition.

Depending on how the organisation is structured, it may be appropriate for other departments to be represented. In the military case these might include doctrine, training policy and logistics for example.

Data Sources and Points of Contact for the Conduct of the TCTNA

Key data sources required for the analysis should be identified, along with key points of contact who can provide access to information/SMEs to inform the analysis.

Data sources may include:

- early human factors analyses
- Ministry of Defence Architectural Framework (MODAF) human and systems views (produced during system design by human factors integration and systems engineering specialists)
- defined organisational performance requirements
- mission essential task lists
- concept documents (Concept of Use (CONUSE), Concept of Employment (CONEMP), Concept of Operations (CONOPS))
- manufacturer's system documentation
- doctrine
- SOPs
- TTPs
- interviews with SMEs
- observations of the task
- existing training documentation

- existing TNA documents
- research reports.

Key points of contact may include:

- training delivery organisations SMEs
- manufacturer's representatives
- organisational training specialists.

Resources

This section should identify the team members and resources that are required to perform the TCTNA and generate the required outputs.

Methodology

This section should describe how the TCTNA methodology is to be applied in order to generate the required outputs. If any deviations from the TCTNA approach are to be made or alternative analysis methods are to be used they should be documented and justified. Key factors that should be considered include:

- *Iterations of the process* – If the TCTNA is associated with an equipment acquisition project, it is likely that the method will be applied iteratively in order to inform successive levels of Business Case as well as to provide a final recommended option.
- *Linkages to other TNAs/TCTNAs* – Where multiple TNAs/TCTNAs are being conducted, careful coordination will afford opportunities for efficiencies in data capture and facilitate the development of efficient training solutions which maximise the use of common training facilities.

Plan and Timescales

A project plan showing the sequencing of activities with timescales and milestones should be provided.

Constraints, Assumptions, Risks and Opportunities

Constraints, Assumptions, Risks, and Opportunities (CARO) related to the timely and effective conduct of the TCTNA process should be identified and recorded. These should be monitored and updated as the TCTNA activity progresses. In addition, CARO related to the training requirement which are known at project inception (such as applicable organisational policy directives) should also be recorded.

Chapter 9
Team/Collective Task Analysis

Overview

The aim of this stage of analysis is to generate a sufficiently detailed description of the team or collective task, as it is performed in the operational environment, to inform the specification of the required training environment and training overlay. Figure 9.1 shows where Team/Collective Task Analysis fits into the TCTNA process sequence. In an ideal situation the task scope and the training audience will be clearly defined at the Project Initiation stage and serve as inputs to this analysis stage. However, at the early stages of the development of a new capability this clarity may not have been achieved and will only become clear through the detailed task analysis process.

The Team/Collective Task Analysis is the foundation upon which the remainder of the analyses are built. It is of critical importance because any errors or omissions which occur at this stage propagate through the rest of the process and are likely to result in a training solution being developed which cannot deliver effective training. Notably:

- If task components are missed then they will not be fed into training objectives and the training overlay will be incomplete.
- If task conditions are not completely captured this will lead to:
  - incomplete specification of the training environment;
  - incomplete definition of the training environment management capabilities that are required to set-up and control appropriate scenarios.
- If assessment criteria and the information requirements to inform such assessment are not captured correctly, data capture requirements for the training environment will not be fully specified.

The outputs from Team/Collective Task Analysis feed into Training Environment Analysis, Training Overlay Analysis and Constraints, Assumptions, Risks and Opportunities Analysis.

Process

The analysis approach is based around the Team Performance Model shown in Figure 9.2 (described in detail in Chapter 3). The team process components and categories of outcomes guide the identification of the types of processes which are being undertaken in the task, and the outputs that are generated. Similarly, the

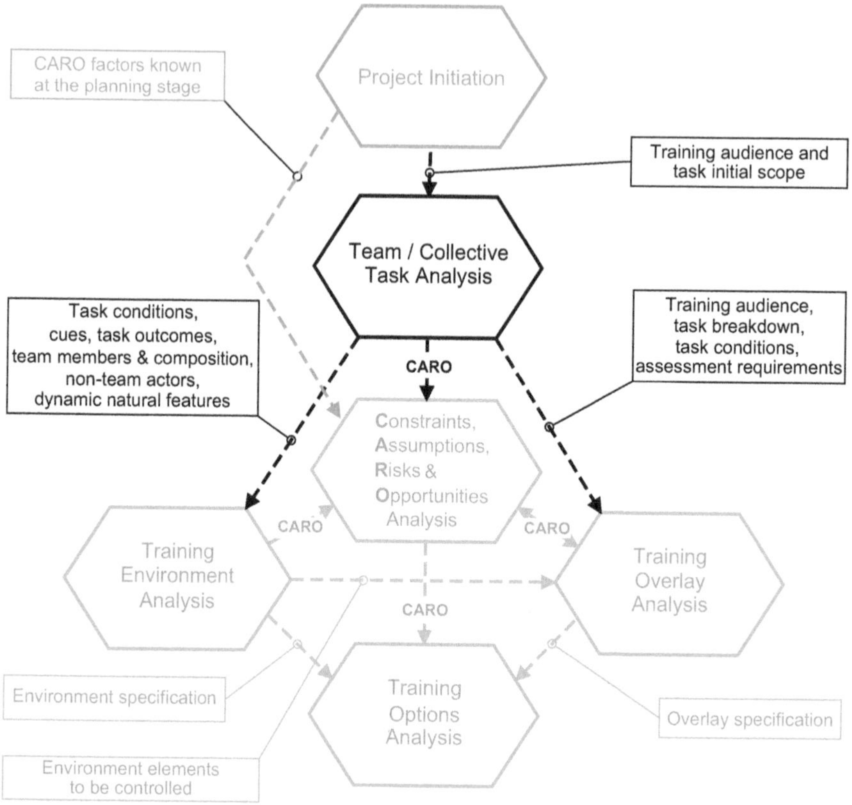

Figure 9.1 TCTNA process sequence – Team/Collective Task Analysis

components of the task environment serve as categories of elements with which the team members interact in carrying out team tasks.

The process steps for Team/Collective Task Analysis are:

1. establish the scope of the task;
2. describe the training audience;
3. identify the range of initial conditions in which the task may be executed;
4. conduct a detailed analysis of the task;
5. identify the teamwork-related KSAs required to support task execution.

Products

The products of this process are:

- *Task scope description* – This provides a description of the scope of the task, identifying the boundary of the task in terms of conditions under

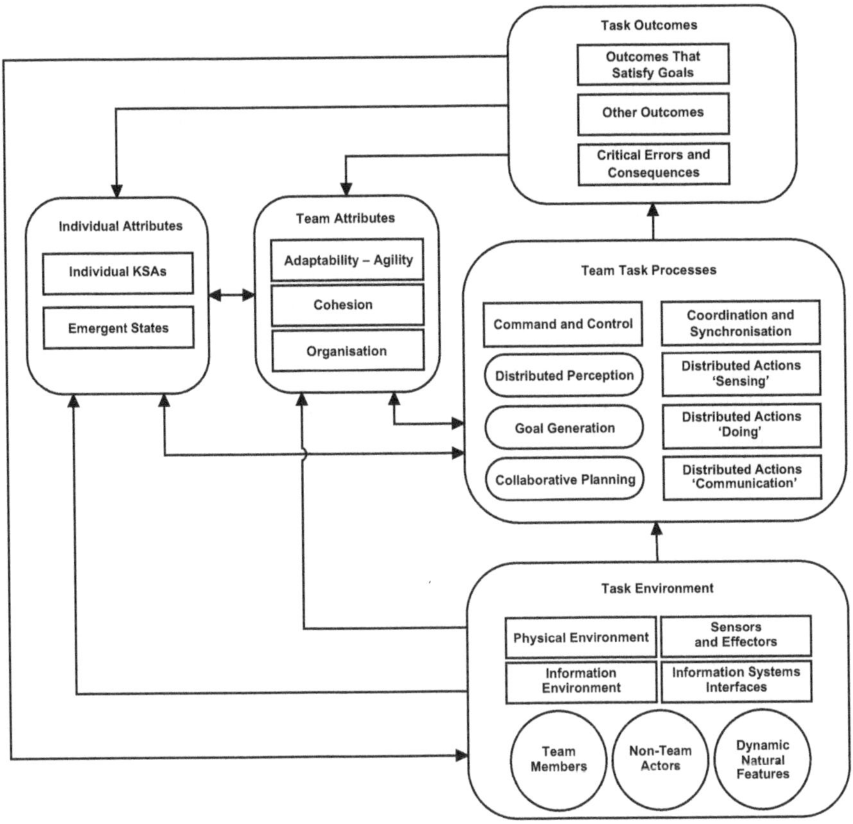

Figure 9.2 The Team Performance Model

which the task is to be executed, the agencies involved in executing the task and the nature of the required outcomes.

- *Team/Collective Organisation Description* – This provides a description of the structure of the organisation and the roles of the components within it. This defines the training audience. This information informs the conduct of the task analysis and may need to be generated with reference to outputs from systems engineering, system safety and human factors integration (HFI) activities.
- *Scenario tables* – Scenario tables provide a comprehensive description of the conditions and context in which the task is executed in the form of a table listing the static and dynamic elements in the physical and information environment, and the events that could occur in a typical scenario. This informs the detailed analysis of the task and the subsequent specification of the training environment in the Training Environment Analysis stage.
- *Task network diagram* – This diagrammatic representation of the task shows how the task is decomposed into sequences of subtasks and shows

the products as intermediate outcomes of each subtask. The construction of the diagram aids the analyst in decomposing the task into its constituent sub-components and displays parallel and concurrent phases of activity within the team, and can be used to facilitate discussions with SMEs in the development and validation of the analysis.

- *Task description table* – This provides a detailed description of each sub-task identified in the task network diagram in terms of its participants, inputs, processes including team interactions, relevant task conditions, systems used by the team (sensors, effectors and information systems interfaces), task outputs and evaluation criteria for both task processes and outputs (with associated data capture requirements). This data informs Training Environment Analysis and forms the basis of the development of training objectives, the assessment strategy and the specification of the training overlay interface to the training environment in the Training Overlay Analysis.
- *Team KSA table* – This captures the underpinning KSAs which the team members require to form appropriate mental models of the task and how the team executes it. This information is used to inform the development of knowledge-related training objectives in the Training Overlay Analysis.

Information Sources

Examples of sources that can be used to inform the task analysis include:

- contingent capability requirements and standards
- mission essential task lists
- Concept of Use (CONUSE), Concept of Employment (CONEMP), Concept of Operations documents (CONOPS)
- manufacturer's system documentation
- doctrine
- SOPs
- TTPs
- interviews with SMEs
- observations of the task
- existing training documentation
- early human factors analyses
- Ministry of Defence Architectural Framework (MODAF) human and system views (produced during system design by HFI and systems engineering specialists).

Information availability will depend on the context in terms of the stage within which the analysis is being conducted. If the analysis is driven by the need to enhance training capability for a platform or capability which is already in existence, it is likely that all of the sources listed above would be available. Where new platforms or capabilities are being developed, then much less information is likely to be available, as doctrine may not have been developed, SOPs are unlikely to exist and the task cannot be observed. In such circumstances greater reliance will have to be placed on discussions with SMEs about how it is anticipated that the platform/capability will be used. SMEs would include both military staff who can give an operational perspective on task execution, but also HFI specialists and systems engineers involved in systems design.

At the earliest stages of acquisition, lack of detailed information will preclude a very detailed task analysis. The task analysis will have to be amplified and revised as more information becomes available throughout the acquisition cycle.

Establishing the Scope of the Task

If the task boundaries are clearly defined then the detailed analysis can start straight away. In such a case the task scope description will simply be a reiteration of the scope definition given in the initiation document. However, if this is not the case, then some high-level analysis of the available documentation, such as the CONEMP and doctrine, along with discussions with SMEs, may be required in order to define the boundaries of the task and the related training audience. This involves a high-level analysis of the task, its goals and the context in which it is performed in order to identify the broad sub-tasks involved and the teams and assets that are required to perform them. This is illustrated in the Carrier Enabled Power Projection Case Study at Appendix B. Consideration of the full scope of a task may well extend outside the boundaries of a single platform. Capturing the wider potential scope of the task helps to mitigate against the generation of training stovepipes, by properly capturing the interoperability requirements inherent in collective tasks. Where a platform or capability has a number of discrete tasks which it must fulfil, a separate analysis of each may be conducted for each task.

Describing the Training Audience

A description of the structure of the team/collective organisation and the function of its sub-components serves two main purposes at this point. It aids in the analysis of the task and also provides the reader of the report with useful information to aid the understanding of the task analysis as it is subsequently presented. Discussions with SMEs and the collection of relevant documents can usefully be conducted at the same time as discussions about the scope of the task in the previous step.

The exact format and nature of this description will vary depending on the scale and complexity of the organisation and the level of detail available at the time the analysis is being conducted (for example, very early in the acquisition cycle the information available may be at a relatively high level). The level of definition that is required is a sufficient breakdown of the organisational structure to identify the teams, sub-teams or individuals that are interacting in the conduct of the task. Consideration needs to be given to functional relationships that span multiple platforms or component groups of an overall organisation. For example, in a maritime Task Group (TG), the AAW function is delivered by individuals and teams across multiple platforms. The AAW Commander may well be the AAW Officer of the Type 45 (the primary AAW platform in the TG). As such he will have a direct responsibility to the TG Commander for the conduct of that function, as well as coming under the chain of command of the Captain of the Type 45. Examination of organisational charts for the HQ structure and for each platform in the TG alone would not reveal this functional relationship.

Potential representations of use in describing the organisation include: narrative descriptions, organisational charts, diagrams showing functional relationships, role description tables and diagrams showing platform workspace layouts. Figure 9.3 shows an organisational chart for the team members involved in force protection of a Type 23 Frigate. Three local environments can be identified: the bridge, the upper deck and the ops room.

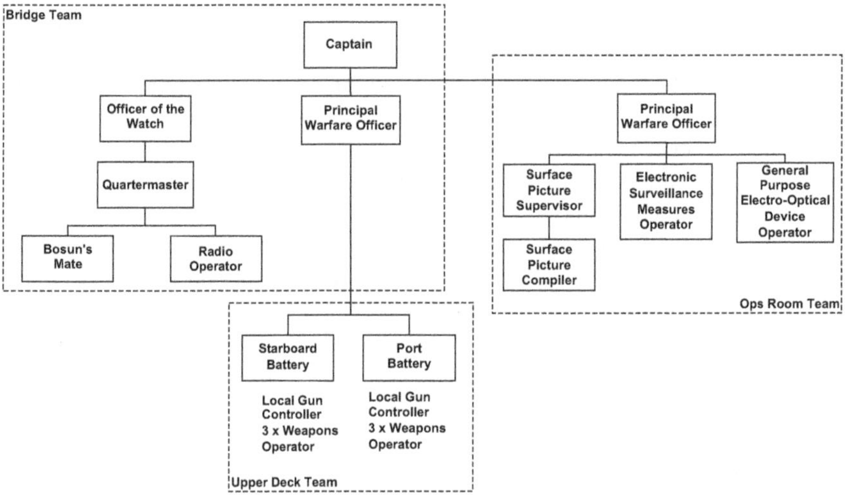

Figure 9.3 Type 23 organisational chart for force protection against an asymmetric threat for a Type 23 Frigate

The roles of the individuals can be described in a team role table as shown in Table 9.1. This may be populated at the individual level or could have descriptions of sub-team roles, depending on the scale and complexity of the organisation that is being described and the level of detail relevant to the subsequent task analysis decomposition.

The anticipated experience level of the team should also be identified, this will be influenced by factors such as the typical rates of staff rotation through posts, whether or not the training is for a completely new system, and if new concepts of operation are being implemented.

Table 9.1 Example team role table entries

Role	Description
Captain	Overall responsibility for the safety of the ship. Provides guidance to the Principal Warfare Officer (PWO) on tactical decision making and application of ROE.
Principal Warfare Officer (PWO)	Tactical command of the ship and integrated use of the sensors and weapons systems. Application of ROE. Issuing weapons control orders.
Weapons Operator	Search for and evaluate threats, prioritise targets. Operate weapons (General Purpose Machine Gun (GPMG), Mini-gun) including loading, firing, re-loading and conducting stoppage drills.

Scenario Description

The purpose of scenario tables is to capture all of the environmental elements that constitute the overarching conditions under which the task may have to be executed, and events that can occur in typical scenarios. Relevant conditions can be identified by considering types of problems that the team have to solve or environmental characteristics which cause team actions to become easier or more difficult to perform. Table 9.2 provides descriptions of a set of categories that can be used to capture relevant information. The physical, information, human and military conditions categories are likely to be applicable in all cases. The political and economic categories may not be applicable at lower tactical levels, but will be of particular importance at the higher operational and strategic levels. Descriptions should identify how the conditions manifest in task execution. The examples given in Table 9.2 in each category are illustrative and as such do not constitute a complete checklist.

The categories can be used to capture the pertinent conditions that would apply in typical scenarios. Table 9.3 provides an example of a table format that could be used for capturing a scenario description. Additional conditions may be identified during the detailed analysis of the task, and the table updated.

Conducting a Detailed Analysis of the Task

Carrying out the detailed analysis of the task is the most complex part of the Team/Collective Task Analysis phase. In this section an overview of the process sequence and outputs is provided followed by guidance on the construction of each of the products.

Table 9.2 Task conditions categories

Category	Description
Political	Political conditions that might impact on task execution. Examples include: • home popular support • UN Security Council Resolution • host government support • host population support • presence of non-state actors • absence of the rule of law.
Economic	Economic conditions that might impact on task execution. Examples include: • economically failing country • presence of starvation and famine • economic interests of indigenous security forces • corruption amongst government and local officials.
Physical	Aspects of the physical environment of relevance to the task. Examples include: • geography (air, land, maritime) • climate • level, nature and density of urban and rural development/infrastructure • transport infrastructure (air, land, maritime) • day/night • pattern of life (air, land, maritime).
Information	Information and communications systems of relevance to the task. Examples include: • coalition and alliance systems such as datalink pictures, intranets, video feeds • internet and broadcast media (radio, TV) • print media • intelligence products.
Human	Aspects of the human environment that might impact on task execution. Examples in a military context may include: • dress, language, religion, culture, beliefs and values of the indigenous population, enemy forces and coalition partners • indigenous population attitude to own forces (hostile, tolerant, indifferent or friendly). In a military context the detail of the human landscape may be broken down into civilian and military using the following categories: *Civilian* Numbers, organisation (if applicable), equipment, actions/activities. *Military* Numbers, organisation, disposition, equipment, capability and doctrine of: Enemy forces: • conventional forces • asymmetric forces. Neutral forces. Friendly forces: • own forces • coalition partners • host nation forces. Team resources including: • sensors and effectors, information systems • consumables: days of supply, fuel, food, ammunition • internal information environments.

Table 9.3 Example scenario table for Maritime Force Protection of a Type 23 Frigate against an asymmetric threat

Scenario	Maritime Force Protection of a Type 23 Frigate against an asymmetric threat, whilst at cruising watch in open waters
Physical	Open waters leading to confined waters such as straits, transits and harbour approaches. Variable sea-state and possibility of poor visibility due to local weather conditions.
Information	Personal role radios and internal command lines on-board. Royal Navy Command Support System chat connection to higher command.
Human	*Civilian* Local shipping, operators of local shipping (not hostile). Possibility of smugglers in fast moving small craft. *Military* Friendly forces: Command headquarters. Neutral forces: None. Enemy forces: Asymmetric threat of non-conventional forces in small, fast craft equipped with small arms, rocket propelled grenades, or operating waterborne improvised explosive devices. Possibility of simultaneous attack from multiple craft. Resources: • radar • electro-optical sensors • GPMG x 4 • Mini-guns x 2 • flares, searchlight and loudhailer.
Events	Small, unidentified, fast craft heads towards ship on collision heading, does not respond to escalation of force measures, turns to follow ship when ship changes heading and speed, opens fire on ship with small arms and rocket propelled grenades. Small, unidentified, fast craft does not respond to escalation of force measures, turns to follow ship when ship changes heading and speed, maintains collision course with ship (waterborne IED). Smugglers in small, fast craft heading towards ship do not respond to escalation of force measures, but maintain their heading. Fishing vessels on collision heading with the ship respond to escalation of force measures.

Process Sequence and Outputs

The detailed task analysis involves the following data capture steps:

1. identify the sequences of intermediate outcomes that need to be achieved to meet the overall goal (these can be thought of as the "steps" or milestones in the task);
2. identify the actions that need to be undertaken to achieve each of these intermediate outcomes;
3. identify the initial conditions that initiate each of the actions and the dynamic conditions that impact on task execution;
4. identify the action inputs in terms of cues from sensing the physical environment and information from the information environment;

5. identify the sensors, effectors and information systems interfaces that the team use to carry out each sub-task;
6. identify the team interactions that need to occur to support sub-task execution;
7. identify the conditions that may impact on team performance during task execution;
8. identify the assessment criteria for the process and outcome for each action;
9. identify the data capture requirements to support assessment against the criteria.

The suggested formats for representing data captured from these steps are the Task Network Diagrams and the Task Description Table. These are illustrated in overview in Figure 9.4. Alternative diagrammatic notations could be used, such as HTA notation, although limitations in the power to convey visually the sequential and parallel nature of tasks need to be considered. Whatever graphical notation is used, all fields in the related Task Description Table need to be completed.

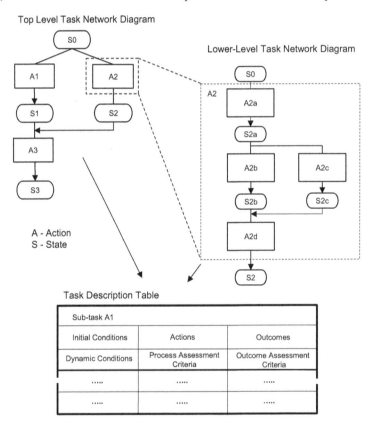

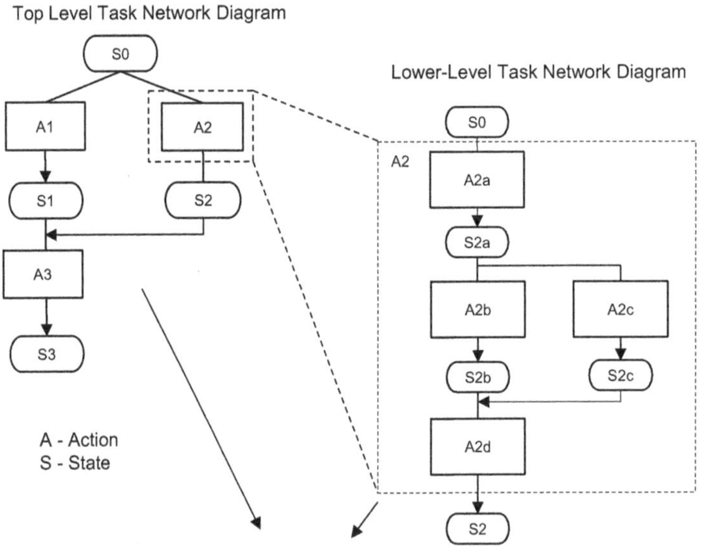

Task Description Table

Sub-task A1		
Initial Conditions	Actions	Outcomes
Dynamic Conditions	Process Assessment Criteria	Outcome Assessment Criteria
.....	.....	.....
.....	.....	.....

Figure 9.4 Overview of Task Network Diagrams and the Task Description Table

The Task Network Diagrams show the sequence of actions undertaken and intermediate outcomes that are generated in task execution. These are generated from data collected from steps 1 to 3. The Task Description Table captures the details of each action and its related outcomes and is populated using the data collected from steps 4 to 9.

Task Network Diagram

Notation

The Task Network Diagram provides a visual representation of the task breakdown using a node and link notation. The symbols used are shown in Table 9.4. Nodes represent states of the physical and information environments and links represent actions that are taken to change these environments.

Table 9.4 Task Network Diagram symbols

Category	Symbol	Description
State		Nodes represent states of the physical and information environment, which can be initial conditions or the results of actions of the team as task outcomes, non-team actors or dynamic natural features (for example wind, tides). For cross-reference in the Task Description Table they are numbered in the format SX.X.
Action		Action links are used to represent actions carried out by the team. For cross-reference in the Task Description Table they are numbered in the format AX.X.

Decomposition of the task into its subcomponents requires the identification of the sequences of actions that the team undertakes to get from the starting state of the environment to the desired outcome, along with the intermediate outcomes that they generate along the way. Based on the Team Task Model described in Chapter 3, the types of actions that the team can undertake and the types of outcomes which they produce are shown in Table 9.5.

It is also useful to explore which of the environmental stressors in the following list may impact on team performance and what events in the environment would bring them about:

- threat/stress
- performance pressure

Table 9.5 Team action types and output types

Team action types	Outcome types
Interactions with the external environment (physical and information): • sensing (actions which transform our understanding of the external world) • actions to transform the physical environment 'doing' • communication with non-team actors. Internal actions / team collaborative activities: • distributed perception to build SA • goal generation • collaborative planning • coordination and synchronisation of action • communication • performance monitoring • coordination of information within the team to develop SA, support goal generation or collaborative planning, or to support synchronisation and coordination of activity.	Generation or modification of team information products (for example the production of a plan). Changes in the external physical environment (for example effects of engaging the enemy). Changes in the external information environment (for example publication of a press release).

- high workload
- time pressure
- adverse physical conditions
- high information load
- incomplete and/or conflicting information
- multiple information sources
- visual overload
- auditory overload
- shifting goals
- ill-structured problems
- rapidly changing and evolving scenarios (dynamic environments, uncertain environments).

Figure 9.5 shows a top-level Task Network Diagram for the Surface Protection against an Asymmetric Threat Task.

The following explanation defines the types of action that were identified in constructing this Task Network Diagram and the nature of the outputs from each action and the logic of the sequencing shown.

The first task (A1.0) is building the surface picture. This is a sensing activity carried out by the ops room team using their sensors and the bridge team conducting a visual search. The output of the task is the unknown contact being reported, shown as output state.

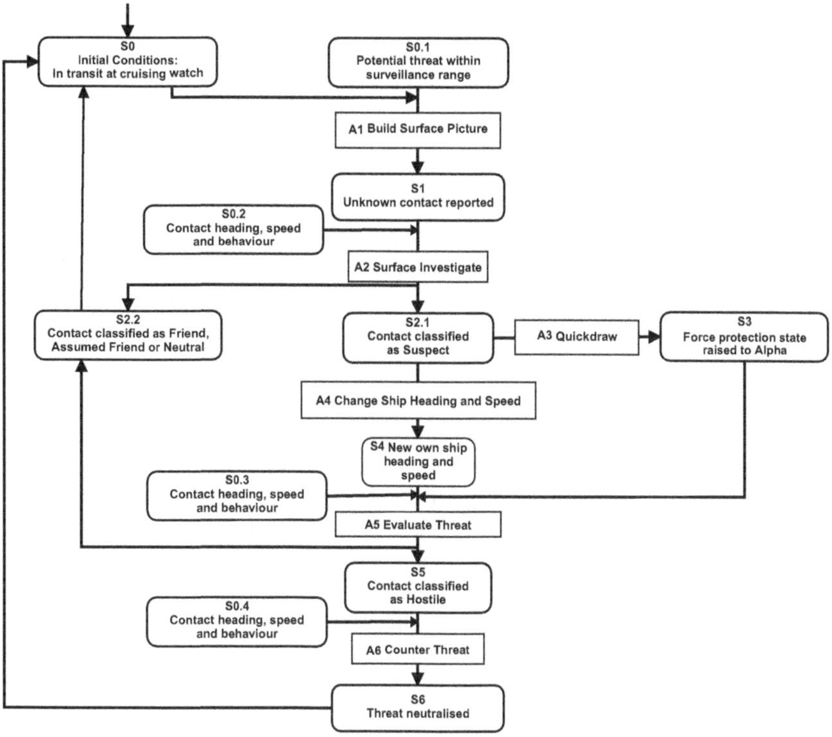

Figure 9.5 Top-level breakdown of the Surface Protection against an Asymmetric Threat Task

Once a contact has been reported the surface investigate action (A2.0) is instigated by the Officer of the Watch (OOW). This is a SOP whereby all the team involved in the search focus on the area where the contact has been reported to see if the contact appears in their sensors and report, in turn, what information they have about the contact. This surface investigate action draws upon environmental state S0.2 'Contact heading speed and behaviour'. The OOW synthesises that information to determine how the contact is to be classified. This set of activities embraces sensing and perception to yield situational awareness with the products being either: the contact being classified as neutral, friend, assumed friend or suspect. Contact classification is broadcast so that all team members know the status of the contact and can anticipate what actions will follow.

If the contact is classified as friend, assumed friend or neutral no further action is taken, indicated by no further actions being linked to that output node. If the contact is classified as suspect, two actions ensue, indicated by the two action links attached to the 'Contact classified as Suspect' output node. The two actions that follow are the raising of the force protection state to Alpha (weapons manned and with ammunition ready for loading) by means of executing the Quickdraw SOP (A3.0) and turning the ship (A4.0). Raising the Force Protection State is an action in the physical environment which involves the Upper Deck Weapons Operators collecting the GPMGs and the Mini-guns, setting them up on the weapons mounts, loading them ready for use and manning the weapons. The Captain would also be called to the bridge. At the same time as the Quickdraw SOP is executed the ship is turned (A4.0). This is done to inform the next action of evaluating the threat (A5.0). The positioning of A3.0 'Quickdraw' and A4.0 'Change Ship Heading and Speed' actions indicates *parallel and independent actions* performed by different teams in response to the same initiating conditions.

Evaluating the threat (A5.0) is a more complex action. It involves the monitoring of the contact following the alteration of the ship's heading change. If the contact turns to maintain a pursuit course then a range of force escalation measures are employed. These can include broadcast of radio warnings, shining a search light at the vessel, firing flares and ultimately firing warning shots. If the contact changes course to remain in pursuit of the ship and then maintains its pursuit course despite the escalation of force measures being employed, then it would be classified as hostile. Equally, if it displayed other hostile intentions, such as the crew appearing to prepare weapons for use, the contact would immediately be classified as hostile. This action is therefore a composite of a number of actions causing transformation in the physical environment, as well as sensing and perception actions to develop situational awareness and planning activity in terms of the sequence of force escalation measures to be used. The heading and behaviour of the contact are key inputs to the monitoring action and are shown as an input state in the diagram.

The final action is to 'Counter Threat' (A6.0) which involves the Captain or PWO issuing weapons control orders to the Weapons Operators to engage the contact, and the Weapons Operators then engaging the contact until it turns away or is neutralised.

Decomposition and stopping rules

Where a complex action has been identified, it is a candidate for further decomposition. The 'Evaluate Threat' action (A5.0) is a case in point. Two options are possible: the existing diagram can be expanded to show more detail, or a separate diagram can be drawn to show the expansion of the action. The decision about whether to expand a diagram at a given level, or to introduce a lower level, is a matter of judgement based on the complexity of the diagram at any given level. If the level of complexity becomes hard to follow, or if there is simply too much detail to fit onto a page, the use of lower-level diagrams should be considered.

A related issue is at what level should the analysis stop? Given that the purpose of the analysis is to understand how the team/components of the collective organisation interact to achieve the task, the decomposition should go no lower than identifying an action by an individual team member. That is to say, individual tasks should not be decomposed as this level of detail is not required. As an example, turning the ship is conducted by the Bosun's Mate at the helm. The set of actions he undertakes to do this, such as turning the wheel and checking the compass for the achievement of the desired heading are not relevant at this level of analysis. It may be appropriate to stop at a higher level, where an action is carried out by multiple team members/sub-teams/teams, provided the description of the process and the outputs, along with the associated assessment criteria has sufficient clarity. Figure 9.6 shows a separate diagram to show the detail of the Evaluate Threat action. Using this approach, the overall task breakdown can be shown as a hierarchy of diagrams. Figure 9.6 also shows how interaction of the actions of the team and the actions of non-team actors can be shown. To capture the notion that the escalation of force measures can influence the behaviour of the contact a link is shown from the Escalation of Force Measures Executed state (S5.3) to a contact action which in turn leads to a further contact heading speed and behaviour state (S0.3.1).

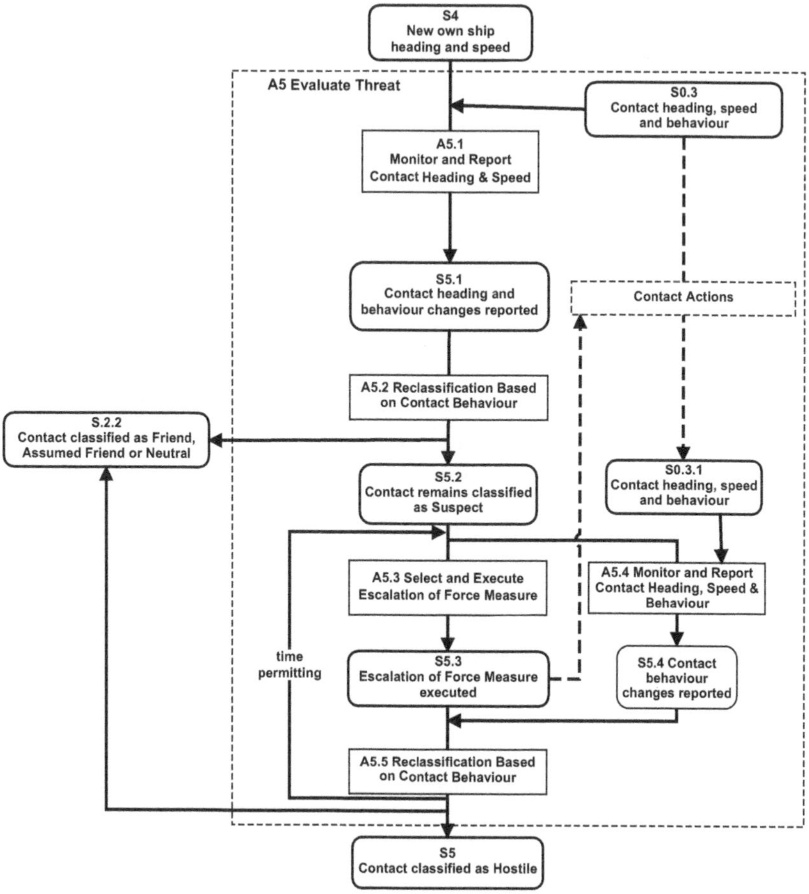

Figure 9.6 Expansion of Evaluate Threat

Task Description Table

The Task Network Diagram provides a clear means of showing the task breakdown, capturing the sequences of actions and their inputs and outputs. However, the detail associated with each action cannot be captured on the diagram. The Task Description Table provides a structured format for capturing the essential detail about task actions and their inputs and outputs. An example format and the entry for action A1.0 'Build Surface Picture' is shown in Table 9.6. The top row in the table entry for an action captures the inputs, process and outputs of the task. Key errors and their consequences are also identified. The bottom row captures the environmental variables that affect the difficulty of the task (and as such need to be replicated in the training environment) and the process and output assessment criteria, along with data capture requirements to inform these assessments.

Table 9.6 Example Task Description Table entries

Action A1.0 Build Surface Picture		
Initial conditions and cues S0: Neutral shipping. Coastline. Maritime zones, channels and lanes. S0.1: Potential threat vessel types.	**Actions (sensing, doing, communication)** *Participants*: Bridge team and ops room team *Process description*: (including sensors, effectors and information system interfaces used). Bridge team conduct visual search, ops room team use radar and Electronic Surveillance Measures (ESM) sensors to conduct radar and ESM search. Contacts reported. *Key team interactions*: Broadcast of the contact report to all team members to maintain situational awareness.	**Outcomes** S1: Contact report giving range, bearing, heading and speed.
Dynamic conditions Volume of shipping in the area; Multiple, simultaneous threats; Sea state; Visibility (fog and rain).	**Process assessment criteria** Comprehensiveness of the search though all arcs using appropriate resources/sensors. **Data capture requirements** Observation and recording of team actions and communication.	**Outcome assessment criteria** Timeliness, completeness accuracy of the report, on the correct circuits. **Data capture requirements** Recording of contact reports, time and originator.

Identify the Teamwork-related KSAs Required to Support Task Execution

In this section we capture the knowledge that all team members require about how the team operates in order to engage effectively in the team tasks. Table 9.7 provides an example of a table capturing team KSAs. Chapter 3 provides detailed descriptions of team KSAs that may be applicable. This material would typically form the content of training sessions conducted before task practice is conducted.

Table 9.7 Team KSA requirements table

Category	Description
Knowledge	
Team/team member roles	Roles of Upper Deck Team (UDT), Ops Team and Bridge Team.
Task models	Self-defence ROE. SOPs: Surface Investigate, Quickdraw, Escalation of Force.
Skills	
Voice procedures	Use of correct phraseology for reporting target sightings.
Attitudes	
Positive team orientation	Display a positive commitment to collaboration to achieve team goals.

Practical Issues in Conducting the Team/Collective Task Analysis

In an ideal situation the scope of the task will be clearly defined, as will the training audience, documentation pertinent to the analysis will be readily available for the team to read into the task and SMEs will have unlimited time to work with the team through the analysis process. In practice these conditions rarely, if ever, exist.

In practice it is suggested that an initial meeting with SMEs is planned (multiple initial meetings if there are SMEs from different areas to be consulted) to discuss the task to get an initial idea of the scope of the task, the training audience, and conditions and an overview of what the task involves. This also provides an opportunity to explore what documentation is available to shed light on the details of the task. It can also be helpful to brief the SMEs on the structure of the analysis and provide examples of the format of the outputs so that they understand what the analyst is trying to achieve and what they can expect to be shown at future meetings for review.

Data and documentation collected at this initial meeting (or meetings) can then be analysed in order to start drafting analysis products which can then be presented and discussed at subsequent meetings. A second face-to-face meeting to discuss the initial draft outputs is helpful as it provides the opportunity to explain the format of the outputs in more depth as well as discussing the accuracy and completeness of the outputs and to identify what additional information is required. Experience from the application of the techniques described in this section, including the development of the case studies, suggests that SMEs are usually able to make sense of the analysis products such that it is possible to circulate documents for review and then follow up with phone calls/video conference discussions to discuss the details, should face-to-face meetings not always be possible (although face-to-face meetings are generally preferable, particularly for complex tasks).

Typically, the analysis process is iterative, with refinements to all of the products occurring as the analysis progresses.

Chapter 10

Constraints, Assumptions, Risks and Opportunities Analysis

Overview

The aim of the Constraints, Assumptions, Risks and Opportunities (CARO) Analysis is to identify the CARO that inform the development of the Training Strategy during Training Overlay Analysis and the identification of viable training solutions during Training Options Analysis. Figure 10.1 shows the interconnections of CARO analysis with the other processes in the TCTNA process sequence. Some CARO, such as applicable training policy, may well be known at the Project Initiation stage. Others will become apparent as each stage of analysis is conducted. Consequently, CARO analysis is a process that runs in parallel with the other analytical processes up to the end of Training Overlay Analysis.

Process

CARO analysis involves the analysis of policy documents and discussions with SMEs from across all areas of capability to determine CARO related to the training audience, the training task, the training environment and the training overlay.

Products

The products of CARO Analysis are:

- *Constraints Table* – list of constraints and consequences for option development
- *Assumptions Table* – list of assumptions and consequences for option development
- *Risk Register* – list of risks, effects and potential mitigations for option development
- *Opportunities Table* – list of opportunities and consequences for option development.

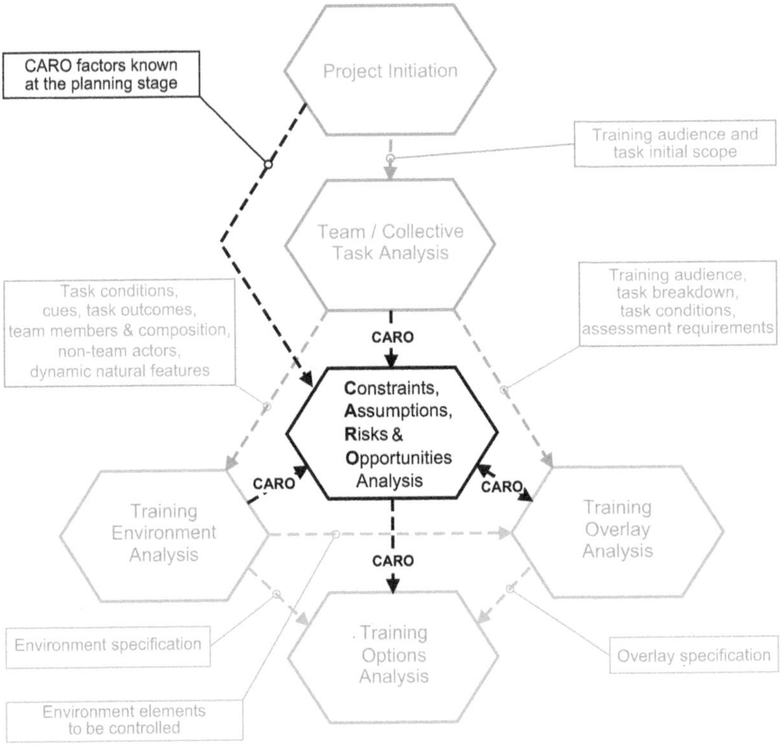

Figure 10.1 TCTNA process sequence – CARO Analysis

Information Sources

Potential sources of information are:

- stakeholders for the TCTNA (provide expertise from the operational domain and the supporting capability areas)
- discussions with staff responsible for training policy
- training policy documents
- discussions with SMEs from training delivery organisations.

Constraints Table

There are a wide variety of constraints which may limit the viability of training options. Considerations include, but are not limited to:

- *Training policy* – There may be policy constraints on how training is conducted, such as the qualifications required for instructors to be able to conduct certain types of training, and the requirement for simulation to be explored as an option.

- *Safety* – Safety can be a significant constraint on the choice of training environment. A typical example would be limitations on live firing. Invariably, weapons effects and other potential hazardous modes of action in the environment (for example high-speed, low-level flying) have to be simulated or conducted in special areas.
- *Cost* – Budget limitations, and budgetary cycles may preclude the selection of training options due to cost (costs can include initial acquisition, on-going maintenance of training systems and the cost of running each training event).
- *Training audience availability* – There may be a limit on how long a training audience is available for training and when that availability falls. The availability of a training audience may necessitate the splitting of the overall collective task into a number of collective sub-tasks. Consideration also needs to be given to the availability of the training audience to participate in exercise planning, which may occur anything from six months to a year in advance of the exercise in which they are due to participate.
- *Resource availability* – Limitations on the availability of resources to generate the required training environment and training overlay, such as training areas, equipment required for training, daily and seasonal weather conditions and suitably experienced role-players, can all limit training options. Similarly a lack of resources to support the trainees' collective task will limit which tasks or sub-tasks can be trained or exercised.
- *Security* – Security considerations, such as the security classification of a simulation system precluding its linkage to unclassified systems, may also constrain training options, as may commercial considerations such as International Traffic in Arms Regulations or certification requirements to link synthetic environments.
- *Interoperability* – Legacy training systems and training management systems may have limited potential for linking to newly developed systems.

A suggested format for capturing constraints is shown in Table 10.1.

Opportunities Table

Opportunities to exploit or develop existing training capabilities and synergies with other training development projects should be identified, as they may have the potential to yield training solutions that are more cost effective. As an example, the development of a synthetic training solution for individual training may require only modest extension in capability to cater for team training. A suggested format for capturing opportunities is shown in Table 10.2.

Table 10.1 Example Constraints Table format

Constraint	Consequence
Weapons effects cannot be trained live	Weapons effects need to be simulated.
A simulator system is classified	The simulator cannot be networked with unclassified simulation systems.
Augmentees are only available for two weeks per year for training	Duration of training with a fully augmented team is limited.
Instructional role-players are only available for one week per year	Interactions with specific non-team actors may be limited. Tasks involving these non-team actors should be identified and prioritised.

Table 10.2 Example Opportunities Table format

Opportunity	Description
Individual weapons training system due for refresh	Team training solution could be based on an extension of the individual weapons training solution provided there are enough weapons stations to cater for team training requirements, suitable communication facilities and adequate red force representation.

Assumptions Table

Any assumptions which impact on the determination of the training strategy and development of options need to be identified. All of these assumptions reference the properties of the collective task, and as a result can only be generated following an analysis of the collective task. Assumptions can typically relate to:

- training target audience
- training throughput
- when training is required to start, finish and how often it may be required, including whether an interim solution is required
- training location, particularly resulting from policy guidance
- how to deliver training, including any policy / concepts and doctrine guidance
- whether training will need to be integrated into existing training solutions.

Risk Register

Risks to the successful completion of the analysis and to the subsequent delivery of an effective training solution also need to be captured. The risk register format used in acquisition projects is shown in Table 10.3. It is suggested that this format

Table 10.3 Example of a Risk Register

Serial	Risk description	Cause	Effect	Probability *	Impact **			Risk owner	Mitigation action	Mitigation cost (£k) ***
					Perf	Cost	Time			
1	Intellectual Property Rights constraints on the release of technical publications for training.	Technical publications are not available in time to inform training analysis and design.	TNA and training design incomplete; immature training solution.	Med	High	High	High	Project team commercial	Project team contracts for technical publications to be made available for training analysis and design.	1,000
2	Availability of SMEs to inform training analysis and design.	SMEs not available in time to inform training analysis and design.	TNA and training design incomplete; immature training solution.	Low	Med	Low	High	TNA Steering Group	TNA Steering Group to direct SME availability, including industry SMEs.	Minimal (for example travel)
3	Availability of doctrine to inform training analysis and design.	Doctrine not available in time to inform training analysis and design.	TNA and training design incomplete; immature training solution.	Med	Med	Low	High	Operational owner	Project team and operations staff to identify when concepts and doctrine will be published and make them available to training analyst and designer.	Minimal

* Probability of occurrence: High / Med / Low.
** Impact to schedule Performance / Cost / Time: High / Med / Low.
*** Mitigation costs are for illustration purposes only, and are not intended to be representative.

is used for compatibility purposes. Categories of risk which should be considered include but are not limited to:

- risk to capability and effects of not having a training solution ready in time to meet the In-Service Date (ISD)
- risk to capability of not having a training solution adequately resourced through life of the capability
- risk in being unable to articulate training requirements in time to inform training solution design
- risk to training requirements articulation, design and development through unavailability and/or immaturity of key information
- risk to personnel structures through inappropriate training and/or lack of throughput capacity.

Chapter 11
Training Environment Analysis

Overview

The aim of Training Environment Analysis is to produce a specification for a training environment that will support all aspects of task execution identified in the Team/Collective Task Analysis. Figure 11.1 shows the position of Training Environment Analysis in the TCTNA process sequence.

Whilst the Team/Collective Task Analysis identifies all of the elements that have to be present in the training environment, the information is spread across the different analysis outputs. Training Environment Analysis produces a consolidated representation of these elements diagrammatically, and adds supporting detail about the precise fidelity requirements for each element. A key point to note is that it is implementation agnostic; that is to say it is not predicated upon any particular implementation for any of the elements. It serves as a neutral specification which holds true for any type of implementation, which could involve a mix of live and synthetic components. It informs the Training Overlay Analysis, as all of the elements will need to be considered from the perspective of training environment management, and Training Option Analysis as the set of environment requirements against which each option will be evaluated for compliance.

Process

The analysis process is based around the Training Environment Model shown in Figure 11.2.

The process steps are:

1. Produce Training Environment Diagram(s) which capture all of the elements that must be represented in the training environment to support task execution.
2. Produce specifications which capture the fidelity requirements for each of the elements in the diagram(s).

Products

The products of this process are:

- *Training Environment Diagram(s)* – These provide a diagrammatic representation of the required training environment in terms of the physical and information environment components and the actors and their interfaces

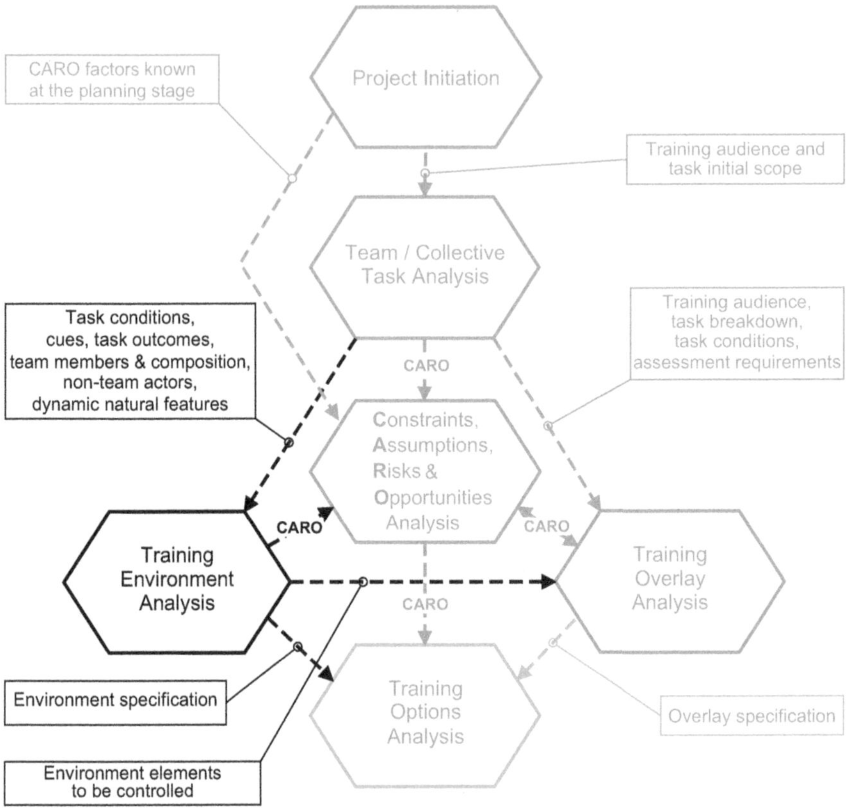

Figure 11.1 TCTNA process sequence – Training Environment Analysis

to these environment components. They facilitate the development of the environment table and aid review of the environment specification with SMEs.

- *Training Environment Specification Tables* – These provide a description of the essential properties of each element in the environment in terms of its fidelity requirements, including behaviour where appropriate. The set of tables produced constitute a complete specification of the required training environment characteristics against which training environment options can be evaluated during Training Options Analysis.

Information Sources

Information sources for this stage of analysis include:

- the Team/Collective Task Analysis
- system documentation
- SME interviews.

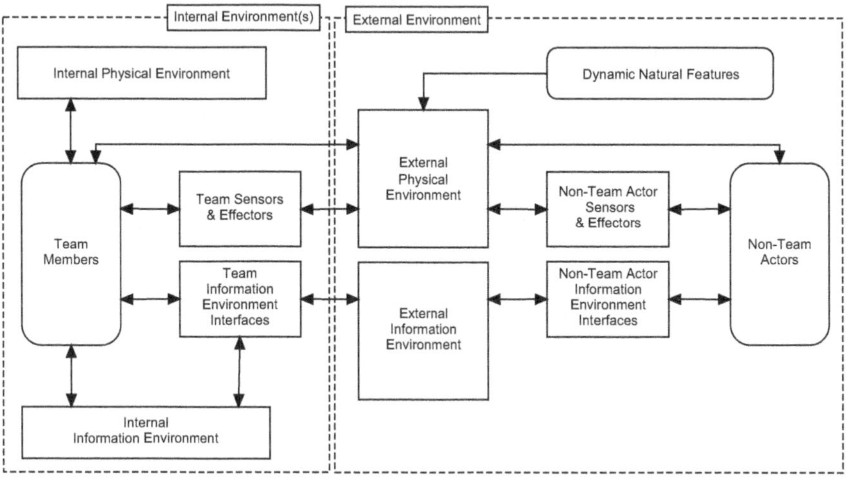

Figure 11.2 Training Environment Model

Approach

The suggested approach for this stage of the analysis is firstly to produce a diagrammatic representation of the environment (a Training Environment Diagram), showing the elements in the environment and the interactions that occur between them. Following this, detailed specifications for each of the elements and the modes of action that connect them are generated. The Training Environment Diagram provides an overview of all of the key features of the environment. Its construction helps the analyst build a picture of the environment and serves as a useful tool for facilitating discussion with SMEs in both the development and the validation of the specifications developed in this stage.

The required physical and information environment elements will have been identified in the task conditions table produced during Team/Collective Task Analysis, along with the non-team actors relevant for the task. The descriptions of sub-task inputs, outputs and the dynamic variables affecting each sub-task, captured in the task description table, provide the information source for identifying the sensors, effectors and information system interfaces and the dynamic properties of the elements in the environment.

Training Environment Diagram

The Training Environment Diagram provides a visual representation of the required elements in the training environment and the interactions between them. An example is shown in Figure 11.3. It is constructed in layers which reflect the layers in the training environment model (shown in Figure 11.3 and described in Chapter 5 previously) and uses the same box and arrow notations. The layers in the diagram can be drawn horizontally or vertically depending on what looks best. In the example in Figure 11.3, they have been drawn horizontally as this was judged to be most readable for the scale of the diagram. Where the environment is more complex, the diagram can be split into sections, for example by internal environment or by having separate diagrams to show direct interactions and interactions supported through sensors, effectors and information environment interfaces. The example in Figure 11.3 opposite is an example of a diagram that represents a section of a training environment, in this case for the Bridge team involved in Surface Protection. It is one of a set of three diagrams that capture the complete training environment for Force Protection training. The complete set can be seen in the Force Protection Case Study in Appendix A.

Training Environment Specification Tables

Having produced a Training Environment Diagram, a specification can be developed for each of the elements identified. In order to devise potential training solutions it is necessary to know not only what elements from the task environment need to be present in the training environment, but also the fidelity requirements for each element. In the specifications, fidelity requirements are broadly considered in terms of physical and functional fidelity, defined as follows:

- *physical fidelity* – the physical attributes of the element, such as look, feel, weight, size and sound
- *functional fidelity* – how the element functions in terms of the responses that it produces to the inputs that it receives.

The description of the training environment has been divided into three sections which cover:

- physical and information environment
- sensors, effectors and information environment interfaces
- non-team actors.

Tables 11.1–11.3 provide descriptions of the information required for each type of environment element and suggested table structures, along with example table entries (*shown in italics*) drawn from the Force Protection Case Study at Appendix A.

Table 11.1 describes the requirements for specifying physical and information environment components along with dynamic natural features. Physical environment elements are specified in terms of physical and functional fidelity. Information environment elements do not have physical fidelity as such, in that information may be expressed through a number of interfaces, so the description focuses on the properties of the system for communication systems and the properties of the data or content held in information systems. The properties

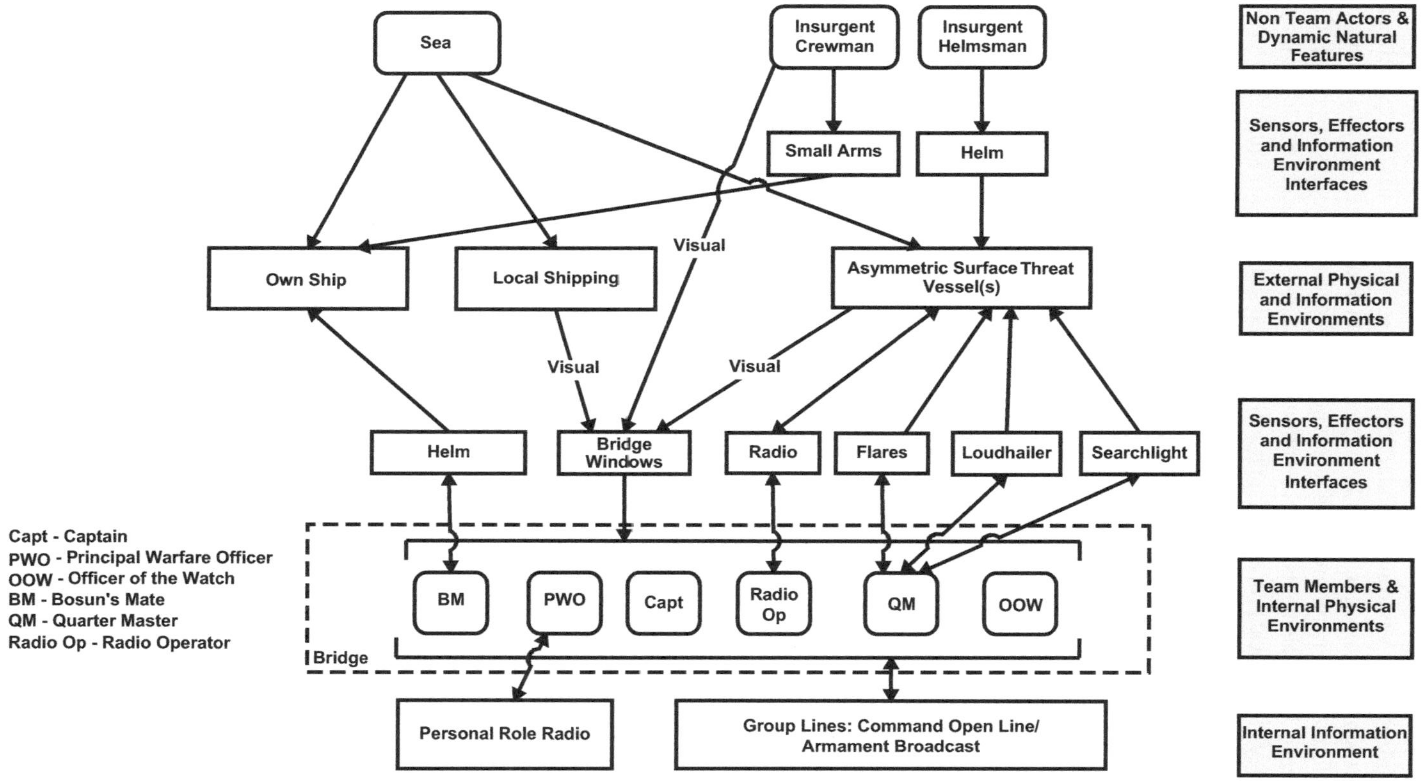
Sea
Insurgent Crewman
Insurgent Helmsman
Non Team Actors & Dynamic Natural Features
Small Arms
Helm
Sensors, Effectors and Information Environment Interfaces
Visual
Own Ship
Local Shipping
Asymmetric Surface Threat Vessel(s)
External Physical and Information Environments
Visual
Visual
Helm
Bridge Windows
Radio
Flares
Loudhailer
Searchlight
Sensors, Effectors and Information Environment Interfaces
Capt - Captain
PWO - Principal Warfare Officer
OOW - Officer of the Watch
BM - Bosun's Mate
QM - Quarter Master
Radio Op - Radio Operator
BM
PWO
Capt
Radio Op
QM
OOW
Bridge
Team Members & Internal Physical Environments
Personal Role Radio
Group Lines: Command Open Line/ Armament Broadcast
Internal Information Environment

Figure 11.3 Example Training Environment Diagram

of the information environment interfaces, such as monitor size or specialised keyboards, are described in the following section. In addition we need to describe local information environment elements such as printed map boards and reference documents. These elements of information which have a physical existence may have relevant physical properties (such as the printed size of map) which also have to be identified, if they are of relevance to the task in question.

Table 11.1 Physical and information environment element specifications

Environment element	Physical fidelity	Functional fidelity
External physical environment elements	Description of the physical attributes of the element of significance for the task being executed: includes shape, size, colour, sound and so on.	Description of the dynamic attributes of the element and its responses to actions from other elements in the environment (for example in a car accident, what sorts of damage would be sustained by the vehicles for given impact velocities?).
Asymmetric surface threat vessel	*Representative physical appearance and size for types of craft considered to be in use by asymmetric forces.*	*Speed, acceleration and rate of turn should be representative for each type of craft considered to be in use by asymmetric forces.*
Dynamic natural environment features	Description of the physical attributes of the element of significance for the task being executed: includes shape, size, colour, sound.	Description of the dynamic attributes of the element and its modes of action on other elements in the environment.
Sea	*Colour, appearance of waves.*	*The waves associated with higher sea states cause vessels to pitch and roll. The movement of asymmetric threat vessels is significant because the weapons operators have to shoot at a moving target. The movement of the own ship is significant because upper deck movement affects the weapons operators' sight picture and the effective arc of the weapons.*

Table 11.1 Physical and information environment element specifications

Environment element	Physical fidelity	Functional fidelity
Internal physical environments	The following need to be captured for each working environment identified: • the number of people to be accommodated • the sensor and effector user interfaces in the workspace • information systems interfaces (internal and external) • other elements required in the workspace (for example desks, consoles) • relative orientation and positioning of team-member workstations if deemed significant • any other physical features of the workspace that are considered relevant for task performance (for example lighting levels, background noise).	Whilst the description of the work environment is mainly concerned with physical attributes, there may be functional elements that need to be captured such as control of lighting levels.
Upper deck	**Team Members**: 3 starboard weapons operators, 3 port weapons operators. **Interfaces**: 4 GPMGs on mounts, 2 Mk 44 Mini-guns on mounts, personal role radio for each weapons operator.	Each weapon position should be at the correct apparent height above sea level. Each weapon position should have a representative field of view, including any restrictions to lines of sight that are appropriate relative to its position on the ship. The weapon operators' angle of view of the sea and surface vessels, and the alignment of the weapon, should move in response to sea state and ship's manoeuvre.

Table 11.1 Physical and information environment element specifications (*concluded*)

Environment element	Physical fidelity	Functional fidelity
Information environments (internal and external)	**Communications systems**: Communications systems such as radio networks, intercom systems, video-teleconferencing facilities need to be identified along with pertinent functional properties (such as range, number of channels available and susceptibility to interference in the case of radio networks). Example: ***Personal Role Radio Network for Upper Deck Crews***: *Representative point to point communications between upper deck weapons operators and Bridge Team members equipped with radios. Communication channel latency and signal/noise characteristics such as noise, garbling, drop out, distortion, over-talk and roundtrip delays should be representative including loss of communications quality where ship superstructure interferes with point-to-point communications.* **Information systems**: The critical features that have to be captured for information systems are the nature of the data that has to be accessible in the information environment. If live sources are not being accessed, the data will have to be created (for example photos, video, map data, text). Typically, information environments are provided as a set of software applications such as the Royal Navy Command Support System or the Army's Battlefield Information System Applications run on Bowman. Access to internet sites through web-browsers may also be included. Each of these systems needs to be identified and the nature of the data which can be sent and received (such as chat, downloading imagery) needs to be specified. Referencing out to technical manuals for these systems can provide supporting detail. The aim here is to capture the key elements that need to be provided for training purposes. Elements that are accessed directly in the internal information environment such as printed maps, telephone directories and printouts from information sources, should also be identified here. Example: ***Royal Navy Command Support System (RN CSS) Chat***: *Replication of RN CSS chat capability giving connectivity between other RN Ships in a Task Group and reach-back to Component Command HQ, with multiple chat windows to support multiple groups.*	

Table 11.2 describes the requirements for specifying sensors, effectors and information environment interfaces. This table has additional columns in order to specify both the functional and physical fidelity requirements of the user interface as well as the modes of action that they support and the associated functional fidelity requirements for these modes of action.

Table 11.2 Sensor, effector and information environment interface specifications

Interfaces	User interface physical fidelity	User interface functional fidelity	Modes of action supported and functional fidelity
Sensors and effectors	Physical properties of the interface to the sensor/effector that must be represented and to what degree of fidelity for task execution.*	Interactions that must be supported.*	Description of the modes of action supported, with associated functional fidelity requirements.
	* There may be more than one type of interface for a given sensor, providing different functionality to different users, in which case multiple interface physical and functional description pairs will be required.		
GPMG + mount	*Accurate representation of the physical attributes of the weapon and mount to include: shape and size, weight, trigger, safety catch, cocking handle, loading mechanism, sight field of view and adjustment, arcs of manoeuvre (lateral and vertical), and firing arc stops.*	*Trigger pressure, sight picture, recoil, manoeuvrability on the mount, tracer visibility.*	*Modes: firing 7.62 rounds and tracer, continuously and in bursts. Fidelity: accurate representation of: rate of fire, range, muzzle velocity, round trajectory, live round/tracer mix, effects of fall of shot on the target.*
Information environment interfaces	**Communications systems**: hardware requirements for communications systems need to be identified such as control panels, speakers, headsets and visual displays, along with critical aspects of physical fidelity for each.	**Communications systems**: functional requirements – such as channel selection, push to transmit, display adjustment – that are critical to the task must be identified.	**Communications systems**: modes of action in terms of information transmission and reception need to be identified (voice, video, data) along with associated functional requirements such as the number of channels that need to be supported.

Table 11.2 Sensor, effector and information environment interface specifications (*concluded*)

Interfaces	User interface physical fidelity	User interface functional fidelity	Modes of action supported and functional fidelity
	Information systems: these are often accessed through off-the shelf PC hardware, in which case the specification needs to capture the hardware required and any significant performance requirements. Where bespoke interfaces are used, the significant details of the interface need to be captured. Each of the applications that make up the information environment and have screen interfaces whose significant properties need to be captured (referencing out to systems manuals is an appropriate approach for capturing detail).	**Information systems**: the key aspects of functionality associated with both the hardware interface and the applications interfaces for data entry and retrieval (referencing out to systems manuals can also be made here).	**Information systems**: the required capabilities for information transfer in and out of each application should be identified here (referencing out to system manuals can be made).
Command line interface on-board ship	*Headsets with microphones. Control panel with switches to select transmit and receive on each channel.*	*Selection of multiple lines to receive (on left/ right ear phones) and a single line to transmit.*	*Transmission to all users with receive selected on a given channel. Reception from all users transmitting on lines which are selected for receive.*

Table 11.3 presents guidance for specifying the requirements for non-team actors. Whilst humans could be characterised under the headings of physical and functional fidelity, more meaningful labels are appearance and behaviour. Under the description of behaviour we also describe the enablers of behaviour – the sensors, effectors and information environment interfaces that non-team actors may use. There is also a column for knowledge and skills. Non-team actors play a critical role in training through their interactions with the environment and the team. A non-team actor without the necessary knowledge and skills is unlikely to be able to provide credible interactions, and as such will jeopardise training effectiveness.

Non-team actors require sensors and effectors to support the interactions that they have with the team. User interface fidelity is less of a concern as they are not part of the training audience, but by the same token completely unfamiliar interfaces would incur a training burden. However, the modes of action with the environment and their associated fidelity are significant because they directly affect the training audience, in the form of cues and responses to which they have to react.

Table 11.3 Non-team actor specifications

	Appearance	Behaviour	Knowledge and skills
Non-team actors	If the actor appears in the physical environment, relevant physical attributes such as manner of dress and language need to be captured.	Non-team actors play a critical role in terms of the interactions with the team and the environment which they have to support. Therefore, the actions which they take and sensors, effectors and information environment interfaces that they use must be defined. This may include cultural aspects of behaviour.	In order to play their role, non-team actors may need specific knowledge and skills, such as tactics and doctrine, voice procedures and related terminology, ability to operate sensors, effectors and information system interfaces.
Insurgent helmsman	*Dress consistent with the local population.*	*Manoeuvring of the insurgent craft in response to escalation of force measures and direct fire from the ship, consistent with known insurgent tactics.*	*Ability to helm the insurgent craft. Knowledge of insurgent tactics in response to escalation of force measures and direct fire.*

Chapter 12
Training Overlay Analysis

Overview

The aim of Training Overlay Analysis is to produce a training overlay specification, against which training options can be developed. Figure 12.1 shows the position of Training Overlay Analysis in the TCTNA process sequence.

The preceding stages in the TCTNA process have been concerned purely with analysis. In Chapter 1 we made the point that TNA is concerned with both analysis and high-level training design, and it is in the Training Overlay Analysis that the design component emerges in the development of the training strategy. As shown in the detailed training overlay model in Figure 12.2 the training strategy serves as part of the specification for the remaining components of the training overlay. In addition to the development of the training strategy, Training Overlay Analysis is also concerned with the specification of the capabilities that the training environment must provide to support training environment management, instruction and exercise management, shown in the high-level team training model displayed in Figure 12.3. It also specifies training staff capability requirements.

Process

Training Overlay Analysis has two stages:

1. development of the training strategy
2. development of detailed training overlay requirements.

Products

The following products are produced by this phase of analysis:

- *Training strategy* – describes the training strategy to be adopted and specifies:
  - training priorities and objectives
  - training audience
  - training design
  - training system capacity
  - training location
  - training provision.

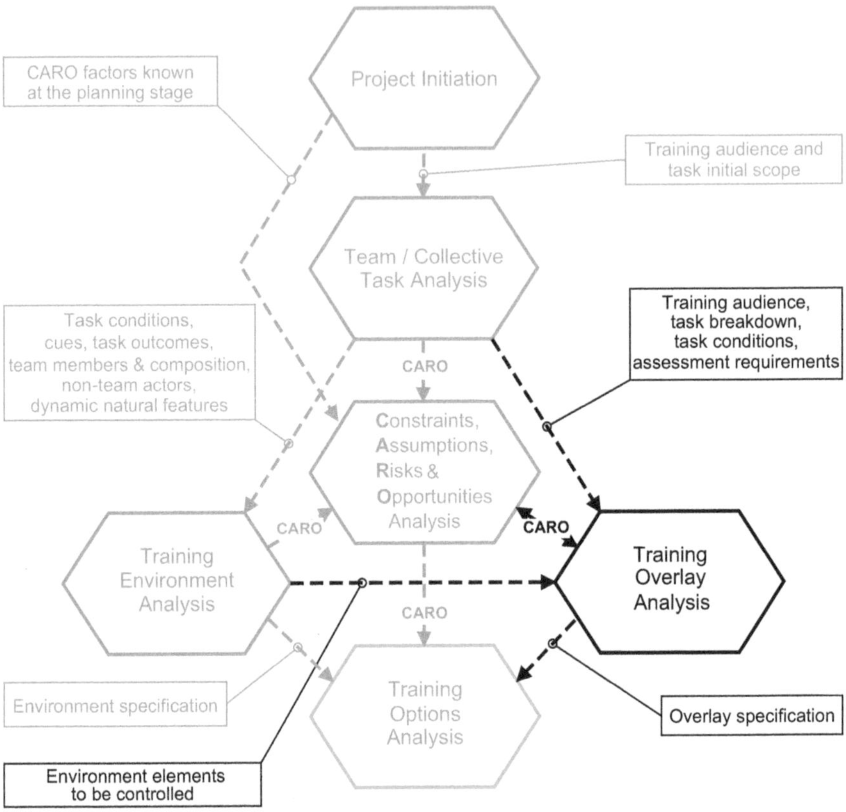

Figure 12.1 TCTNA process sequence – Training Overlay Analysis

- *Detailed training overlay requirements* – describe:
  - training staff capability requirements
  - the specification of the training overlay interfaces to the training environment.

Information Sources

Information sources for this stage of analysis include:

- the Team/Collective Task Analysis
- the Training Environment Analysis
- the CARO Analysis
- training delivery SMEs
- training systems suppliers.

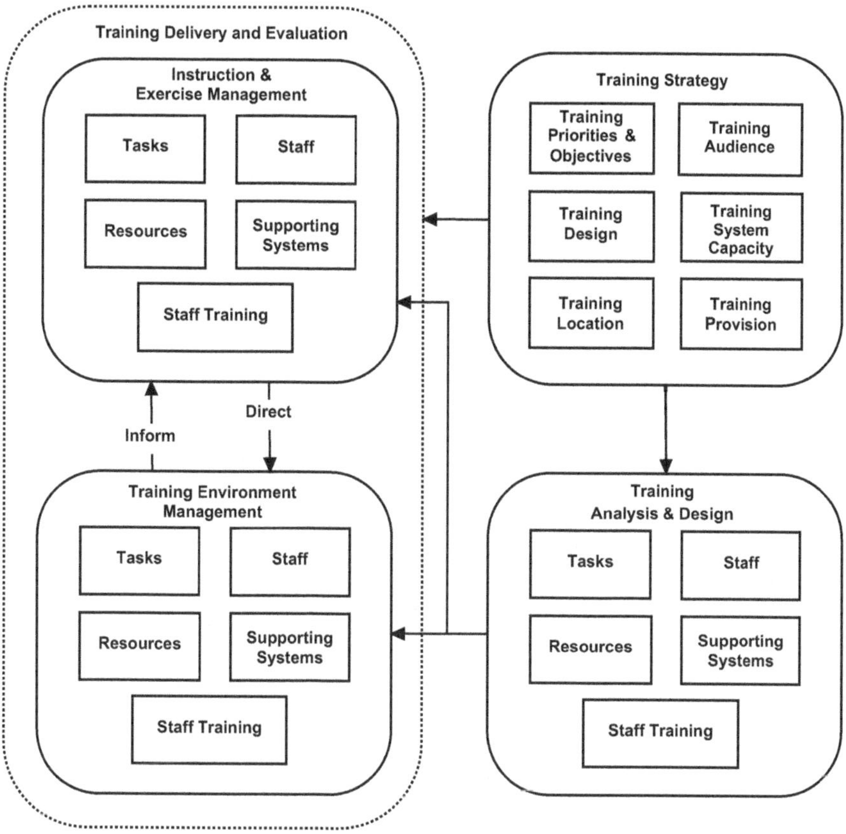

Figure 12.2 Detailed training overlay model

Determining the Training Strategy

The training strategy defines how training is to be conducted. If it is not already defined as an input to the TCTNA as an assumption then it has to be developed as part of the TCTNA process. The strategy has to take account of all of the constraints, assumptions, risks and opportunities that have been identified in the CARO Analysis.

Training Priorities

One of the considerations that have to be addressed, particularly at the early stages of the acquisition cycle, is the 'do nothing' option. The suggested approach for addressing this question is to conduct a risk analysis of not training the task or its subcomponents. An example of a risk analysis table for this purpose is shown in Table 12.1. It is based on an assessment of critical errors and consequences which have been captured in the Task Description Table. Risk is determined based on

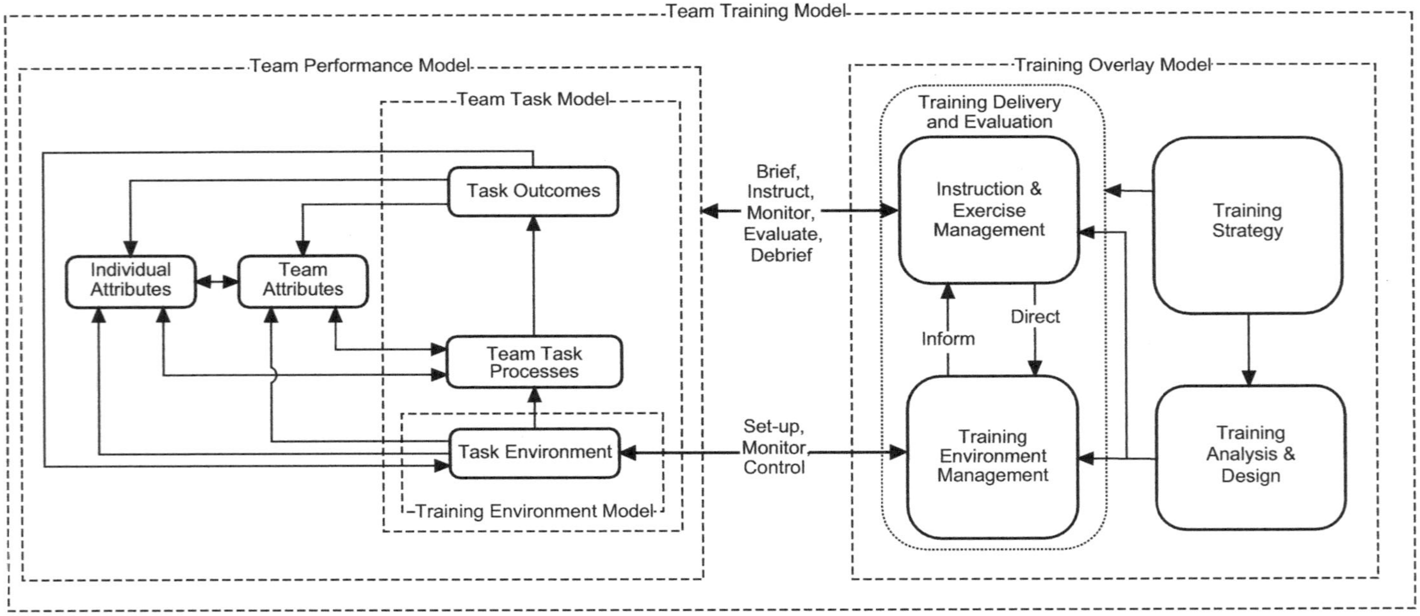

Figure 12.3 High-level team training model

Table 12.1 Training priority risk analysis table

Task	Critical error	Consequences	L* (H, M, L)	S** (H, M, L)	R*** (H, M, L)	Train (Y, N)
1.0 Protect ship from an asymmetric waterborne threat	Ineffective search	Late detection of potential threat	L	M	L	N
	Contact not correctly categorised as suspect	Upper deck weapons not deployed in time to counter the threat – damage to ship, threat to life	L	H	M	Y
	Contact not correctly categorised as hostile	Contact able to get within range to attack the ship – damage to ship, threat to life	L	H	M	Y
	Inaccurate fire from upper deck weapons	Contact able to attack the ship – damage to ship, threat to life	H	H	H	Y

* L = Likelihood, ** S = Severity, *** R = Risk

the likelihood of an error occurring and the severity consequences. In the example given, as there are risks that have been assessed as medium or high, training is recommended as a mitigation.

Where there are multiple tasks that the team or collective organisation has to undertake, priorities could be assigned based on the level of risk. These priorities could inform the relative emphasis to be placed on each activity in terms of training time and frequency.

Training Objectives

Having established what tasks are to be trained, training objectives can be developed for each component of training. Training objectives are derived from the Task Description Tables developed in the Team/Collective Task Analysis.

It is suggested that training objectives are written in the standard format of performance, conditions and standards. However, given that a critical feature of team and collective training is to train the integration of team members into a team and teams into collective organisations, it is recommended that:

- training objective performance statements capture the critical team interactions that are required for successful achievement of the task action being performed;

- the standards should include assessment criteria for these interactions;
- objectives should be developed for supporting team knowledge requirements identified within the Team/Collective Task Analysis.

Examples of objectives which meet these criteria are shown in Table 12.2.

Table 12.2 Objectives table

Overarching conditions applicable to all training objectives:
With the ship in transit, at cruising watch state, with all existing contacts classified, in an area where there is a potential asymmetric surface threat. (Detailed description provided by the training environment model.)

Performance	Conditions	Standards
Task Execution		
1.0 Carry out the Surface Investigate procedure ensuring that: a. team member responses are synchronised so that information is not lost and the process is not delayed by over-talk; b. the resulting contact classification is broadcast to the Bridge, Upper Deck Weapons and Ops Room teams to maintain team situational awareness.	Given a reported contact which is a potential threat.	Contact correctly classified. Team member responses synchronised in accordance with the Surface Investigate SOP. Classification broadcast to all appropriate teams.
Supporting KSAs		
1.0 Explain the Surface Investigate SOP	Given an example of a contact report and the data available to each team member involved in searching for contacts.	In accordance with the Surface Investigate SOP.

Training Audience

Once the training priorities have been established, it is possible to make a definitive statement about the composition of the training audience. This may be different from that originally defined at Project Initiation, as a result of a greater understanding of who exactly is involved in the tasks to be trained, and the results of prioritisation decisions. For example, when considering the training requirement for infantry battalions across an entire army, it may be decided that only a limited subset of ground forces are provided with arctic and mountain warfare training based on the perceived likelihood of the requirement for operations in that environment.

Training Design

With the training objectives defined and the training audience established, high-level training design can be carried out. We suggest that it covers:

- training structure and sequencing
- training and assessment methods
- training duration
- training environment selection.

Training structure and sequencing

The training strategy needs to define how training is to be structured. In the simplest case, the whole team or collective organisation would conduct all of its training together in one environment in the same training event. However, depending on the scale and nature of the task, the scale and nature of the team and all of the constraints that may apply, this may be neither achievable nor desirable. Therefore, consideration needs to be given to whether or not it is necessary or appropriate to decompose the overall training requirement into sub-components, with training the whole organisation together being one of those sub-components. The structuring of training should take account of:

- sub-tasks which do not require the whole team
- sub-tasks which have different environmental requirements
- sub-tasks which are prone to skill fade and require additional practice
- sub-tasks that are separated in time
- Sub-tasks that are conducted in parallel which would benefit from separate practice.

Training methods

Having established the structure and sequencing of training, training and assessment methods can be considered. Methods need to be matched to the nature of the task and its coordination requirements, and the experience level of the teams to be trained. This information can be drawn from the Team/Collective Task Analysis. Consideration also has to be given to the measures that are to be used. These can be derived from the process and outcome assessment criteria identified in the Task Description Tables developed in the Team/Collective Task Analysis. The associated provision of feedback to the training audience is of critical importance in the training process. In team and collective training events formative and summative assessment is typically based on observation of task performance with feedback being provided thorough feedback sessions in the form of AARs. There are a number of factors to be considered in the AAR process including:

- location
- levels at which AAR are delivered
- duration
- frequency
- time allocated to fix faults identified in the AAR before exercise play continues
- systems to support AAR.

These factors influence training duration, resource requirements and staff requirements.

Another key component which has to be identified at this stage is whether there will be a requirement for training analysis and design activity prior to each training event.

Training duration

In an ideal situation the duration of training would be sufficiently long for all aspects of the task to be trained to be repeated until the desired level of competence is achieved. For example, a military battle staff will typically spend a period of time developing an operational plan at the end of which it publishes an operation order. This is then followed by a period of time conducting the operation, during which it is engaged in both the command and control of its subordinates executing the operation whilst simultaneously producing detailed revisions of the plan as the operation unfolds. SME judgement may be that the training ideal would be a five-day planning period followed by 10 days of execution, giving a total duration of 15 days for the exercise. In practice, exercise duration is likely to be a compromise between the training ideal and constraints such as training audience availability, training environment availability and cost. Exercise duration could be constrained to being 10 days in duration for example. Training duration informs the calculations for training system capacity.

Training environment selection

The selection of training environments is informed by the training structure, training methods and the features of the task environment that have to be supported (as defined by the Training Environment Specification developed in the Training Environment Analysis). It also has to take account of the pertinent CARO, such as the availability of live training environments and extant synthetic training environments; training environments being considered for related individual, team and collective training components; and training policy. If the training structure is such that the training audience is split into sub-groups for some elements of training, or the training is split by sub-task, there may be different training environment requirements for each sub-component of training (a sub-component may only require a subset of training environment elements). This is shown graphically in Figure 12.4 for a set of sub-tasks that have different training environment requirements.

An example of such a sub-division occurs in flight crew training. The procedural elements of system operation and instrument flying do not require outside world visual cues or motion cues, so this element of training is often conducted in flight training devices which replicate the aircraft systems but do not have a visual system or motion platform. Most flight simulator manufacturers offer this sort of training device. On the other hand, separate training environments may not be a cost effective solution. For example, when an in-house system for training Royal Air Force TriStar aircrew was developed, it was identified that utilisation of the full flight simulator, which had a motion platform and visual system, would be sufficiently light that the additional cost of a flight training device was not justified. Consequently, procedural training was conducted in the full flight simulator.

At this stage the broad capability requirements can be specified, though these may be refined through the process of Training Options Analysis at the next stage of the TCTNA process.

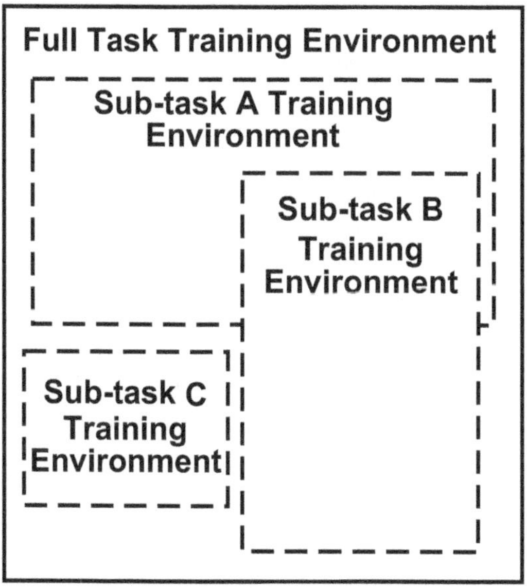

Figure 12.4 Sub-task training environment requirements

Training System Capacity

The training throughput has to be determined for the training solution. This is a function of the number of teams/collective organisations that have to be trained and the rate at which they have to be trained. This needs to take account of:

- initial surge training requirements (such as conversion of teams onto a new platform)
- steady state training requirements, including refresher training requirements.

Training Provision

Training provision may be made wholly from internal resources, or contracted out in whole or in part. This may be directed by organisational policy. If not, a preferred option may be stated at this stage, or left open until training options have been considered in more detail.

Detailed Training Overlay Requirements

Having established the training strategy, detailed training overlay requirements to support its implementation can be established. Whilst the full details of training overlay provision (as shown in the training overlay model in Figure 12.2) cannot be determined until training options are developed, a number of significant requirements can be specified at this stage. These are the training staff capability requirements and the training overlay interface to the training environment.

Training Staff Capability Requirements

Whilst the numbers of staff required to deliver each component of the training strategy cannot be established until the exact implementation is known (that is, when an option is defined in detail), at this stage it is possible to determine the likely mix of staff with different types of qualifications and experience that will be required to deliver the range of instructor tasks required for training analysis, design and delivery. For example, in the Maritime Force Protection example the following categories of training staff would be required:

- Lieutenant Commander (Lt Cdr) Staff Warfare Officer (Above Water Warfare specialist) to plan the exercise and brief, monitor and debrief the PWOs and the Captain
- Lt Cdr Staff Warfare Officer (Navigation specialist) to brief, monitor and debrief the Bridge Team (with the exception of the PWO and Captain)
- Warrant Officer (WO) / Chief Petty Officer (CPO) Above Water Tactics Instructor to brief, monitor and debrief the Ops Room Team ratings
- WO/CPO Above Water Warfare (Gunner) to brief, monitor and debrief the Upper Deck Team.

Having training staff with appropriate qualifications and levels of experience is central to the delivery of effective training.

Training Overlay Interfaces to the Training Environment

There are three key aspects of the training overlay interface to the training environment that have to be specified:

- the features that support training environment management
- the features that support instruction and exercise management
- additional role-player requirements for part-task training.

The development of these requirements is informed by the Training Environment Specification (which details all of the elements in the training environment) and the outputs of the Team/Collective Task Analysis (which include the data capture requirements for assessment and the nature of the scenarios in which the task is conducted). Ensuring that task scenarios can be replicated is critical if task fidelity in the training system is to be achieved. Table 12.3 shows a format that can be used to specify training overlay interface requirements with some examples shown from the Maritime Force Protection case study.

Table 12.3 Training overlay interface requirements table – example entries

Training environment element	Training overlay requirements to interface to the training environment	
	Environment management	**Instruction and exercise management**
Training audience	N/A	Brief, monitor, record, and debrief performance. Replay of actions for debrief.
Asymmetric surface threat	Set up of single or multiple contacts, control of course, speed and actions. View track and current position.	Recording of course and actions for debrief. View track and current position.
Sea state	Set the sea state for the exercise.	N/A
Upper deck weapons	Set/reset and monitor ammunition levels.	Recording of hits on target for debrief.

The totality of these features should provide the necessary capability to construct and deliver appropriate training scenarios. However, this requirement can be reinforced by including a scenario specification. An example is shown in Table 12.4, using the format for capturing scenarios that was used in the Team/ Collective Task Analysis. If the requirement for multiple training environments has been identified in order to cater for training sub-tasks separately, there may be a need to define scenario subsets appropriate to the sub-tasks for use in these separate environments.

Table 12.4 Scenario specification

Scenario	Maritime Force Protection of a Type 23 Frigate against an asymmetric threat, whilst at cruising watch in open waters
Physical	Open waters leading to confined waters such as straits, transits and harbour approaches. Variable sea-state and possibility of poor visibility due to local weather conditions.
Information	Personal role radios and internal command lines on-board. Royal Navy Command Support System chat connection to higher command.
Human	*Civilian* Local shipping, possibility of smugglers in fast-moving, small craft. Operators of local shipping (not hostile). *Military* Friendly forces: Command Headquarters. Neutral forces: None Enemy forces: Asymmetric threat of non-conventional forces in small, fast craft equipped with small arms, rocket propelled grenades or operating waterborne improvised explosive devices. Possibility of simultaneous attack from multiple craft. Resources: • radar • electro-optical sensors • General Purpose Machine Guns (GPMG) x 4 • Miniguns x 2 • flares, searchlight and loudhailer.
Events	Small, unidentified, fast craft heads towards ship on collision heading, does not respond to escalation of force measures, turns to follow ship when ship changes heading and speed, opens fire on ship with small arms and rocket propelled grenades (asymmetric threat). Multiple small, unidentified craft heading to the ship as above. These may be: • on the same approach vector at the same time • on different approach vectors at the same time • on different approach vectors at different times. Small, unidentified, fast craft, does not respond to escalation of force measures, turns to follow ship when ship changes heading and speed, maintains collision course with ship (waterborne improvised explosive device). Smugglers in small, fast craft heading towards ship, do not respond to escalation of force measures, but maintain their heading. Fishing vessels on collision heading with the ship respond to escalation of force measures.

Where the training strategy includes sub components of the whole task to be trained separately, there will be a requirement for role players additional to those specified as Non-Team Actors in the Training Environment Analysis. This is illustrated in Figure 12.5 for the training of a whole force composed of a Command Element and Force Elements. Force Element training will require an additional Higher Control (HICON) role-player team to represent the Command Element. Similarly, Command Element training will require a LOCON role-player team to represent the Force Elements. These additional requirements should be specified at this point, using the same format as that for Non-Team Actors in the environment specification developed in the Training Environment Analysis.

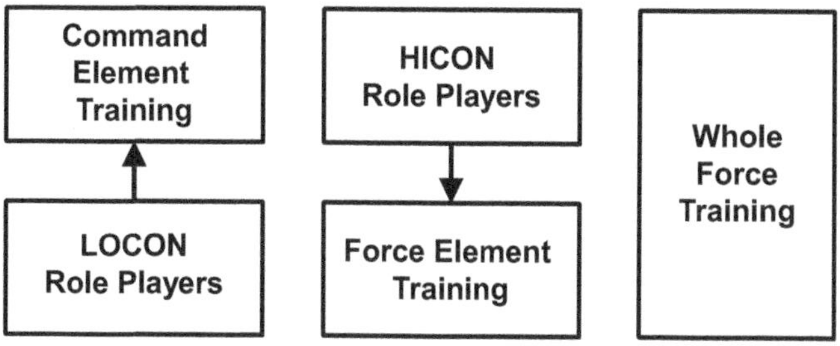

Figure 12.5 Additional role-player requirements for part task training

Chapter 13
Training Options Analysis

Overview

The aim of Training Options Analysis is to identify a set of options to deliver the required training in accordance with the training requirements defined by Collective Task Analysis, Training Environment Analysis and Training Overlay Analysis, taking account of CARO that have been identified. These options are then evaluated against an agreed set of criteria and a recommendation is made about which option(s) should be pursued.

Training Options Analysis is the last process in the TCTNA process sequence, as shown in Figure 13.1.

Process

Training Options Analysis comprises the following steps:

1. Development of option evaluation criteria
2. Training Option Definition, which involves:
    a. identification of options
    b. option description
    c. option costing.
3. Training Option Evaluation

Products

- *Selection criteria* – definition of the criteria to be used in evaluating options such as cost, development time and risk
- *Option descriptions* – detailed description of each option which explains how the training environment and training overlay are to be instantiated, including manning requirements
- *Option evaluation table(s)* – an assessment of each option against the specifications developed in Training Environment Analysis and Training Overlay Analysis, and against the criteria developed in selection criteria definition
- *Training option recommendation* – rank-ordered recommendations of which option(s) to pursue.

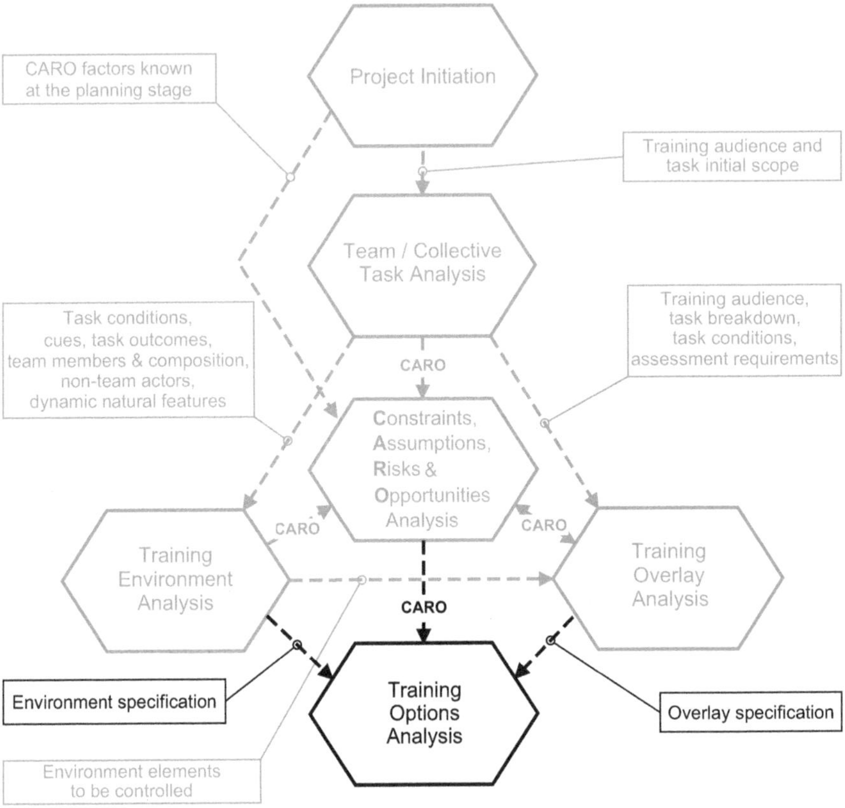

Figure 13.1 TCTNA process sequence – Training Options Analysis

Information Sources

Information sources for this stage of analysis include the following:

- Training Environment Analysis
- Training Overlay Analysis
- CARO Analysis
- current training system technical documentation describing system capabilities
- training delivery SMEs
- industry training systems SMEs.

Development of Option Evaluation Criteria

In order to evaluate putative options a set of evaluation criteria have to be developed. Criteria may be categorised as essential and desirable. Criteria should be based on the solution to meet or address through life factors such as:

- performance
  - coverage of the training objectives
  - provision of the required range of conditions in the training environment
  - support for the training overlay functions.
- costs
- ability to support the required training throughput (surge and steady state)
- flexibility (for example adaptability to meet future changes)
- interoperability (for example potential for a synthetic training system to be connected to other synthetic training systems)
- development time relative to key dates such as ISD
- training requirements for staff to operate and maintain the training solution.

Additional criteria may be derived from the constraints, assumptions, risks and opportunities that have been derived in the CARO Analysis. Key stakeholders should be consulted in the process of developing the criteria and the criteria should be agreed by the key stakeholders.

Training Option Definition

Identification of Options

Training options may include:

- the use of existing training capability
- the adaptation or extension of existing training capability
- the development of new capability
- the use of another nation's training capability.

Identifying suitable training options can be a challenging process, and may require consultations with industrial training systems providers to understand the current 'art of the possible'. The CARO that have been identified in the preceding stages of analysis must be considered when exploring possible options.

Where the organisation already has similar or related training capabilities, a useful starting point can be to evaluate the capabilities of these extant systems to support the required training, particularly in terms of the capabilities of the training environments available. Table 13.1 shows a tabular format that can be used for recording the capability of such extant systems. The left hand column lists all of the training environment elements that are required, as detailed in the training environment specification. Ticks and crosses have been used to denote if the required element is fully supported or not supported at all. A text explanation has been provided where an element is only partially supported. These evaluations are based upon consideration of the requirements for each element as specified in both the training environment specification and the training overlay specification. Shading has also been used to highlight areas of non-compliance (a traffic-light

Table 13.1 Example option evaluation table for the suitability of existing systems for Maritime Force Protection training

Environment Elements	Existing systems (evaluation of fidelity and overlay requirements of the training environment)				
	Live with Ship's Rigid Inflatable Boat (RIB) as the asymmetric threat	Live with contracted RIBs as the asymmetric threat	Maritime synthetic training system (MaST)	Close range weapons trainer (simulator)	Bridge trainer
Sea	✓	✓	x	✓	✓
Insurgent crewman	✓	✓	x	✓	✓
Insurgent helmsman	✓	✓	x	✓	✓
Local shipping	x	x	✓	✓	✓
Own ship	✓	✓	✓	✓	✓
Asymmetric threat	Limited number	✓	✓	✓	✓
Asymmetric threat small arms	Critically limited – no weapons effects	Critically limited – no weapons effects	x	✓	x
Radar	✓	✓	✓	x	x
ESM	✓	✓	✓	x	x
GPEOD	✓	✓	✓	x	x
Helm	✓	✓	✓	x	✓
Bridge windows	✓	✓		x	✓
Radio	✓	✓	✓	x	x
GPMG	Critically limited – no weapons effects	Critically limited – no weapons effects	x	Limited – only 2 GPMGs	x

Key:

✓	Fully meets the requirement
text	Partially meets the requirement
x	Does not meet the requirement / Not supported

colour-coding system could be used to good effect here with red for non-compliant, green for fully compliant and amber for partially compliant).

The existing systems listed in Table 13.1, whilst loosely based on extant systems in the Royal Navy, are purely illustrative and do not constitute an accurate or complete representation of real life systems. Equally, the training objectives shown are hypothetical based upon the example analysis conducted in the Case Study at Appendix A, using the process described in Chapter 12. However, the table does show how the attributes of systems can be collated and compared to inform option generation.

The advantage of producing a table of this type is that it makes it easy to identify training capability gaps and potential options for combining extant capabilities to address the new training requirement. In the example in Table 13.1 it can be seen that in principle a combination of the Maritime Synthetic Training System and the Close Range Weapons Trainer could address all of the requirements, with the exception of the number of GPMGs required. The feasibility of actually connecting these training systems together, along with their availability and location would also need to be considered. This is addressed in the detailed development of training options described below.

Option Description

Each potential training option needs to be described in sufficient detail for it to be costed and then evaluated. Table 13.2 provides some suggested headings for such descriptions.

Headings 3–7 are based on the detail of the training overlay model. Headings 8–14 are based on the supporting capability areas which are likely to be involved in the implementation of the training system (we have used the UK MoD Defence Lines of Development dimensions in the table, but these can be replaced with those of the owning organisation as appropriate).

Option Costing

Cost is a critical factor in evaluating training options. Table 13.3 gives examples of sources of cost broken into capital costs, which are one-off costs associated with the acquisition of the training capability, and annual through-life support costs.

Training Option Evaluation

Comparison of Options

Having identified the option evaluation criteria and developed a set of training options, the options can then be evaluated against these criteria. Table 13.4 provides an example of a format that could be used for recording the results of the

Table 13.2 Headings for option descriptions

Heading	Description
1. Overview	Introductory description of the option.
2. Training environment provision	Detailed description of how the required training environment is to be provided and how it will support the required training delivery and training environment management functions. Its location should also be specified.
3. Staff tasks	Details of the tasks to be carried out in analysis and design, instruction and exercise management, and training environment management (which drive staff numbers).
4. Staff requirements	Details of the numbers, qualification and experience of training and supporting staff (including role-players) required to operate, maintain and support the training solution.
5. Supporting systems	Details of the supporting systems required to support analysis and design and training delivery and evaluation.
6. Resources	Details of the resources required to support analysis and design and training delivery and evaluation.
7. Staff training requirements	Training requirements for training staff and supporting staff.
8. Training linkages	Description of how the training will be integrated with other components in the individual-team-collective training continuum.
9. Equipment	Equipment requirements for the training solution.
10. Infrastructure	Infrastructure requirements and changes necessary to accommodate the training solution.
11. Information	Description of how training information is managed, requirements for procurement of data (for example geographic data, manufacturers' system data).
12. Organisation	The responsibilities of organisations to procure, resource, manage and prioritise use of the training solution.
13. Logistics	Logistics requirements: such as transportation (including of personnel), maintenance, operation of the training system.
14. Interoperability	Description of the interoperability of the training system with other training systems.

Table 13.3 Potential sources of whole life costs

Capital cost items	Annual through-life support costs
Training media	Live and workplace training
Integration into existing training solutions	Instructors
Training support systems	Train the trainer courses
First of class training	Training support staff
Reference documentation	Training administrators
Training design	Travel and subsistence
New or refurbished training infrastructure	Consumables and utilities
IT infrastructure	Training design
Risk mitigation	Training publications
	Facilities management

Table 13.4 Option evaluation table

	Options		
Evaluation criteria	**Live with ship's RIB as the asymmetric threat**	**Live + remote controlled boat**	**Enhanced close range weapons trainer + MaST**
1. Availability	✓	✓	High at Devonport
2. Accessibility	✓	✓	High at Devonport
3. Cost	Low	Medium	High
Training objectives	**Training objective coverage**		
TO 1 Search for threat	✓	✓	✓
TO 5 Counter threat	No weapons effects	No weapons effects from threat	✓
Environment elements	**Evaluation of environment and overlay requirements**		
Sea	✓	✓	✓
Insurgent crewman	✓	x	✓
Insurgent helmsman	✓	x	✓
Local shipping	x	x	✓
Own ship	✓	✓	✓
Asymmetric threat	Limited	1/3 scale	✓
Asymmetric threat small arms	Critically limited – no weapons effects	x	✓
GPMG mode of action	x No fall of shot	✓	✓
GPMG overlay reqts	x No hit indication	✓	✓
Mk44 HMI	✓	✓	✓
MK44 mode of action	x No fall of shot	✓	✓
Mk44 overlay reqts	x No hit indication	✓	✓

Key:

✓	Fully meets the requirement
text	Partially meets the requirement
x	Does not meet the requirement

evaluation. It is a simple extension of Table 13.1, with the addition of the agreed evaluation criteria and the training objectives. Ticks and crosses have been used in the table to indicate full compliance and non-compliance respectively, with text descriptions if an option is only partially compliant. Shading or colour coding (such as traffic-light coding) can also be used to advantage to highlight areas of compliance and non-compliance. More detail about option compliance can be provided in accompanying notes if needed. If a more formal decision analysis process is required, numerical ratings for each criterion can be used and these can be weighted if some criteria are deemed to be more important than others.

Training Option Recommendations

Once the evaluation of the options has been completed, a rank-ordered set of recommended options can be proposed. A supporting justification for the rank ordering can be derived from the compliance statements in the option evaluation table and any accompanying notes.

Appendix A
Maritime Force Protection Case Study

Introduction

This case study is designed to show the detailed application of the analysis components of the TCTNA methodology. All elements of the method are shown with the exception of the derivation of detailed costings.

The case study is focused on the training required for ships' personnel to protect the ship against the threat from asymmetric forces using high-speed, small craft such as Rigid Inflatable Boats (RIBs) with powerful engines. Typically, they would be armed with small arms and rocket propelled grenades (RPGs). They may also employ water-borne improvised explosive devices (IEDs), which are the same craft but packed with explosives set up to detonate on impact with the ship. The case study focuses on how a Type 23 Frigate would employ its close range weapons to counter such a threat. Type 23 Frigates are armed with four GPMGs and two Mk44 Miniguns for countering such threats as shown in Figure A1.

On a Type 23 Frigate, both a GPMG and a Minigun are located on each side of the bridge on the bridge wings. In addition, two further GPMGs are located at the stern of the ship, on either side of the flight deck. The task involves staff in the operations room manning sensors, on the bridge carrying out visual lookout and controlling the ship and on the upper deck manning the weapons. The Captain and the PWO on the bridge direct the tactical engagement.

A full TCTNA of Force Protection would take into account the requirements for all vessel types, in particular Type 45 and Type 26, and would also consider the more complex case of attacks by swarms (up to 20 or more) of larger Fast Inshore Attack Craft. However, this would have made the case study unnecessarily large for our purposes. Whilst the consideration of only one vessel type is an artificial constraint, the case study demonstrates the use of all of the analytical tools and representations in the TCTNA method.

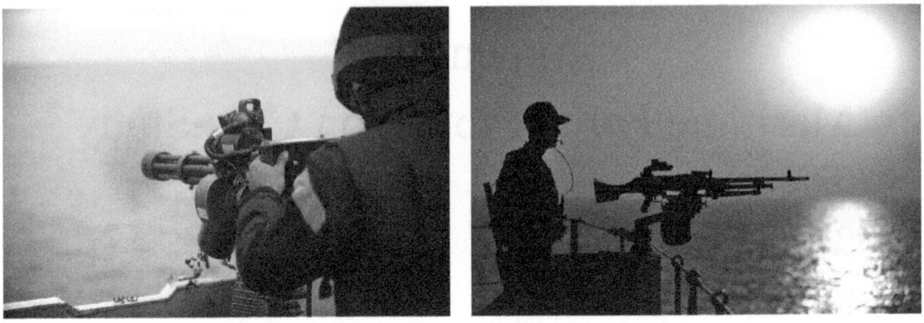

Figure A1 GPMG (a, left) and Mk44 Minigun (b, right) mounted on a Type 23 Frigate

Maritime Force Protection training is carried out within Basic Operational Sea Training (BOST) and Directed Continuation Training (DCT) and when under way to operational deployments where an asymmetric threat is possible. BOST and DCT are conducted by Flag Officer Sea Training (FOST) staff based at Her Majesty's Naval Base (HMNB) Devonport in South West England. BOST is conducted whenever a ship comes out of refit or is newly commissioned. DCT is conducted periodically after BOST until the ship goes back into refit.

Team/Collective Training Analysis

Task Scope

Scope is limited to the ship in transit in cruising watches in open or confined waters where there is a potential threat of terrorist elements operating small craft in small numbers. These may be armed with small arms, rocket propelled grenades or waterborne improvised explosive devices.

Team Organisation Description

There are three teams involved: the Upper Deck Team (UDT), the Bridge Team and the Ops Room Team. The organisation of these teams is shown in Figure A2. Team member roles are described in Table A1.

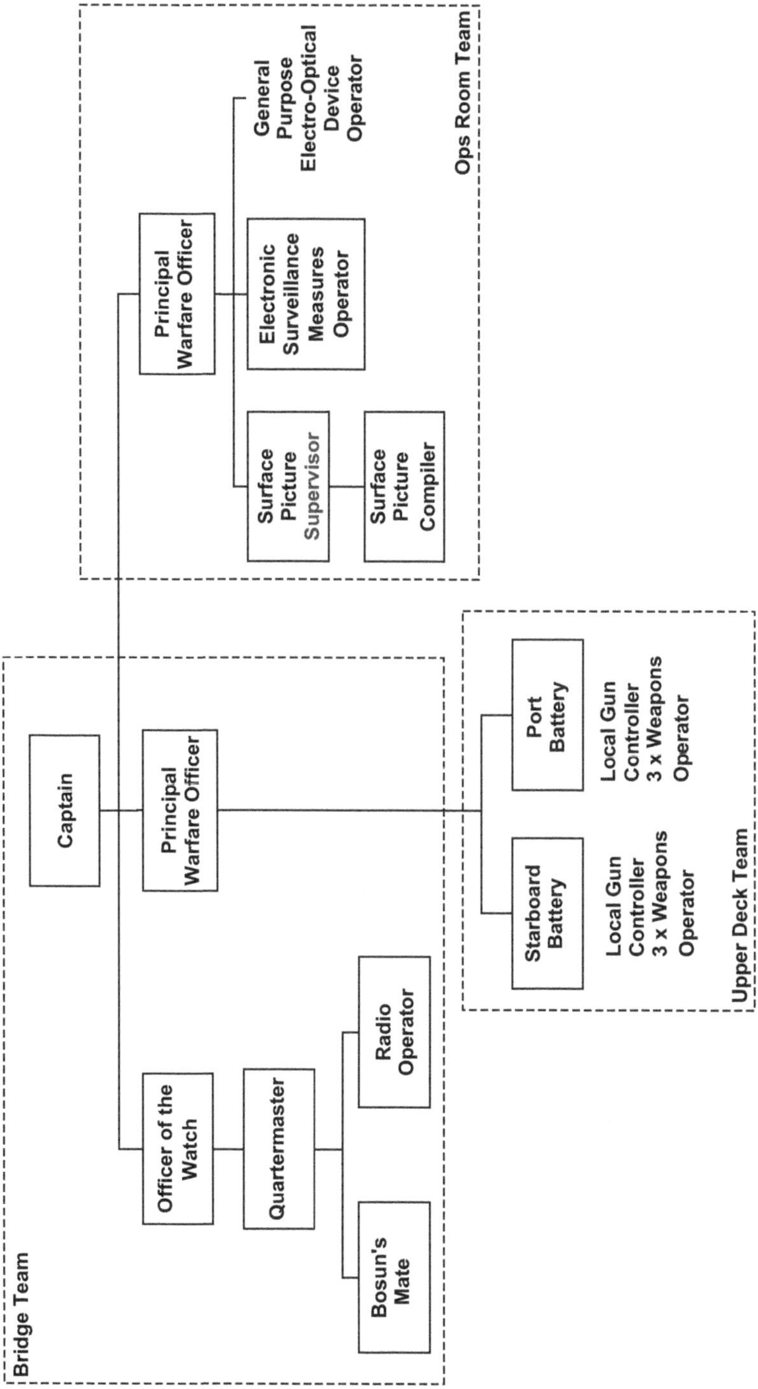

Figure A2 Organisational chart

Table A1 Team roles

Role	Description
Captain	Command of the ship. Guidance to the PWO on tactical decision making and application of ROE.
PWO	Tactical control of the ship and integrated use of the sensors and weapons systems. Application of ROE. Issuing weapons control orders.
OOW	Responsible for the safe conduct of the ship, controlling the ship and initiating force protection measures in the absence of the Captain and PWO.
Quartermaster (QM)	Supervision of the Bosun's Mate and Radio Operator, escalation of force measure (searchlight and flares), look out.
Bosun's Mate (BM)	Responsible for controlling heading and speed of the ship as directed by the OOW.
Radio Operator	Radio operation and lookout can be one and the same.
Local Gun Controller (LGC)	Relaying of weapons control orders to weapons operators and supervision of their actions in carrying them out. Prioritisation of targets. Lookout.
Weapons Operator	Search for and evaluate threats, prioritise targets. Operate weapons (GPMG, Minigun) including loading, firing, reloading and conducting stoppage drills.
Surface Picture Compiler (SPC)	Compilation of surface picture, radar search.
Surface Picture Supervisor (SPS)	Supervision of compilation of the surface picture.
Electronic Surveillance Measures (ESM) Operator	ESM search.
General Purpose Electro Optical Device Operator	Visual search using the General Purpose Electro Optical Device (GPEOD).

Scenario Description

Table A2 Scenario Table

Scenario	Maritime Force Protection of a Type 23 Frigate against an asymmetric threat, whilst at cruising watch in open waters.
Human	Operators of local shipping (not hostile).
Physical	Open waters leading to confined waters such as straits, transits and harbour approaches. Variable sea-state and possibility of poor visibility due to local weather conditions.
Information	Personal role radios and internal command lines on-board. Royal Navy Command Support System chat connection to higher command.
Civilian	Local shipping, possibility of smugglers in fast-moving, small craft.
Military	Friendly forces: Command Headquarters. Neutral forces: None. Enemy forces: Asymmetric threat of non-conventional forces in small, fast craft equipped with small arms, rocket propelled grenades, or operating waterborne improvised explosive devices. Possibility of simultaneous attack from multiple craft. Resources: • radar • electro-optical sensors • GPMG x 4 • Miniguns x2 • flares, searchlight and loudhailer.
Events	Small, unidentified, fast craft heads towards ship on collision heading, does not respond to escalation of force measures, turns to follow ship when ship changes heading and speed, opens fire on ship with small arms and rocket propelled grenades. Small, unidentified, fast craft does not respond to escalation of force measures, turns to follow ship when ship changes heading and speed, maintains collision course with ship (waterborne IED). Smugglers in small, fast craft heading towards ship, do not respond to escalation of force measures, but maintain their heading. Fishing vessels on collision heading with the ship respond to escalation of force measures.

Task Network Diagram

Figures A3 and A4 provide a decomposition of the surface protection task. Threat evaluation (task A5.0 in Figure A3) was considered to be sufficiently complex that it merited further decomposition as shown in Figure A4. Task A6 Counter Threat could also have been decomposed further, but in this instance it was decided that sufficient detail could be adequately captured in the Task Description Table (Table A3), and so it was not merited.

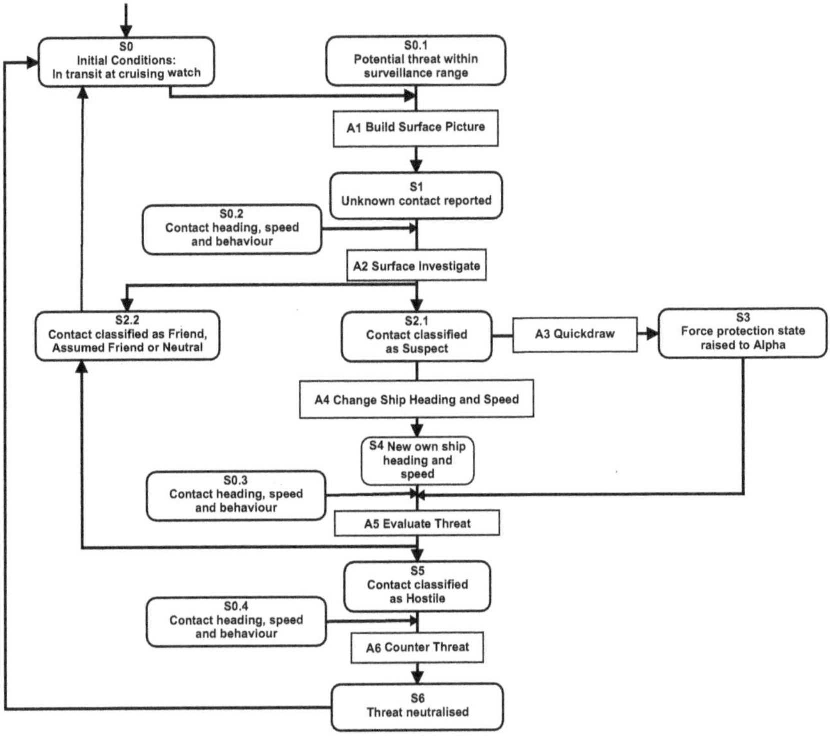

Figure A3 Top-level breakdown of the Surface Protection against an Asymmetric Threat Task

The surface protection task was found to be in part procedural and in part adaptive. Procedural elements included tasks A2 Surface Investigate and A3 Quickdraw, for which there are established SOPs. Neutralising the threat requires the application of tactical principles regarding manoeuvring the ship and the

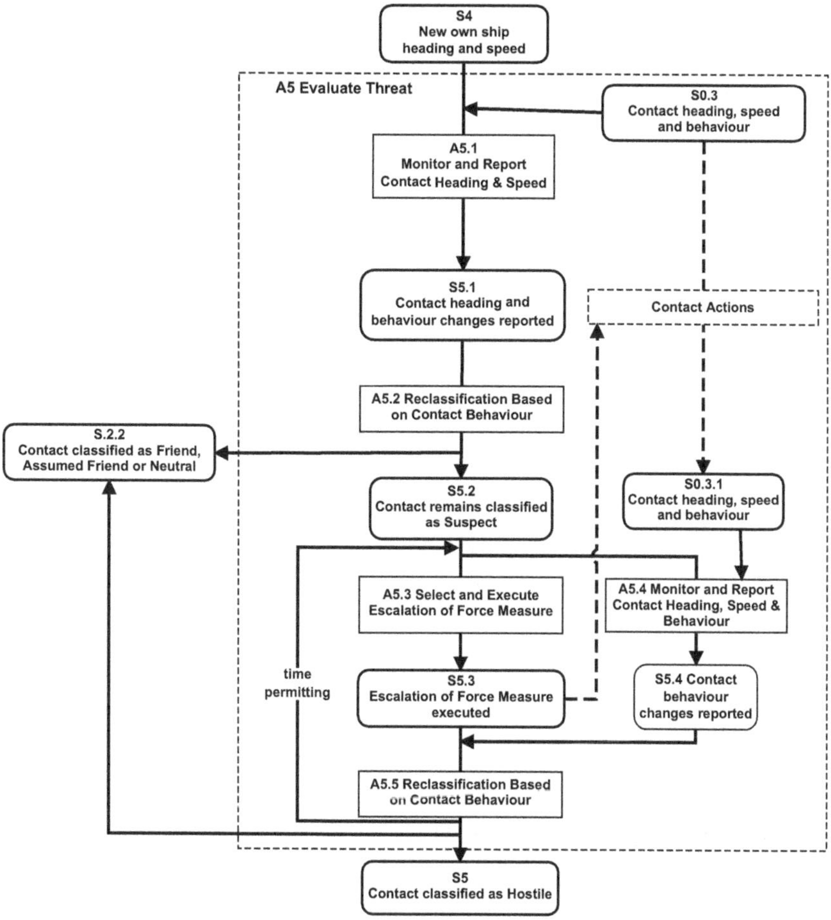

Figure A4 Expansion of the Evaluate Threat task

direction of available firepower, which becomes more complex the greater the number of threats that there are, and was therefore considered to be an adaptive task. Furthermore, it is possible that the multiple threats could arise:

- on the same approach vector at the same time
- on different approach vectors at the same time
- on different approach vectors at different times.

The team would have to demonstrate adaptive coordination in order to deal with these threat situations.

Table A3 Task Description Table

A1 Build Surface Picture		
Initial conditions and cues S0 Neutral shipping; Coastline; Maritime zones, channels and lanes. S0.1 Potential threat.	**Actions (sensing, doing, communication)** *Participants:* Bridge Team and Ops Room Team. *Process description (including modes of action):* Bridge Team conduct visual search, Ops Room Team use radar, GPEOD, and ESM sensors to conduct radar and ESM search. Contacts reported by broadcast to members of all teams. *Key team interactions:* Broadcast of the contact to all team members to maintain situational awareness.	**Outcomes** S1 Contact report giving range, bearing, heading and speed.
Dynamic conditions Volume of shipping in the area; Multiple, simultaneous threats; Sea state; Visibility.	**Process assessment criteria** Comprehensiveness of the search though all arcs. **Data capture requirements** Observation and recording of team actions and communication.	**Outcome assessment criteria** Timeliness, completeness accuracy of the report, on the correct circuits. **Data capture requirements** Recording of contact reports, time and originator.
A2 Surface Investigate		
Initial conditions and cues S0.2 Contact heading, range speed and behaviour.	**Actions (sensing, doing, communication)** *Participants:* Bridge Team and Ops Room Team. *Process description (including modes of action):* Each team member reports the information they have about the contact in accordance with the Surface Investigate SOP,	**Outcomes** S2.1 Contact classified as suspect. S2.2 Contact classified as friend, assumed friend or neutral.

Table A3 Task Description Table

A2 Surface Investigate (cont)		
	which specifies the sequence of reporting. OOW synthesises the information and determines the classification. SPC manually enters contact if not seen on radar and adds classification. *Key team interactions:* Coordinated reporting ensuring no over-talk, broadcast of classification so all team members have the required SA about the contact.	
Dynamic conditions Volume of shipping in the area; Multiple, simultaneous threats; Sea state; Visibility (fog, rain).	**Process assessment criteria** Comprehensiveness of the search though all arcs. **Data capture requirements** Observation and recording of team actions and communication.	**Outcome assessment criteria** Timeliness, completeness accuracy of the report, on the correct circuits. **Data capture requirements** Recording of contact reports, time and originator. Surface picture.
A3 Quickdraw		
Initial conditions and cues S2.1 Contact classified as Suspect.	**Actions (sensing, doing, communication)** *Participants:* OOW, Captain, UDT. *Process description (including modes of action):* Quickdraw SOP called (broadcast) by OOW and Captain called to bridge if not present. UDT collects weapons, installs them on mounts and arms them (Force Protection State Alpha). Captain goes to bridge. *Key team interactions:* UDT reports when actions complete so that the Bridge Team are aware of the new state being reached.	**Outcomes** S3 Force protection state at Alpha.

Table A3 Task Description Table

A3 Quickdraw (cont)		
Dynamic conditions Volume of shipping in the area; Multiple, simultaneous threats; Sea state; Visibility.	**Process assessment criteria** Comprehensiveness of the search though all arcs. **Data capture requirements** Observation and recording of team actions and communication.	**Outcome assessment criteria** Timeliness, completeness accuracy of the report, on the correct circuits. **Data capture requirements** Recording of contact reports, time and originator.
A4 Change Ship Heading and Speed		
Initial conditions and cues S0 Neutral shipping; Coastline; Maritime zones, channels and lanes. S2.1 Contact classified as suspect.	**Actions (sensing, doing, communication)** *Participants:* OOW, Bosun's Mate. *Process description (including modes of action):* OOW orders heading and speed change, Bosun's Mate changes heading and speed at the helm. *Key team interactions:* Checking of revised heading and speed by OOW.	**Outcomes** S4 Own ship new heading and speed.
Dynamic conditions Sea state.	**Process assessment criteria** Correct orders and checking sequence. **Data capture requirements** Observation of the team and communication.	**Outcome assessment criteria** Appropriate magnitude of heading and speed change ensuring weapons are taken out of effect by the degree of heel during the turn. **Data capture requirements** Recording of heading and speed change.
A5 Evaluate Threat		
A5.1 Monitor and Report Contact Heading and Speed		
Initial conditions and cues S0.3 Contact heading, speed and behaviour.	**Actions (sensing, doing, communication)** *Participants:* Bridge Team and Ops Room Team. *Process description (including modes of action):*	**Outcomes** S5.1 Changes in contact, heading, speed and behaviour reported.

Table A3 Task Description Table

A5.1 Monitor and Report Contact Heading and Speed (cont)		
	Bridge Team monitor contact visually and report changes. Ops Room Team use radar, GPEOD and ESM sensors and report changes. *Key team interactions:* Broadcast changes in contact behaviour to all team members to maintain situational awareness	
Dynamic conditions Volume of shipping in the area; Multiple, simultaneous threats; Sea state; Visibility (fog, rain).	**Process assessment criteria** Effectiveness of monitoring (arcs covered). **Data capture requirements** Observation and recording of team actions and communication.	**Outcome assessment criteria** Timeliness, completeness, accuracy of the report, on the correct circuits. **Data capture requirements** Recording of contact reports, time and originator.
A5.2 Reclassification Based on Contact Behaviour		
Initial conditions and cues S5.1 Contact heading, speed and behaviour.	**Actions (sensing, doing, communication)** *Participants:* PWO/ Captain. *Process description (including modes of action):* Evaluation of reports of contact behaviour and broadcast of the reclassification. *Key team interactions:* Broadcast changes in contact behaviour to all team members to maintain situational awareness.	**Outcomes** S2.2 Contact classified as Friend, Assumed Friend or Neutral. S5.2 Contact remains classified as Suspect.
Dynamic conditions Volume of shipping in the area; Multiple, simultaneous threats; Sea state; Visibility.	**Process assessment criteria** Effectiveness of monitoring (arcs covered). **Data capture requirements** Observation and recording of team actions and communication.	**Outcome assessment criteria** Correctness of classification. **Data capture requirements** Recording of classification and contact data.

Table A3 Task Description Table

A5.3 Select and Execute Escalation of Force Measures		
Initial conditions and cues S0 Neutral shipping; Coastline; Maritime zones, channels and lanes. S5.2 Contact remains classified as Suspect.	**Actions (sensing, doing, communication)** *Participants:* Bridge Team, Ops Team and UDT. *Process description (including modes of action):* Captain/PWO selects and orders Escalation of Force measure (radio broadcast, flares, searchlight, audio warning by loudhailer, warning shots). *Key team interactions:* Timely reporting of changes in contact behaviour to inform decision making about further use of escalation measures and contact classification.	**Outcomes** S5.3 Escalation of Force measure executed.
Dynamic conditions Volume of shipping in the area; Multiple, simultaneous threats; Sea state; Visibility (fog, rain).	**Process assessment criteria** Correct choice of escalation measure sequence and timeliness of use, correctness of deployment. **Data capture requirements** Observation and recording of team actions and communication.	**Outcome assessment criteria** Timeliness, completeness accuracy of the report, on the correct circuits. **Data capture requirements** Recording of contact reports, time and originator.
A5.4 Monitor and Report Contact Heading and Speed		
Initial conditions and cues S0.3.1 Contact heading, speed and behaviour.	**Actions (sensing, doing, communication)** *Participants:* Bridge Team and Ops Room Team. *Process description (including modes of action):* Bridge Team monitor contact visually and report changes, Ops Room Team use radar, GPEOD and ESM sensors and report changes.	**Outcomes** S5.4 Changes in contact, heading, speed and behaviour reported.

Table A3 Task Description Table

A5.4 Monitor and Report Contact Heading and Speed (cont)		
	Key team interactions: Broadcast changes in contact behaviour to all team members to maintain situational awareness.	
Dynamic conditions Volume of shipping in the area; Multiple, simultaneous threats; Sea state; Visibility.	**Process assessment criteria** Effectiveness of monitoring (arcs covered). **Data capture requirements** Observation and recording of team actions and communication.	**Outcome assessment criteria** Timeliness, completeness, accuracy of the report, on the correct circuits. **Data capture requirements** Recording of contact reports, time and originator.
A5.5 Reclassify Contact		
Initial conditions and cues S0 Neutral shipping; Coastline; Maritime zones, channels and lanes. S5.4 Changes in contact heading, speed and behaviour reported.	**Actions (sensing, doing, communication)** *Participants:* PWO/ Captain/SPC. *Process description (including modes of action):* Decision on classification based on contact data reported. *Key team interactions:* Broadcast of the classification to all team members to maintain situational awareness.	**Outcomes** S2.2 Contact classified as Friend, Assumed Friend or Neutral. S5 Contact classified as Hostile.
Dynamic conditions Volume of shipping in the area; Multiple, simultaneous threats; Sea state; Visibility (fog, rain).	**Process assessment criteria** – **Data capture requirements** Observation and recording of team actions and communication.	**Outcome assessment criteria** Correctness of classification. **Data capture requirements** Recording of classification.
A6 Neutralise Threat		
Initial conditions and cues S0 Neutral shipping; Coastline; Maritime zones, channels and lanes.	**Actions (sensing, doing, communication)** *Participants:* Bridge Team, UDT, PWO. *Process description (including modes of action):*	**Outcomes** S6 Threat neutralised.

Table A3 Task Description Table (*concluded*)

A6 Neutralise Threat (cont)		
S5 Contact classified as Hostile. S0.4 Contact behaviour heading and speed.	Use of GPMG and Minigun by the UDT to fire aimed shots under collective self-defence ROE, as ordered by the Captain/PWO. Coordinated movement of the ship as required, directed by the PWO/Captain. UDT can instigate firing under individual self-defence ROE. *Key team interactions:* Monitoring of weapons controllers by Local Gun Commander.	
Dynamic conditions Volume of shipping in the area; Multiple, simultaneous threats; Sea state; Visibility.	**Process assessment criteria** Correct orders and actions sequence. Effective monitoring of UDT. Correct tactical use of fire and manoeuvre. **Data capture requirements** Observation and recording of team actions and communication.	**Outcome assessment criteria** Effective neutralisation of threat. **Data capture requirements** Recording of weapons effects.

Team KSA Requirements

Team KSA requirements are shown in Table A4.

Table A4 Team KSA Requirements Table

Category	Description
Knowledge	
Team/team member roles	Roles of UDT, Ops Team and Bridge Team.
Task models	Self-defence ROE, SOPs: Surface Investigate, Quickdraw, Escalation of Force.
Skills	
Voice procedures	Use of correct phraseology for reporting target sightings.
Attitudes	
Positive team orientation	Display a positive commitment to collaboration to achieve team goals.

Constraints, Assumptions, Risks and Opportunities Analysis

Table A5 Constraints

Constraint	Consequence
Live weapons cannot be used against manned targets.	Weapons effects have to be simulated.
Current Close Range Weapons training system and Bridge Simulator located at Collingwood.	Not easily accessible to crews undertaking Basic Operational Sea Training in Devonport.
MaST, Bridge Trainer and Close Range Weapons Trainer primarily used for individual training.	Very limited availability for team training.

Table A6 Opportunities

Opportunity	Description
Close Range Weapons Trainer at Collingwood	Weapons trainer offers positions for GPMG and Minigun (2 each) and position for PWO – weapons effects shown.
MaST	Available at Devonport – offers ops room simulation.
Bridge Trainer	Available at Collingwood – provides bridge simulation.

Training Environment Analysis

Training Environment Diagrams

Three training environment diagrams are shown in Figures A5 to A7, one for each of the Bridge, Ops Room and Upper Deck Teams. This split was chosen because there was too much detail for the entire environment representation to be shown on a single page.

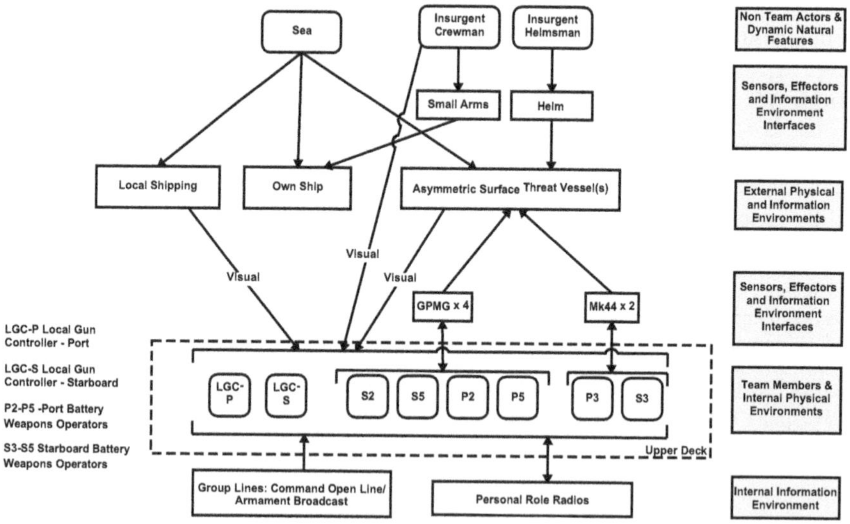

Figure A5 Upper Deck Training Environment Diagram

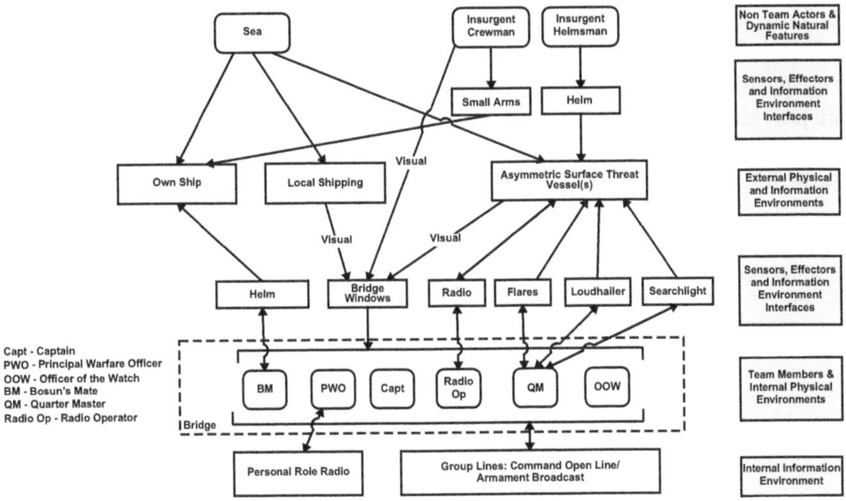

Figure A6 Bridge Training Environment Diagram

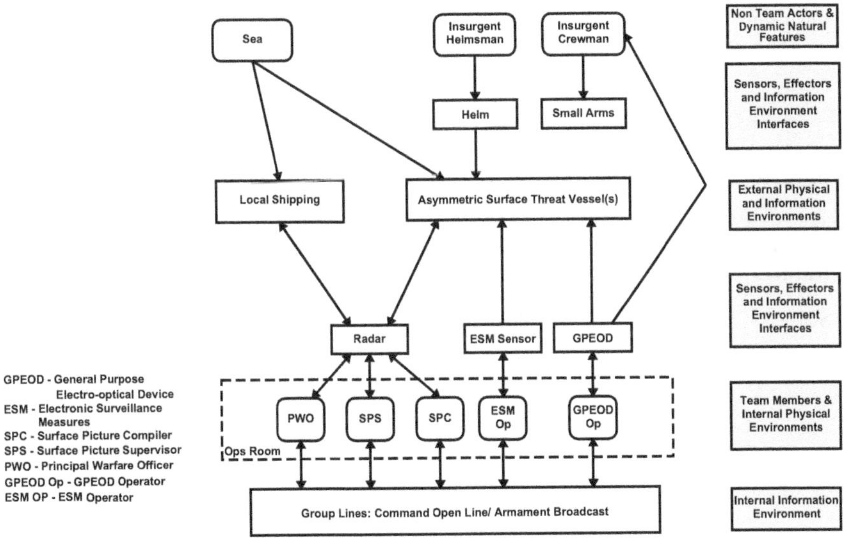

Figure A7 Ops Room Training Environment Diagram

Training Environment Specification Tables

The specifications for all of the elements shown in the set of Environment Diagrams are provided in Tables A7 to A13.

Table A7 External Physical Environment Specifications

External physical environment elements	Physical fidelity	Functional fidelity
Asymmetric surface threat vessel	Representative physical appearance and size for types of craft considered to be in use by asymmetric forces.	Speed, acceleration and rate of turn should be representative for each type of craft considered to be in use by asymmetric forces.
Neutral shipping	Representative appearance for shipping in the area.	Steer representative courses moving at representative speeds and rates of turn.
Own ship	N/A	Representative speed range, acceleration and rate of turn and tactical turn diameter.

Table A8 Dynamic Natural Features Specifications

Dynamic natural features	Physical fidelity	Functional fidelity
Sea	Colour, appearance of waves.	The waves associated with higher sea states cause vessels to pitch and roll. The movement of asymmetric threat vessels is significant because the weapons operators have to shoot at a moving target. The movement of the own ship is significant because upper deck movement affects the weapons operators' sight picture and the effective arc of the weapons.

Table A9 Internal Physical Environment Specifications

Internal physical environments	Physical fidelity	Functional fidelity
Upper deck	*Team Members:* 3 starboard weapons operators, 3 port weapons operators. *Interfaces:* 4 GPMGs on mounts, 2 Mk 44 Miniguns on mounts, personal role radio for each weapons operator.	Each weapon position should be at the correct apparent height above sea level. Each weapon position should have a representative field of view, including any restrictions to lines of sight that are appropriate relative to its position on the ship. The weapon operators' angle of view of the sea and surface vessels, and the alignment of the weapon, should move in response to sea state and ship's manoeuvre. The relative position of the weapons positions with respect to each other is significant, as their arcs of fire may or may not overlap and the operators have different views of the same visual scene.
Ops room	*Team Members:* 5 *Interfaces:* 3 – radar, GPEOD and ESM.	None
Bridge	*Team members:* 5 *Interfaces:* 4 – helm, 5 comms control panels, radio, bridge windows.	None

Table A10 Information Environment Specifications

Information environments (internal and external)	
Communications	
Personal role radio	Representative point to point communications between upper deck weapons operators and Bridge Team PWO. Communication channel latency and signal/noise characteristics such as noise, garbling, drop out, distortion, over talk and roundtrip delays should be representative including loss of communications quality where ship superstructure interferes with point-to-point communications.
Group lines	Command Open Line, Armament Broadcast. Intercom capability, broadcast of Command Line and Armament Broadcast to Upper Deck Team.

Table A11 Sensors and Effectors Specifications

Sensors and effectors	Interface physical fidelity	Interface functional fidelity	Mode of action/ functional fidelity
GPMG + mount	Accurate representation of the physical attributes of the weapon and mount to include: shape and size, weight, trigger, safety catch, cocking handle, loading mechanism, sight field of view and adjustment, arcs of manoeuvre (lateral and vertical) and firing arc stops.	Trigger pressure, sight picture, recoil, manoeuvrability on the mount, tracer visibility.	Modes: firing 7.62 rounds and tracer, continuously and in bursts. Fidelity: accurate representation of: rate of fire, range, muzzle velocity, round trajectory, live round / tracer mix, effects of fall of shot on the target.
Mk44 Minigun + mount	Accurate representation of the physical attributes of the weapon and mount to include: shape and size, weight, trigger, safety catch, cocking handle, loading mechanism, sight field of view and adjustment, arcs of manoeuvre (lateral and vertical) and firing arc stops.	Trigger pressure, sight picture, recoil, manoeuvrability on the mount, tracer visibility.	Modes: firing 7.62 rounds and tracer, continuously and in bursts Fidelity: accurate representation of: rate of fire, range, muzzle velocity, round trajectory, live round / tracer mix, effects of fall of shot on the target.
Helm	Representative wheel, throttles and speed/ heading displays.		Representative rate of turn and acceleration response to angle of wheel inputs and throttle quadrant position.
Bridge windows	Representative size and orientation field of view.	Representative resolution.	N/A
Loudhailer	Press to broadcast.	Broadcast when selected	Representative audible range.
Searchlight	Searchlight direction control.	Control direction of the searchlight.	Representative range and intensity.

Table A11 Sensors and Effectors Specifications (*concluded*)

Sensors and effectors	Interface physical fidelity	Interface functional fidelity	Mode of action/ functional fidelity
Flares	Flare firing capability.	Control of direction and firing of the flare.	Representative range and intensity of flare.
ESM	Selection of data pages for emitters detected.	Data displayed in the correct format.	Representative detection capability of radio and radar signals.
GPEOD	Joystick control of direction, zoom control, selection of thermal image / optical image.	Representative resolution of display and appearance of optical and thermal images.	Representative optical and thermal sensitivity.
Radar	Representative controls for operating the radar and entering tracks. Representative display resolution.	Radar range display changes appropriate to selections. Manually entered tracks display appropriately.	Representative range and resolution.

Table A12 Information Environment Interface Specifications

Information environment interfaces	Interface physical fidelity	Interface functional fidelity	Mode of action/ functional fidelity
Group Line Interface on-board ship	Headsets with microphones. Control panel with switches to select transmit and receive on each channel. Speakers for broadcast to upper deck.	Selection of multiple lines to receive (on left/ right ear phones) and a single line to transmit.	Transmission to all users with receive selected on a given channel. Reception from all users transmitting on lines which are selected for receive. Broadcast on upper deck.
Radio	Control panel switches to select frequency and to transmit.	Selection of transmit and receive.	Representative range.

Table A13 Non-Team Actor Specifications

Non-team actors	Appearance	Behaviour	Knowledge and skills
Insurgent helmsman	Dress consistent with the local population.	Manoeuvring of the insurgent craft in response to escalation of force measures and direct fire from the ship, consistent with known insurgent tactics.	Ability to helm the insurgent craft. Knowledge of insurgent tactics in response to escalation of force measures and direct fire.
Insurgent crewman	Dress consistent with the local population.	Firing at the ship when within range.	Knowledge of insurgent tactics. Weapons handling skills.

Training Overlay Analysis

Training Strategy

Training priorities

Risk assessment: Training is required for this task based on the analysis in Table A14.

Table A14 Training Priority Risk Analysis Table

Task	Critical error	Likelihood (H, M, L)	Consequences	Severity (H, M, L)	Risk (H, M, L)	Training required as mitigation (Y, N)
1.0 Protect ship from an asymmetric waterborne threat	Ineffective search	L	Late detection of potential threat	M	L	N
	Contact not correctly categorised as Suspect	L	Upper deck weapons not deployed in time to counter the threat – damage to ship, threat to life	H	M	Y
	Contact not correctly categorised as Hostile	L	Contact able to get within range to attack the ship – damage to ship, threat to life	H	M	Y
	Inaccurate fire from upper deck weapons	H	Contact able to attack the ship – damage to ship, threat to life	H	H	Y

Training objectives

Table A15 Training objectives

Performance	Conditions	Standards
Task execution		
Search for threat ensuring: Coordinated visual search through all arcs	Potential threat in search area	Timely report of contact Coverage of visual search arcs monitored on Bridge
Surface Investigate ensuring: Coordinated report of information from each person involved in search	Contact reported	Correct classification of contact Well-coordinated reporting in correct sequence
3.0 Raise Force Protection State	Contact classified as Suspect	Timely mounting and arming of weapons
4.0 Evaluate Threat	Contact classified as Suspect	Appropriate use of manoeuvre and selection of escalation of force measures
5.0 Counter Threat ensuring: Compliance with weapons control orders monitored by local gun controller	Contact exhibiting hostile intent/ behaviour	Lethal force used with correct application of self-defence ROE Correct weapons control orders issued and complied with Appropriate monitoring of weapons operators by local gun controller
Supporting KSAs		
Knowledge		
Surface Investigate and Quickdraw SOPs	Given a scenario	Accurate description of SOP sequence
Self-defence ROE	Given a scenario	Correct identification of circumstances when use of lethal force for self-defence is legal
Attitudes		
Team orientation	In a scenario	Collaborate with other team members to achieve team goals

Training audience
The Training Audience is the Upper Deck Team, the Bridge Team and the Operations Room Team.

Training design

Training structure and sequencing Training to be conducted as a single task (not trained in parts). Training is required during BOST and DCT. Review of team KSA, SOPs, ROE and tactics will be undertaken prior to task practice. Refresher training is also required in transit to deployments and on deployments, where there is a perceived threat.

Training methods The review of team KSA, SOPs, ROE and tactics will be conducted as an instructor facilitated discussion. This will be followed by full task practice consisting of a range of scenarios. An Events Based Approach to training will be adopted with a progression of tasks of increasing diversity and complexity designed to engender adaptive responses.

Training duration Allowing for briefing and debriefing, along with scenario presentations it is estimated that a training event would last for half a day.

Training environment selection Training *en route* will be live. Training during BOST/DCT could be implemented in either the live environment or a synthetic environment, provided all aspects of the training environment specification can be met.

Training system capacity requirements
Based on 18 events per year the capacity requirement would be nine days of availability per year. A full training throughput calculation would need to take account of all vessel types that require training (T23, T26 and T45). For the purposes of this case study it is estimated that for BOST and DCT 18 events per year would be required to cater for all vessel types.

Training provision
Live training is to be provided from within organisational resources. A synthetic training solution could be contractor owned and operated, but with military instructional staff.

Training Overlay Requirements for the Training Environment

Training staff requirements
The following categories of training staff would be required:

- Lieutenant Commander (Lt Cdr) Staff Warfare Officer (Above Water Warfare specialist) to plan the exercise and brief, monitor and debrief the PWOs and the Captain

- Lt Cdr Staff Warfare Officer (navigation specialist) to brief, monitor and debrief the Bridge Team (with the exception of the PWO and Captain)
- WO/CPO Above Water Tactics Instructor to brief, monitor and debrief the Ops Room Team ratings
- WO/CPO Above Water Warfare (Gunner) to brief, monitor and debrief the Upper Deck Team.

Training overlay interfaces to the training environment
The training overlay requirements for the training environment are shown in Table A16. The scenario requirements are as described in Table A17.

Table A16 Training overlay requirements for the training environment

Training environment element	Training overlay requirements	
	Environment management	**Training delivery**
Training audience	N/A	Brief, monitor, record and debrief performance Replay of actions for debrief
Physical environment elements		
Asymmetric surface threat	Set up of single or multiple contacts, control of course, speed and actions	Recording of course and actions for debrief
Neutral shipping	Configuration of course and speed for each ship	Recording of track for AAR
Own ship	Configuration of start position, heading and speed	Recording of track for AAR
Dynamic natural features		
Sea state	Set the sea state for the exercise	N/A
Sensors		
Radar	N/A	Capture of ESM output for AAR
ESM	N/A	Capture of radar display images for AAR
GPEOD	N/A	Capture of GPEOD display images for AAR

Table A16 Training overlay requirements for the training environment (*concluded*)

Effectors		
Upper deck weapons (Minigun and GPMG)	N/A	Recording of hits on target for AAR
Helm	N/A	
Information environment – communications systems		
Radio	N/A	Recording of voice comms and replay for AAR
Group lines	N/A	Recording of voice comms and replay for AAR

Table A17 Scenario requirements

Scenario	Maritime Force Protection of a Type 23 Frigate against an asymmetric threat, whilst at cruising watch in open waters.
Human	Operators of local shipping (not hostile).
Physical	Open waters leading to confined waters such as straits, transits and harbour approaches. Variable sea-state and possibility of poor visibility due to local weather conditions.
Information	Personal role radios and internal command lines on-board. Royal Navy Command Support System chat connection to higher command.
Civilian	Local shipping, possibility of smugglers in fast-moving, small craft.
Military	Friendly forces: Command Headquarters. Neutral forces: None. Enemy forces: Asymmetric threat of non-conventional forces in small, fast craft equipped with small arms, rocket propelled grenades, or operating waterborne improvised explosive devices. Possibility of simultaneous attack from multiple craft. Resources: • radar • electro-optical sensors • GPMG x 4 • Miniguns x2 • flares, searchlight and loudhailer.

Table A17 Scenario requirements (*concluded*)

Events	Small, unidentified, fast craft heads towards ship on collision heading, does not respond to escalation of force measures, turns to follow ship when ship changes heading and speed, opens fire on ship with small arms and rocket propelled grenades (asymmetric threat). Multiple small, unidentified craft heading to the ship as above. These may be: • on the same approach vector at the same time • on different approach vectors at the same time • on different approach vectors at different times. Small, unidentified, fast craft does not respond to escalation of force measures, turns to follow ship when ship changes heading and speed, maintains collision course with ship (waterborne IED). Smugglers in small, fast craft heading towards ship do not respond to escalation of force measures, but maintain their heading. Fishing vessels on collision heading with the ship respond to escalation of force measures.

Training Options Analysis

Existing Capabilities

The current provision includes the use of contracted Rigid Inflatable Boats (RIBs) and the ship's own RIB to act as potential threats whilst training at sea. Contracted RIBs are used during BOST and during Exercise Joint Venture. In transit, the ship's own RIB would be used. MaST, the Close Range Weapons Trainer and the Bridge Trainer at HMS Collingwood, although not currently linked, offer training environments for the Ops Room Team, the Upper Deck Team and the Bridge Team respectively.

A comparison of these systems for use in Force Protection Training is shown in Table A18. The following coding is used in the table:

✓	Fully meets the requirement
text	Partially meets the requirement
x	Does not meet the requirement / not supported

Both of the currently employed systems (using contracted RIBs or the ship's RIB) are critically limited by the fact that weapons effects are not represented. Furthermore, as there are no tracer or fall of shot indications for the GPMGs or Miniguns, the operators cannot bring them to bear on the target using the doctrinally recommended technique.

Taken together, MaST, the Bridge Trainer and the Close Range Weapons training system offer coverage of the majority of the requirements, with the exception that the Close Range Weapons Trainer only has provision for two

GPMGs. Potentially, they could be networked together. However, availability of the connected system is likely to be a significant issue as the primary use for all of the systems is individual training. Furthermore, the most likely opportunity for crews to have time to undertake training on the system is when they are undergoing BOST or DCT, which is conducted at Devonport. Therefore, accessibility would be an issue.

Table A18 Option evaluation table for existing systems

Environment elements	Existing systems (evaluation of environment and overlay requirements)				
	Live with ship's RIB as the asymmetric threat	Live contracted RIBs as the asymmetric threat	Maritime Synthetic Training System (MaST)	Close Range Weapons Trainer	Bridge Trainer
Sea	✓	✓	x	✓	✓
Insurgent crewman	✓	✓	x	✓	✓
Insurgent helmsman	✓	✓	x	✓	✓
Local shipping	x	x	✓	✓	✓
Own ship	✓	✓	✓	✓	✓
Asymmetric threat	Limited	Limited	✓	✓	✓
Asymmetric threat Small Arms	Critically limited – no weapons effects	Critically limited – no weapons effects	x	✓	✓
Radar	✓	✓	✓	x	x
ESM	✓	✓	✓	x	x
GPEOD	✓	✓	✓	x	x
Helm	✓	✓	✓	x	✓
Bridge windows	✓	✓	x	x	✓
Radio	✓	✓	✓	x	x
GPMG	Critically limited – no weapons effects	Critically limited – no weapons effects	x	Limited – only 2 GPMGs	x

Table A18 Option evaluation table for existing systems (*concluded*)

Environment elements	Existing systems (evaluation of environment and overlay requirements)				
	Live with ship's RIB as the asymmetric threat	**Live contracted RIBs as the asymmetric threat**	**Maritime Synthetic Training System (MaST)**	**Close Range Weapons Trainer**	**Bridge Trainer**
Mk44	Critically limited – no weapons effects	Critically limited – no weapons effects	x	✓	x
Bridge visual	✓	✓	x	x	✓
Ops room	✓	✓	✓	x	x
Bridge	✓	✓	x	Limited – PWO only	✓
Upper deck	✓	✓	x	Limited – no type specific representation of visual surroundings	x
Group lines	✓	✓	✓	x	x
Personal role radio	x	x	✓	x	✓

Option Descriptions

Option 1. Live with ship's RIB as the asymmetric threat

Overview Live training using the ship's own resources.

Training environment provision Ship's RIB driven as suspect craft. Personal weapons with blank firing attachments used as insurgent weapons. Blanks for upper deck weapons. Training to be conducted in a maritime training area used for BOST/DCT. Pre-cursor team KSA training to be classroom based at FOST/or in a briefing space on-board the ship.

Staff tasks

- exercise planning
- exercise control

- KSA classroom training
- brief, monitor, debrief Bridge, Ops Room and Upper Deck Teams.

Staff requirements For training during BOST/DCT the following staff would be required:

- 1 x Lt Cdr Staff Warfare Officer (Above Water Warfare specialist) to plan the exercise; brief, monitor and debrief the PWOs and the Captain; lead the KSA training session prior to task practice
- 1 x Lt Cdr Staff Warfare Officer (navigation specialist) to brief, monitor and debrief the Bridge Team (with the exception of the PWO and Captain)
- 1 x WO/CPO Above Water Tactics Instructor to brief, monitor and debrief the Ops Room Team ratings
- 1 x WO/CPO Above Water Warfare (Gunner) to brief, monitor and debrief the Upper Deck Team
- 2 crew members to crew the RIB and act as insurgents.

For training *en route*:

- PWO to lead the KSA training and plan the exercise
- Executive Officer/ Offwatch PWO to brief and debrief the exercise, and monitor the Bridge Team
- 2 x Off-watch warfare team senior rates to monitor and debrief the Ops Room Team and the Upper Deck Team.

Supporting systems N/A.

Resources Classroom for KSA training at FOST.

Staff training requirements N/A.

Training linkages Training will be scheduled within BOST/DCT.

Equipment Ship's RIB.

Infrastructure N/A.

Information Written assessments during BOST will be held in the extant training administration system at FOST.

Organisation Training during BOST/DCT will be scheduled by FOST staff within the BOST/DCT training programme. Training *en route* will be scheduled by the Captain of the ship.

Logistics Fuel for the RIB, blank ammunition for the GPMGs and Miniguns.

Interoperability N/A.

Option 2. Live with remote-controlled boats as the asymmetric threat

Overview Live training using remote-controlled boats as the asymmetric threat, conducted on a live firing range, or in clear open waters *en route*.

Training environment provision Remote-controlled boat (one-third scale) as the asymmetric threat. Live rounds for upper deck weapons. Training to be conducted in a maritime live firing range used for BOST/DCT or in open waters clear of other shipping beyond the range of the close range weapons (GPMG and Minigun) when *en route*.

Staff tasks

- exercise planning
- exercise control
- operate the remote-controlled boats
- KSA classroom training
- brief, monitor, debrief Bridge, Ops Room and Upper Deck Teams.

Staff requirements For training during BOST/DCT the following staff would be required:

- 1 x Lt Cdr Staff Warfare Officer (Above Water Warfare specialist) to plan the exercise and brief, monitor and debrief the PWOs and the Captain
- 1 x Lt Cdr Staff Warfare Officer (navigation specialist) to brief, monitor and debrief the bridge team (with the exception of the PWO and Captain)
- 1 x WO/CPO Above Water Tactics Instructor to brief, monitor and debrief the Ops Room Team ratings
- 1 x WO/CPO Above Water Warfare (Gunner) to brief, monitor and debrief the Upper Deck Team
- 2 crew members to operate the remote-controlled boats.

Supporting systems N/A.

Resources N/A.

Staff training requirements Operation of the remote-controlled boat system.

Training linkages Training will be scheduled within BOST/DCT.

Equipment Remote-controlled boats and control system.

Infrastructure N/A.

Information Written assessments during BOST will be held in the extant training administration system at FOST.

Organisation

- Navy Command, Maritime Training Acquisition Organisation will be responsible for acquisition of the remote-controlled boat system.
- Training during BOST/DCT will be scheduled by FOST staff within the BOST/DCT training programme. Training *en route* will be scheduled by the Captain of the ship.

Logistics

- Delivery of remote-controlled boat system to ship by lorry.
- Fuel for the remote-controlled boats, live ammunition for the GPMGs and Miniguns.

Interoperability N/A.

Option 3. Enhanced Close Range Weapons Trainer + Maritime Composite Training System

Overview Training during BOST/DCT provided by an enhanced version of the Close Range Weapons Trainer, currently installed in the dockyard at Portsmouth, to be located at Devonport and connected to the MaST installation at Devonport. The Ops Room Team would be accommodated in the MaST with the Bridge and Upper Deck Teams accommodated in the enhanced Close Range Weapons Trainer.

Training environment provision MaST connected to an Enhanced Close Range Weapons Trainer, providing a synthetic environment for the training.

Staff tasks

- exercise planning
- exercise control
- operate MaST and the Close Range Weapons Trainer – to include setting up scenarios, initiating events, recording key activities as directed by the instructional staff, facilitating replay of events for AAR
- KSA classroom training
- brief, monitor, debrief Bridge, Ops Room and Upper Deck Teams.

Staff requirements For training during BOST/DCT the following staff would be required:

- 1 x Lt Cdr Staff Warfare Officer (Above Water Warfare specialist) to plan the exercise and brief, monitor and debrief the PWOs and the Captain
- 1 x Lt Cdr Staff Warfare Officer (navigation specialist) to brief, monitor and debrief the Bridge Team (with the exception of the PWO and Captain)
- 1 x WO/CPO Above Water Tactics Instructor to brief, monitor and debrief the Ops Room Team ratings
- 1 x WO/CPO Above Water Warfare (Gunner) to brief, monitor and debrief the Upper Deck Team
- 2 x Operators for the MAST and Enhanced Close Range Weapons Trainer.

Supporting systems Debrief system to enable replay of events for AAR.

Resources Briefing/debriefing facility.

Staff training requirements

- operation of the combined MaST /Close Range Weapons Trainer system
- operation of the debrief system.

Training linkages Training will be scheduled within BOST/DCT.

Equipment Enhanced Close Range Weapons Trainer.

Infrastructure Building at HMNB Devonport to accommodate the Enhanced Close Range Weapons Trainer, preferably next to / near to the MaST installation.

Information Written assessments during BOST will be held in the extant training administration system at FOST.

Organisation

- Navy Command, Maritime Training Acquisition Organisation will be responsible for acquisition of the Enhanced Close Range Weapons Trainer, its connection to the MaST and any enhancements required to MaST.
- Training during BOST/DCT will be scheduled by FOST staff within the BOST/DCT training programme.

Logistics N/A.

Interoperability The Enhanced Close Range Weapons Trainer will be designed to connect with MaST.

Costs

The estimates of costs for the three options are:

Option 1 – running costs of RIB and blank ammunition cost – low
Option 2 – moderate acquisition cost for remote-controlled boats
Option 3 – very high relative to the other options. Requires acquisition of training system and integration with MaST, infrastructure costs for accommodating it and annual staff and maintenance costs to run it.

(A full cost calculation was beyond the scope of the preparation of this case study.)

Option Evaluation Criteria

The following evaluation criteria were identified:

- availability
- accessibility
- cost.

Option Evaluation

Table A19 provides an evaluation of the three selected options against the three option evaluation criteria, the coverage of the training objectives and compliance with the training environment and overlay requirements.

The following coding is used in the table:

✓	Fully meets the requirement
text	Partially meets the requirement
x	Does not meet the requirement / not supported

The use of a remote-controlled boat which can withstand live fire hits addresses only half of the weapons effects problem. However, since it is not manned, crew actions cannot be observed. Furthermore, as the available remotely operated craft are one-third scale they present an unrepresentatively small target. This may present issues with visual detection and range estimation, as well as being significantly harder to hit (having one ninth of the surface area of a full-size craft). It may also have a negative training effect if the estimation of how far ahead of the craft to fire is estimated in terms of boat lengths.

Linking an enhanced Close Range Weapons Trainer, with representative weapons positions for all of the Upper Deck Team and a Bridge, to MaST at Devonport would provide the capability to train Force Protection effectively, as weapons effects could be provided both ways. It would only be available at

Devonport, but would address training requirements during BOST and DCT. It would also provide the required capability for training Force Protection against Fast Attack Craft (FAC) and Fast Inshore Attack Craft threats (FIAC). Whilst it is a costly option, it is the only option that provides sufficient fidelity of weapons effects to facilitate tactical training beyond the procedural level.

Live training with the ship's RIB as the threat provides a no-additional-cost option for training in transit.

Recommended Option

Live training with own ship's RIB is recommended as the solution for training in transit.

An enhanced Close Range Weapons Trainer located at Devonport would be the recommended solution for training of crews whilst at BOST, as it is the only solution that offers full tactical training rather than procedural training. However, this would be contingent on the outcome of an equivalent study to look at the requirement for training against a FAC/FIAC threat.

Table A19 Option evaluation table

	Options		
Evaluation criteria	**Live with ship's RIB as the asymmetric threat**	**Live + remote-controlled boat**	**Enhanced Close Range Weapons Trainer + MaST**
1. Availability	✓	✓	High at Devonport
2. Accessibility	✓	✓	At Devonport
3. Cost	✓		
Training objectives	**Training objective coverage**		
TO1 Search for Threat	✓	✓	✓
TO2 Surface Investigate	✓	✓	✓
TO3 Raise Force Protection State	✓	✓	✓
TO4 Evaluate Threat	✓	✓	✓
TO5 Counter Threat	No weapons effects	No weapons effects from threat	✓

Table A19 Option evaluation table (*concluded*)

Environment elements	Evaluation of environment and overlay requirements		
Sea			✓
Insurgent crewman	✓	x	✓
Insurgent helmsman	✓	x	✓
Local shipping	x	x	✓
Own ship	✓	✓	✓
Asymmetric threat	Limited	1/3 scale	✓
Asymmetric threat Small Arms	Critically limited – no weapons effects	x	✓
Radar	✓	✓	✓
GPEOD	✓	✓	✓
ESM	✓	✓	✓
Helm	✓	✓	✓
Bridge windows	✓	✓	✓
Radio	✓	✓	✓
GPMG HMI	✓	✓	
GPMG mode of action	No fall of shot		✓
GPMG overlay reqts	No hit indication	✓	✓
Mk44 HMI	✓	✓	✓
Mk44 mode of action	No fall of shot	✓	✓
Mk44 overlay reqts	No hit indication	✓	✓
Bridge visual	✓	✓	✓
Ops room	✓	✓	✓
Bridge	✓	✓	✓
Upper deck	✓	✓	✓
Group lines	✓	✓	✓
Personal role radio	✓	✓	✓

Appendix B
Carrier Enabled Power Projection Case Study

Introduction

The purpose of this case study is to:

- illustrate the application of the TCTNA approach to a large-scale task at the early stages of a project where much of the information and doctrine has yet to mature;
- illustrate how the TCTNA approach can yield useful information when applied in a short timescale as a 'first look';
- demonstrate how the methodology and the use of the analytical tools can be tailored to suit the context of application.

Carrier Enabled Power Projection (CEPP) is a relatively new capability construct which is in the early stages of maturity. It is based around the employment of the Queen Elizabeth Class (QEC) aircraft carrier to serve as a platform from which a variety of air assets may be launched, including attack helicopters, support helicopters for moving troops and matériel, helicopters carrying sensors, and Lightning II (LTNG II) fixed wing aircraft. The concept also embraces a multinational dimension, with a task group being formed with maritime and air assets from multiple nations.

The introduction into service of each of these platforms would be supported by a set of TNAs conducted at the platform level, team and individual levels. These would yield recommendations for training solutions at each of these levels. A full TCTNA would need to be coordinated with these lower level TNAs to ensure that the most effective training solution was developed to address all levels of training, allowing for integration of training system components for the higher levels of training. Furthermore, many of the assets that could be involved in CEPP (such as the Battlestaff, and rotary wing assets) already exist and have training solutions in place based on their current employment. Consideration would have to be given as to what extensions to their extant training systems would be required to cater for preparation for CEPP tasks, as well as their potential integration into a CEPP training solution. The multinational dimension adds further complexity, requiring collaboration and coordination between nations.

This case study was conducted during a one-day workshop with SMEs to discuss generic carrier operations through studying the BR1806 British Maritime Doctrine (Royal Navy 2004), followed by six days of effort allocated to the examination of the available data and writing up the analysis. With such a limitation on time, it was only feasible to look at the highest level of training within the case study. Significant additional work would be required to ensure that a detailed solution achieved the necessary level of integration with lower levels of training.

The TCTNA structure has been followed but not all of the detailed components have been implemented (such as the analysis of detailed fidelity requirements) as they were not appropriate or feasible for such a high-level analysis. The intention is to show how the structure can be used to guide high-level analysis, rather than slavishly followed as a procedure.

This case study should be treated as purely illustrative; any opinions and suggestions are those of the authors and do not reflect UK MOD policy.

Team/Collective Task Analysis

Task Scope and Definition

CEPP is defined by the Single Statement of Need as follows:

> An integrated and sustainable joint capability, interoperable with NATO that enables the projection of UK Carrier Strike and Littoral Manoeuvre power, as well as delivering Humanitarian Assistance and Defence Diplomacy. (Ministry of Defence 2013: 3-20, note 32.)

To conduct a high-level collective task analysis for CEPP to describe this capability relevant British maritime doctrine must be examined, to aid the identification of the task boundary.

A number of maritime doctrinal concepts are relevant to establishing the boundary of the CEPP task. *BR1806 British Maritime Doctrine*, Chapter 3, covers principles governing the use of maritime power and defines and inter-relates the key concepts of power projection, sea control and maritime manoeuvre. It makes the following statements about power projection:

> Maritime Power Projection is the threat or use of maritime combat capabilities at global range to achieve effects in support of national policy objectives; usually to influence events on land directly. (p. 45)

> The ability to conduct Power Projection relies on our maritime forces being able to exploit the sea for our own advantage while denying its use to a potential rival or enemy. (p. 41)

Traditionally this ability to exploit the sea was described as 'command of the sea', however a local and more limited form of command of the sea is known as 'sea control', which is limited in time and space for what is actually necessary for a given operation. BR1806 makes the following statements about sea control:

> Sea control is the condition in which one has freedom of action to use the sea for one's own purposes in specified areas and for specified periods of time and where necessary, to deny or limit its use to the enemy. There is likely to be a requirement for sea control across the spectrum of conflict. (p. 41)

> Sea control comprises control of the surface and sub-surface environments and of the airspace above the area of control. (p. 42)

> Whenever the freedom of action of the maritime force is challenged and, in particular as it approaches the area of operations, there will be a requirement to establish levels of sea control that will be sufficient to ensure its protection and to enable subsequent operations. (p. 150)

From a review of the concepts, Sea Control is an enabler and necessary condition for Power Projection operations to occur, whether these Power Projection operations are Carrier Strike or Littoral Manoeuvre. Littoral Manoeuvre is not formally defined in BR1806, however the term Maritime Manoeuvre is defined as:

> The ability to use the unique access provided by the sea to apply force or influence at a time or place of political choice. (p. 49)

Carrier Strike was not explicitly referenced in BR1806, but as a form of Power Projection it comprises an element of Maritime Force Projection. The relationship of Maritime Force Projection, Maritime Projection and Sea Control is illustrated in Figure B1. As Littoral Manoeuvre is a similar concept to amphibious operations we have subsumed amphibious operations into Littoral Manoeuvre to avoid duplication of concepts.

Non-combatant Evacuation Operations (NEOs) fall into the scope for the CEPP task due to the inclusion of Humanitarian Assistance in the CEPP Single Statement of Need. Tomahawk Land Attack Missiles (TLAM) and Naval Gunfire Support (NGS), while constituting types of Maritime Force Projection, are already trained and operated by the relevant platforms and so would not fall into the scope for CEPP, but may well be tasks conducted in parallel.

Maritime Air Support supports Sea Control, which itself is an enabler for the core Maritime Power Projection sub-tasks of Carrier Strike and Littoral Manoeuvre. Within Maritime Air Support by organic air, BR1806 identifies the operations of air defence of the battlespace (Counter Air), Anti-Surface Force and Combat Support Air Operations (BR1806, p. 162).

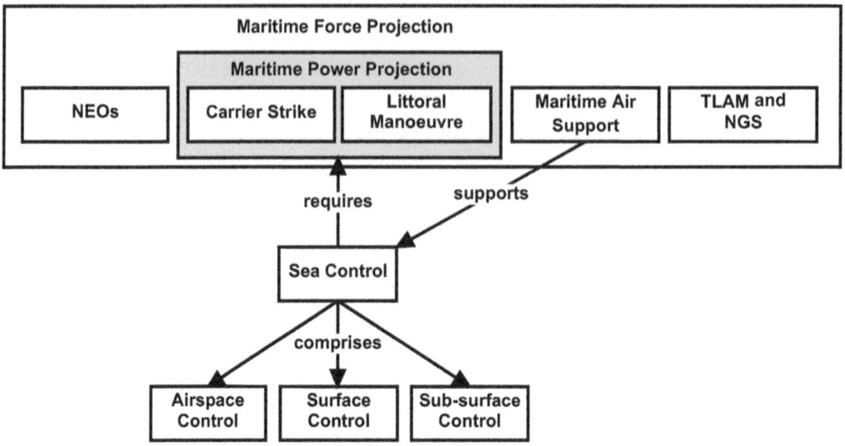

Figure B1 Maritime Power Projection and Sea Control

Air Maritime Coordination (AMC) and its associated Procedures (Air Maritime Coordination Procedures AMC(P)) cover air support for all maritime warfare areas: Anti-Air Warfare (AAW), Anti-Surface Warfare (ASuW) and Anti-Submarine Warfare (ASW). AMC in support of ASuW includes Air Operations in Anti-Surface Warfare (AOASuW) (also known as Anti-Surface Force Air Operations). AOASuW includes Surface Surveillance Coordination, Air Interdiction of Maritime Surface Targets and Maritime Air Support.

The *Strategic Defence and Security Review* (Cabinet Office 2010) reduced the scale of the carrier strike capability based on the following statement:

> We cannot now foresee circumstances in which the UK would require the scale of strike capability previously planned. We are unlikely to face adversaries in large-scale air combat. We are far more likely to engage in precision operations, which may need to overcome sophisticated air defence capabilities. (p. 23)

While large-scale air combat may be unlikely in the future, it is a possibility that in a future conflict the QEC could be targeted with aircraft-launched stand-off weapons, as an alternative to engaging a fifth-generation multi-role aircraft in air-to-air combat. While only one of the eight British aircraft carriers sunk in the Second World War was sunk by air attack, the US and Japan lost a total of 18 aircraft carriers to air attack in the same period. Maritime Air Support acts as an enabler to Sea Control which is a requirement for Power Projection, and so should fall within the remit of CEPP. The alternative would be to eliminate Sea Control from the CEPP task, which would be nonsensical given that while Sea Control is an enabling stage for the Power Projection tasks, it also is a concurrent activity for Power Projection. The concurrency of Sea Control and Power Projection is such that while Sea Control could be seen as defining the context for Power Projection it is actually much more than this. There may be a very real tension between the requirements for conducting both Sea Control and Power Projection simultaneously with finite assets, within the supporting/supported componency relationships in a joint operation. Power Projection operations may necessitate placing the QEC and its escorts in a position where they may potentially be attackable, so remaining out of range of enemy stand-off weapons and or remaining undetected may not be a guaranteed context of CEPP.

In a similar form the tasks of defending the QEC as the centre of the task group from the other major threat types such as submarines (which torpedoed and sank a total of 16 aircraft carriers in the Second World War) would seem to fall into the scope of performing the CEPP task. If the carrier task group doesn't manage to perform Sea Control adequately and protect the QEC and its critical enablers (support vessels and vessels enabling CEPP/Littoral Manoeuvre) then there is no power to project.

In summary it is suggested that the CEPP collective task boundary is defined as shown in Figure B2.

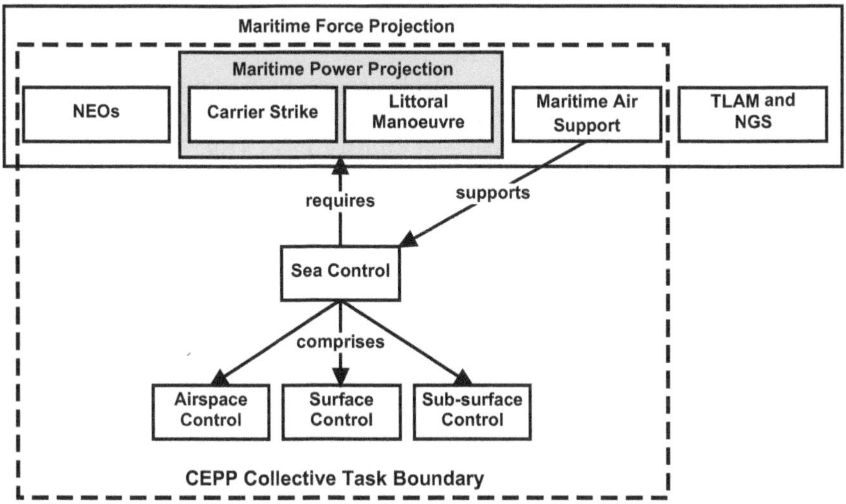

Figure B2 CEPP collective task boundary

One could argue that Sea Control is part of the RN day-to-day activity, and indeed it is. However, the demands of Sea Control associated with the QEC as the hub of the carrier group are much more complex and operate on a much wider scale, and the threats and capabilities faced may be commensurately greater as well.

The need for multi-platform collective training against a credible threat is driven by:

- the scale and complexity of the carrier group which is delivering CEPP

- the Joint C2 construct, the number and type of vessels in the carrier group
- the requirement for AMC and AMC(P)
- the requirement for layered defence including theatre ballistic missile defence
- the size of the Surveillance Area and Classification, Identification, Engagement and Assessment areas needed to protect the Carrier group Vital Area
- the number of high value units to be defended (potentially QEC and two Royal Fleet Auxiliaries as a minimum, most likely more vessels in the Littoral Manoeuvre flavour of CEPP)
- the operation of both fixed wing and rotary wing blue air
- the requirement to perform Sea Control and Power Projection concurrently.

Conditions

Identification of the conditions of the CEPP task will involve analysis of the defence planning assumptions and MoD Studies and Assumptions Group (SAG) scenarios that are relevant. The content of these scenarios will furnish necessary detail on relevant conditions that exist.

Task conditions should reflect a suitable opponent reflecting that 'The ability of our maritime forces to conduct high intensity warfighting confirms upon them their effectiveness across the full range of Military Tasks' (BR1806, p. 58).

Training Audience Description / CEPP Organisation

Having suggested a collective task boundary for CEPP we now need to identify what other partners and which platforms might be involved in CEPP to aid identification of the potential training audience.

CEPP as a UK task

A number of Task Group (TG) configurations that could exploit the Power Projection capabilities of the QEC are possible. These include Carrier Strike, Littoral Manoeuvre and supporting Special Forces (SF) operations. These could exploit the use of rotary as well as fixed wing assets being deployed from the carrier.

CEPP as a combined task

The context for CEPP would be either as a single nation or as part of a coalition which might include NATO allies such as the US, France and Italy. CEPP collective tasks may therefore be combined with US or NATO allies. The main permutations within this are:

- Sea Control and Replenishment Combined Operations
  - UK CEPP task group complemented with US/NATO escort ships, or CEPP escorts (especially Type 45) working with other US/NATO carrier groups.

  - UK CEPP task group supported by US/NATO inorganic air and/or Maritime Patrol Aircraft.
- Power Projection Combined Operations
  - UK CEPP task group complemented with NATO amphibious and landing force units (for example Dutch Marines or US Marines).
  - UK CEPP task group complemented with F35B LTNG II aircraft (US Marine Corps – 340 ac in total on order, 50 available in 2014/2015) or Italian Navy (22 ac to be delivered between 2014 and 2022).
  - UK CEPP task group complemented with NATO SF.
- Combined Carrier Operations
  - QEC combined with US, French and Italian carriers (this may be combined with all or any of the permutations above).

The Statement of Intent (SOI) signed on 5th January 2012 between the US DoD and UK MoD proposed enhanced cooperation on carrier operations and Maritime Power Projection. Part of the work in the concepts and doctrine area involves the development and extension of current (NATO) TTPs and concepts, in areas such as Intelligence, Surveillance, Target Acquisition and Reconnaissance (ISTAR), AMC(P) and Joint Targeting. The SOI covers generation, training, operation and sustainability with the aim of seeking maximum interoperability and synergy.

CEPP as a joint task

The C2 in all CEPP missions is joint in flavour and exists at the operational level of war, so CEPP training will involve close cooperation and alignment between all services involved. The opportunity is to drive collective training for CEPP, 'top-down' as a fully joint activity, aligned to Defence priorities and to exploit the benefits of combined training opportunities in conjunction with other deployed activity (Defence Engagement, Defence Diplomacy, Operational requirements).

Commander Carrier Strike Group (COMCSG) and Commander UK Task Group (COMUKTG) are assumed to be afloat on the QEC in an operational scenario. The Joint Force Air Component Command (JFACC) will be involved if the LTNG II is embarked, and may be involved in the Littoral Manoeuvre flavour of CEPP if inorganic air is a requirement. Lead Commando Group or other high readiness land forces will be required for the Littoral Manoeuvre flavour of CEPP as will the assets from Joint Helicopter Command.

Summary of Platforms and Organisations Potentially Involved in CEPP

Figure B3 is an initial attempt to represent the potential interactions and interfaces within CEPP at their broadest possible level to summarise entities and interactions involved.

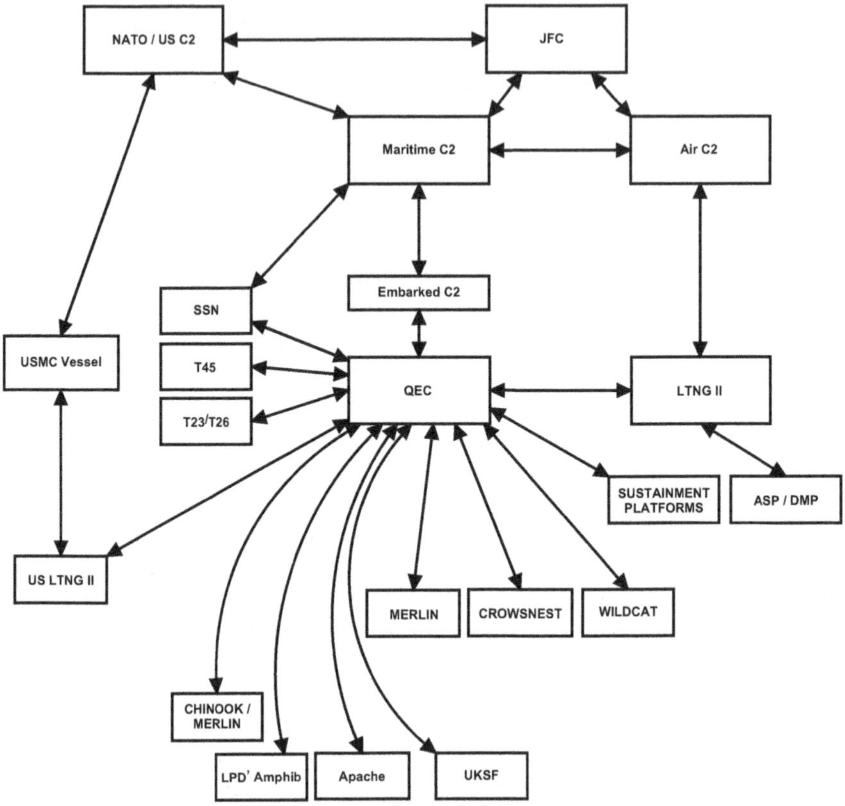

Figure B3 CEPP actors and linkages, high-level view

High-Level Task Analysis

British Maritime Doctrine identifies seven stages of the maritime contribution to a joint campaign identified in (BR1806, p. 147):

1. identification of a crisis
2. force generation
3. deployment
4. Sea Control Operations
5. Maritime Force Projection (includes NEOs, Amphibious Operations, Maritime Air Support and Surface and Sub-surface land attack (currently TLAM and NGS), and Carrier Strike
6. Sustainment Operations
7. recovery/redeployment.

Clearly some of the stages overlap and run concurrently – for example Sea Control Operations must run concurrently with Maritime Force Projection and Sustainment Operations. This is illustrated in Figure B4.

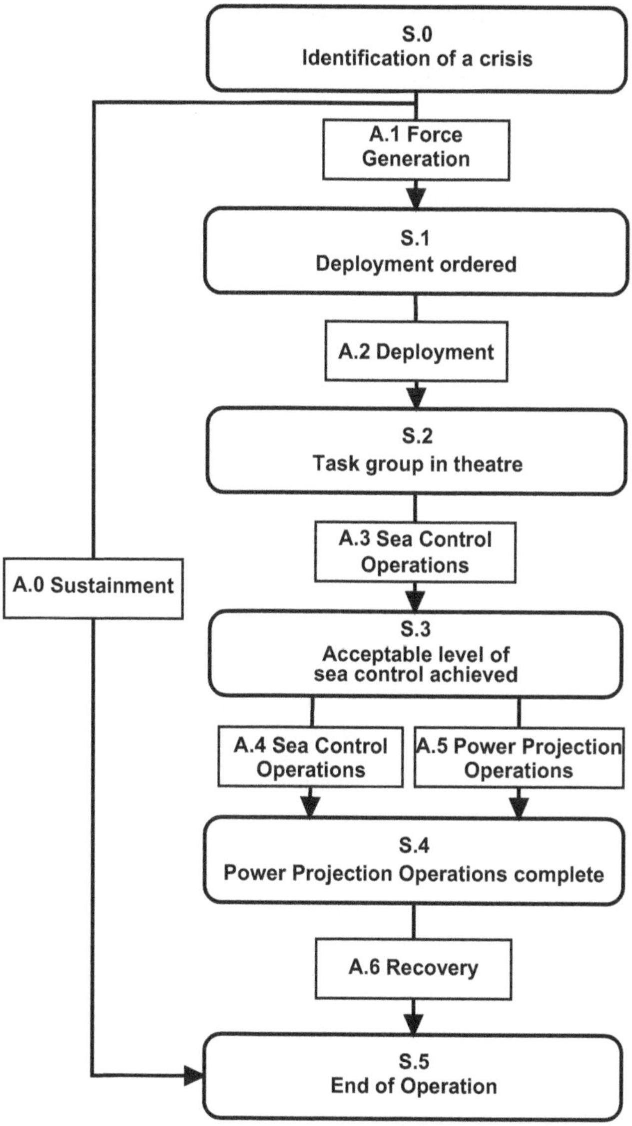

Figure B4 Task network diagram for CEPP

Limits on available time and information precluded the development of a Task Description Table.

Constraints, Assumptions, Opportunities and Risks Analysis

Table B1 Constraints table

Constraint	Consequence
QEC platform and crew availability	With the QEC at sea for significant proportions of the year and the requirement for Fleet Time Support Periods each year, opportunities for collective training may be limited. There would be limited opportunities to exploit a synthetic training solution based on 'ship alongside' connected into a synthetic environment simulating ship systems.
Training audience and platform availability	The wide variety and large numbers of assets and crews required to deliver full CEPP capability will make the scheduling of training for the full capability challenging.
LTNG II flying hours	Likely to be limited, placing significant constraints on live training.
LTNG II mission profiles	Many LTNG II missions cannot be flown live due to security.
Co-location of training audience	If the training audience is not co-located, the representation of blue actors within a unified battlespace will have to be delivered synthetically.
Synthetic Delivery	Delivery of a synthetic environment to at-sea deployed assets will require bandwidth from a suitable transmitter that has coverage to all platforms engaged in training.
Training Staff	Some training staff will need to be co-located with trainees undergoing collective training and exercising; where platforms are at sea deployed or on long transits to theatre, training staff will have to be embarked.

Table B2 Risks table

Risk	Description
Lack of training audience availability	A large number of players are needed to represent both blue and red command elements and force elements. If elements are not available these may have to be represented and played as 'white' by instructional management. This lowers the value of training as there are lessons to be learnt and training value to be had from both sides of a mutual interaction. For example having to represent an agency such as the JFACC through instructional management is both a large resource burden, but also a missed training opportunity for the JFACC to work with the maritime component.
Lack of availability of non-UK platforms and organisations	As many of the critical team members and organisations are non-naval and non-UK, any delivery of collective training can be jeopardised by a pull-out of command or force elements which are not under the UK chain of command.

Training Environment Analysis

CEPP is delivered by the coordination and synchronisation of actions by a multiplicity of actors and agencies that transcend either what is provided by QEC alone, or even the maritime component alone. Coordination and synchronisation of action relies on a shared plan, and a shared plan is predicated on both an alignment of goals and an alignment of situational awareness. The alignment and distribution of plans, goals and SA is through communication whether face-to-face or mediated thorough technology. C2 as functions are delivered through communication and it is the actions, interactions and interfaces between the elements discussed above (and the task environment) as they are distributed through the collective organisation that one is trying to develop and exercise.

From a QEC-centric view, the CEPP training environment is an environment which exists both internal and external to the QEC, and relies heavily on interactions that are cross-component (joint) and cross-nationality (combined). For a task of this scale which involves a range of teams that occupy very different work spaces we would suggest generating Training Environment Diagrams for each high-level set of actions of the overall CEPP task.

Detailed analysis of the required training environment cannot be undertaken until the range of potential missions/contingencies in which CEPP will be engaged has been defined. Figure B5 shows a hypothetical Training Environment Diagram for a Non-combatant Evacuation Operation.

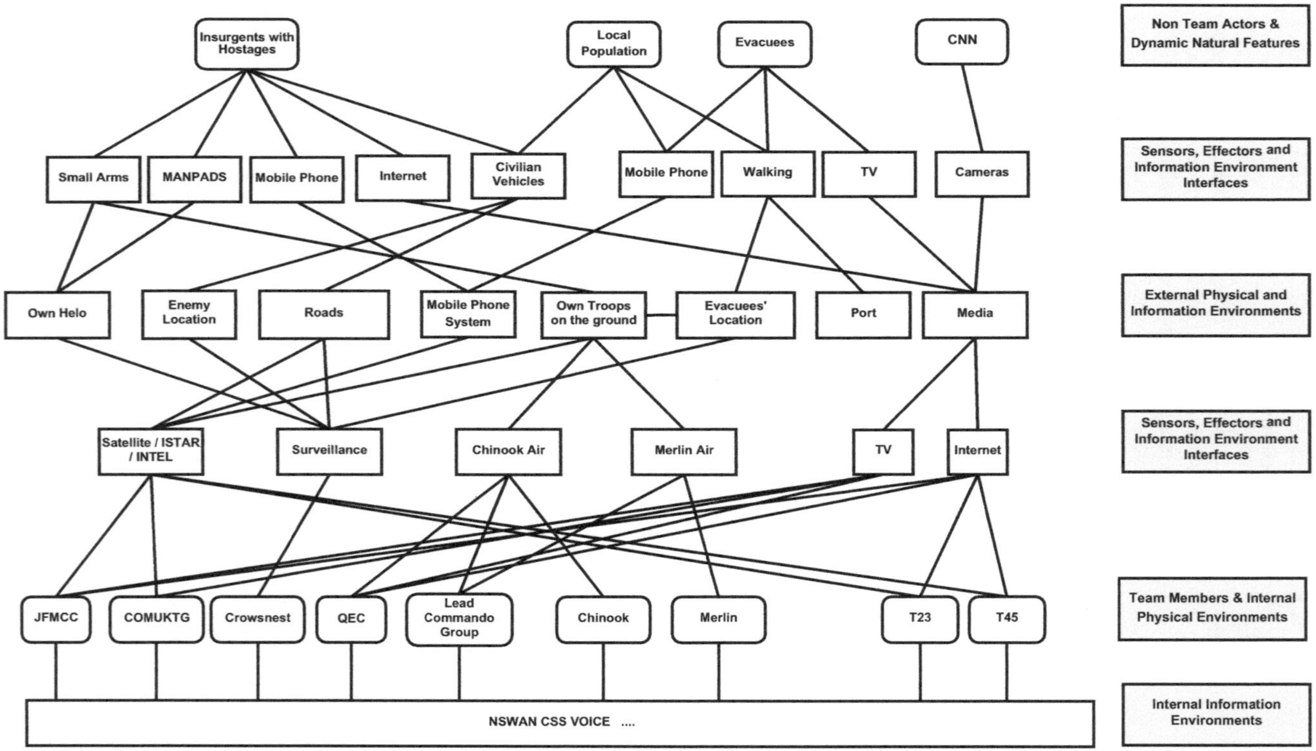

Figure B5 CEPP Training Environment Diagram

Training Overlay Analysis

Training Strategy

Given the large number of actors, agencies and platforms that may be involved in CEPP in its various forms, the likelihood of being able to bring all of the possible assets involved together at the same time for long enough and often enough to achieve all required training is estimated to be less than modest. Therefore, an appropriate strategy would appear to be to divide the training requirement up into a number of parts or task domains.

A plausible task domain can be considered to be a group of elements that meet the following criteria:

- They closely interact with each other in the conduct of the overall task and as such will require training effort to be expended to achieve the required degree of integration.
- Their interactions with other elements in the CEPP domain could satisfactorily be supported by the use of role-players and supporting systems.

Given the number and range of assets involved, the number of permutations for partitioning into domains is large. One sample approach which should be taken as illustrative is shown in Figure B6.

1. UK Joint C2 Battlestaff training – Operational to Tactical level C2 training to stress and develop cross-component interoperability for COMCSG/COMUKTG, LTNG II, QEC ops room. Ideally would involve JFMCC and JFACC as training audience and Joint Forces Command (JFC).
2. Combined C2 training:
   a. UK – US C2 training
   b. UK – NATO C2 training
   c. Force Generation, CEPP reconfiguration training (embarkation of LTNG II, Afloat Spares Pack, Deployed Mission Pack and Personnel), Flying Operations and at-sea sustainment training. This could be extended to included Combined Interoperability Training (the above with partner nations to test interoperability) [not shown on the diagram for clarity].
3. Sea Control training of the QEC task group:
   a. joint
   b. combined.
4. CEPP Littoral Manoeuvre training (embarkation of equipment and personnel, deployment, air and surface manoeuvre):
   a. Joint Littoral Manoeuvre training
   b. Combined Interoperability Training (with US/NATO).

5. CEPP Carrier Strike training:
   a. joint embarked air/maritime training with mixed fixed wing (FW) and rotary wing (RW) assets
   b. combined embarked air / maritime training with mixed FW and RW assets.

Where the dashed lines of a task domain cross over an actor or agency this indicates that the entity may be included or not depending on availability. For example not including LTNG II in Collective Task Domain 3 (Sea Control) means one is performing Sea Control without the benefit of FW organic air support. The lack of availability of LTNG II / LTNG II crews in this context means that there are entire collective tasks such as Defensive Counter Air (DCA) as part of maritime AAW which cannot be trained.

Once one has identified the task domains of interest (which includes a broad definition of training audience subsets), it is possible to make some statements about what sorts of environmental characteristics are relevant for each domain. This in turn enables consideration of how the training overlay as the set of instructional functions might be represented.

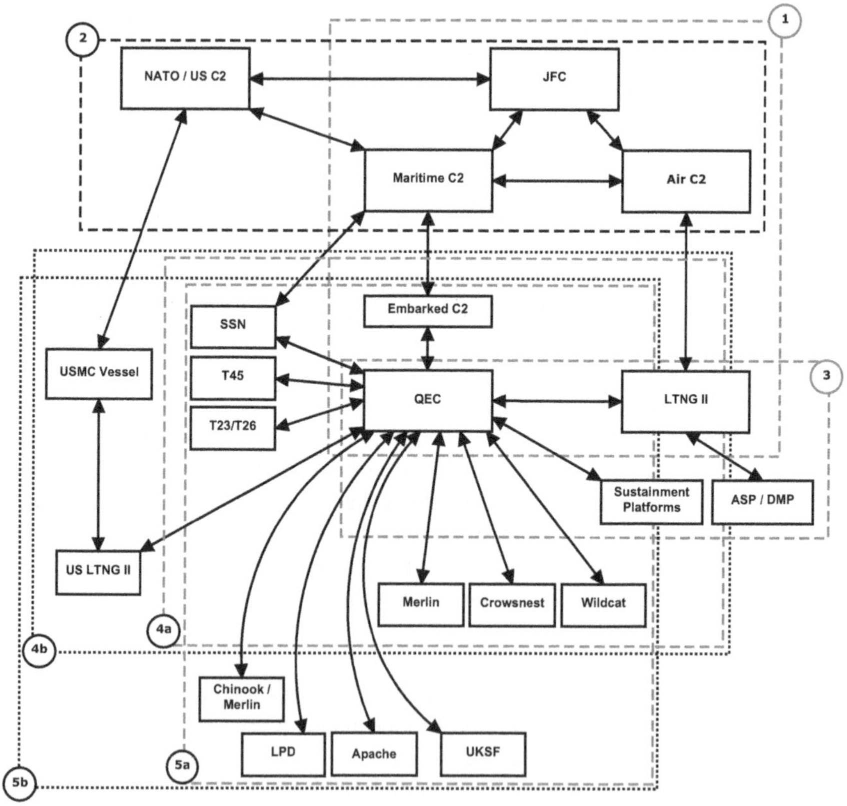

Figure B6 CEPP represented as five interlocking collective task domains

Training Options Analysis

The following analysis considers each of the task domains in turn, providing an overview of the key factors that would need to be considered.

UK Joint C2 Battlestaff Training

The focus of this training event is the integration of the various C2 components involved in CEPP.

Training audience
JFC, Maritime C2 chain, Air C2 chain, Embarked C2, QEC command team, LTNG II command team.

External environment
An evolving and complex crisis/contingency situation, higher command direction, intelligence, legal and political agencies, problem/enemy characteristics.

Internal environment
Situation assessment products, plans, orders, management of resources and internal constraints. Internal systems for communication and information generation, storage and retrieval, including QEC information systems used by the Air Management Organisation on board.

Critical considerations
Location of QEC command team and LTNG II command team for training (embarked or not):

- If these teams are embarked the location (time zone) of the QEC will be a consideration for synchronising with teams working on GMT. Network bandwidth may also be a consideration.
- If these teams are not embarked then relevant shipborne planning and management systems for QEC and LTNG II will need to be replicated or real systems used if they are portable.
- If QEC command team does not use shipborne systems for this training, replication of systems should be as complete as possible or additional training on this element covered within collective task domains 4, 5 or 6.

Environmental instantiation

1. External Physical Environment – At this level of scale the external physical environment is experienced indirectly through information systems available to the teams as the external information environment.
2. External Information Environment – Constructive elements, role played.

3. Internal Physical Environment – Mixture of 'live' workspaces, real equipment with access to required networks at appropriate levels of security. QEC and LTNG II will either use real systems on board which will require a synthetic environmental feed to wherever the QEC is located, or will use real systems located elsewhere which will require a synthetic environmental feed. Where external environment elements feed into internal embarked systems, and alter internally represented information, a means of generating this input must be represented. There is a need to be able to generate injects into these systems, just as injects are generated in the QEC combat system. As an example, software systems reporting LTNG II status may automatically feed into Air Management Organisation software; here there may be a need to generate a certain number of aircraft and make one unserviceable to cause dynamic re-planning to occur. The instructional staff must have a way of informing the training audience that an aircraft is unavailable. In this example there is a need to LOCON the status of the embarked aircraft.

Combined C2 Training

In this collective task domain CEPP is one of a number of tasks that may be undertaken, so the training at this level focuses on the management and application of CEPP with international partners and agencies, alongside other high-level combined/joint tasks.

Audience
JFC, Maritime C2 chain, Air C2 chain, NATO, US, France, Italy.

External environment
An exercise at this level is likely to be a US or NATO led construct with similar properties to task domain 1, but a bigger/more complex problem, over a wider area with a greater number of actors and agencies.

Internal environment
From a UK perspective similar to task domain 1; however, systems and network classification (for example NATO secret vs UK secret) and appropriate interoperability of systems may generate issues. These issues may overspill into the delivery of the training event itself through networking systems together (ITAR, systems certification and so on).

Critical considerations
NATO non-article 5 situations may not allow some flavours of CEPP to be conducted. UK integration into NATO events will be dependent of command element availability, and whether those units are assigned as NATO forces.

Environmental instantiation

1. External Physical Environment – At this level of scale the external physical environment is experienced indirectly through information systems available to the teams as the external information environment.
2. External Information Environment – In a NATO context this will involve NATO systems which will need to be sourced in appropriate numbers.
3. Internal Physical Environment – Mixture of 'live' workspaces, real equipment with access to required networks at appropriate levels of security.

Force Generation, CEPP Reconfiguration Training, Flying Operations and At-Sea Sustainment

The focus of this training event is the integration of LTNG II with real equipment and personnel at sea.

Training audience
Teams involved in Force Generation of QEC, LTNG II Squadron staff and support, Sustainment Platforms.

External environment
The QEC and embarked LTNG II aircraft at sea, loading the Afloat Spares Pack and the Deployed Mission Pack onto QEC, opportunities for live flying, on-board aircraft maintenance and replenishment of QEC and LTNG II by replacement/support ships.

Internal environment
The whole QEC and associated systems.

Critical considerations
Location and availability of QEC and LTNG II.

Environmental instantiation
With the exception of the command teams generating the QEC and LTNG II, most of this training can only be conducted live on board.

Sea Control QEC TG

The focus of this is maritime sea control training for the QEC TG (QEC and escorts)

Training audience
Embarked C2, QEC, LTNG II, Sustainment Platforms, Merlin, Crowsnest, Wildcat, Nuclear-powered Attack Submarines (SSNs), Type 45, Type 23, Type 26.

External environment

Maritime multi-threat environment, submarines, aircraft, surface threats including FIAC, missiles including anti-ship ballistic missiles, blue surface, subsurface and RW platforms.

Internal environment

Communication and information systems that link and are shared between maritime and embarked air platforms to build and share SA, Plans and Command Direction.

Critical considerations

- The scale of a representative TG needing to be represented as compared to availability of maritime platforms and crews.
- The secrecy of LTNG II systems and associated simulators may make networking air and maritime simulation systems very difficult.
- Harmonising LTNG II and maritime training may be difficult.

Environmental instantiation

1. External Physical Environment – Most of these threats can only be represented synthetically with the exception of ASW / enemy submarines, which might be better represented live (if live assets are available) due to potential complexities in synthetic environment representation. LTNG II will be unable to display some capabilities live in peacetime so may be unable to participate in some live training.
2. External Information Environment – Operations rooms and ships' combat systems must be represented with appropriate manning states.
3. Internal Physical Environment – QEC command areas, ships' operations rooms and associated FW and RW operating environments.

Further Points for Consideration in Options Analysis

If maritime platforms at sea are to be used, the key difficulties are delivery of the synthetic environment required to all players (which include RW and FW aircraft), and being in similar time zones. Clearly, delivery of bandwidth to carry a synthetic threat to submarines is a hurdle, for ASW one would need to integrate ASW helicopters into the synthetic representation (or do it live, which for ASW might be preferable). For this reason ASW of all warfare areas is best suited to live delivery providing one can generate enough submarines to play both sides, and enough surface combatants and High Value Units for them to attempt to defend.

Defensive surface warfare training (for example massed waves of missile equipped FAC and FIAC) would benefit from inclusion of aircraft with an ASuW role (Wildcat). Here threat platforms cannot really be economically represented live to the point of firing anti-ship missiles (which cannot be represented live at all).

Anti-air warfare can really only credibly represent threat synthetically – here integration of T45, E3D and Crowsnest with QEC would be important to exploit complementary capabilities while minimising individual platform vulnerabilities. LTNG II in a DCA role to integrate into theatre AAW may only be capable of being flown synthetically.

In all of these types of maritime warfare, the delivery of a suitable synthetic environment to blue platforms at sea and having all required units available at the same time to engage in training is the issue. The air platforms which are vital for all forms of maritime warfare are difficult to integrate into 'at-sea' training and exercising, as are the submarines (unless training is conducted live). Inbuilt training systems in air platforms which are non-flight deck based might be linkable into a synthetically delivered training environment; however, the ship to helicopter interaction would not be fully represented.

The alternatives to 'on-board, at-sea synthetic training' are:

1. The use of shore-based virtual simulators, which could work if all TG platform crews were available for training at the same time. Bandwidth would not be an issue, and the integration of ships' RW assets would be much easier (again if crews are available).
2. Live training accepting deficiencies in threat representation which are critical for meaningful task representation.

In considering the connection of virtual training environments for different platforms or the inclusion of live assets in a training event, the training benefit gained verses the cost to achieve it must be considered. In considering training benefit, the issue of training audience primacy has to be taken into account. For example, inclusion of SSNs could be achieved by the deployment of an SSN to participate in a live exercise, by linking an extant SSN simulator to MaST or by enhancing MaST to support SSN training. If the SSN crew is the primary training audience for the training event planned, then, presuming there is significant training benefit for the SSN crew, it should be possible to construct a robust case for one of the options to be pursued. However, if the SSN crew is not the primary training audience, but is supporting another component, the training benefit which they receive needs to be scrutinized carefully. Given that, in training at this level, it is the integration of components and their interactions which are the focus of training, it may be possible to achieve the desired training effect by using a suitably experienced role-player to provide the SSN interactions with the rest of the training audience.

In summary, the integration of training environments for each platform and command element is complex. An argument could be advanced that it is too complex to address, and therefore the solution should be focused at a platform and command element level, with the roles of other platforms and command elements being provided through the use of role-players or CGF. However, whilst such a solution would facilitate training at a platform or command element level, it would not cause the teams that have to communicate with each other and coordinate their tasks to actually interact with each other, which are generally understood to be key in the development of an understanding of how each other operates and the development of trust. In short, if you want to be certain that jigsaw puzzle pieces do actually fit together to make a picture, you have to get them out of the box at the same time and check they do actually fit. Whilst the entire CEPP jigsaw puzzle is complex and has a number of permutations, not trying to fit any of the pieces together could be construed as unwise.

References

Cabinet Office (2010) *Securing Britain in an Age of Uncertainty: The Strategic Defence and Security Review*. London: The Stationery Office.

Ministry of Defence (2013) *Joint Doctrine Publication 0-30: UK Air and Space Doctrine*. Swindon: Development, Concepts and Doctrine Centre, Ministry of Defence.

Royal Navy (2004) *BR 1806 British Maritime Doctrine*, 3rd edn. London: The Stationary Office.

Index